INTERNET PROTOCOL-BASED EMERGENCY SERVICES

INTERNET PROTOCOL-BASED EMERGENCY SERVICES

Editors

Hannes Tschofenig
Nokia Siemens Networks, Finland

Henning Schulzrinne
Columbia University, USA

This edition first published 2013

Registered office
John Wiley & Sons Ltd, The Atrium, Southern Gate, Chichester, West Sussex, PO19 8SQ, United Kingdom

For details of our global editorial offices, for customer services and for information about how to apply for permission to reuse the copyright material in this book please see our website at www.wiley.com.

Library of Congress Cataloging-in-Publication Data

Tschofenig, Hannes.
Internet protocol-based emergency services / Hannes Tschofenig.
pages cm
Includes bibliographical references and index.
ISBN 978-0-470-68976-9 (cloth)
1. Emergency communication systems. 2. Computer network protocols. 3. Internet. 4. Public safety radio service. I. Title.

TK6570.P8S38 2014
004.67′8–dc23 2013005450

A catalogue record for this book is available from the British Library.

ISBN: 978-0-470-68976-9

Typeset in 10/12pt Times by Aptara Inc., New Delhi, India

1 2013

To my wife Verena and my daughter Elena. Verena encouraged me to work on this project and motivated me when I got stuck due to all my other obligations.

Hannes Tschofenig

Contents

List of Figures

List of Tables

List of Contributors

Bernard Aboba, Skype

Anand Akundi, Telcordia

Carla Anderson, e-Copernicus

Gabor Bajko, Nokia

Ray Bellis, Nominet

Chantal Bonardi, European Telecommunications Standards Institute (ETSI)

Margit Brandl, Nokia Siemens Networks

Guy Caron, Text written in personal capacity; works for Canadian telecommunication operator

John Chiaramonte, Booz Allen Hamilton

Martin Dawson, CommScope

John Elwell, Siemens Enterprise

Gunnar Hellström, Omnitor

Jean-Pierre Henninot, Ministère de l'Economie de l'Industrie et de l'Emploi (MEIE)

Hannu Hietalahti, Nokia

Roger Hixson, National Emergency Number Association (NENA)

Jan Kåll, Nokia Siemens Networks

Jong Yul Kim, Columbia University

Dirk Kröselberg, Siemens

Jim Kyle, University of Bristol

Samuel Laurinkari, Nokia Siemens Networks

Christopher Libertelli, Skype

Gary Machado, European Emergency Number Association (EENA)

John Martin, AuPix Limited

Yacine Rebahi, Fraunhofer FOKUS

Carl Reed, Open Geospatial Consortium (OGC)

Brian Rosen, NeuStar

Jakob Schlyter, Kirei AB

Byron Smith, L.R. Kimball

Wonsang Song, Columbia University

Martin Thomson, Skype

Khiem Tran, CommScope

Gordon Vanauken, L.R. Kimball

Ed Wells, Text written on behalf of CLDXF Work Group (WG), Data Structures Subcommittee, Core Services Committee, National Emergency Number Association (NENA)

James Winterbottom, CommScope

Preface

The Internet has changed our society in astonishing ways during the past 20 years. It has enabled companies as well as individuals to communicate information easily to a worldwide audience. It has also influenced the way we interact with each other.

The topic of this book is a result of this change: emergency services is a feature of the telecommunication system and is therefore tightly bound to the evolution of the underlying communication infrastructure.

As communication moves to the Internet Protocol (IP)-based infrastructure, there is the question of how emergency services should be offered in this new communication environment. From a technology point of view, it has the potential to offer more functionality at a lower cost. This book describes the completed efforts as well as ongoing developments to make IP-based communication systems offer emergency services functionality.

Readers

This book assumes that the reader has a minimum level of familiarity with how the Internet and telecommunication infrastructure works. In particular, some understanding of the Internet Protocol and the Session Initiation Protocol, which is at the heart of the communication architecture, is assumed. Interested readers will find many resources on the Web and in various referenced books and specifications for further reading.

This book is directed toward the following groups:

- *Research Community.* For researchers and students, this book provides a quick entry to the large world of emergency services. Even though the content of the book does not focus on identifying new research areas, a student will be able to capture the state of the art, find two research/pilot projects described in this book, discover information on ongoing work and the organizations and communities where this work is taking place, and see where gaps still exist at the time of writing.
- *Standards Developing Experts.* A standards expert working in one organization is, in our experience, very often unfamiliar with the work done in other standards developing organizations. A lack of understanding of the regulatory environment, ongoing deployment efforts, or broader organizational landscape is not uncommon either. As such, this book will help anyone working on standards to better understand the multiple facets of this topic.
- *Product Architects.* Those who must develop new products or update their existing products to support regulatory mandates or to support new services for their customers face the

difficult decision about what standards to read, support, and implement. This book covers the broad range of standards in the emergency services space and answers many high-level questions. This book, however, does not provide the details for implementers to start coding. The referenced specifications contain the authoritative text, though, with much greater detail and valuable examples.
- *Regulatory Community.* Regulators from all over the world are observing the evolution of telecommunication, which is a heavily regulated area, and are wondering about how the desire of companies and their customers for global communication over the Internet is changing their role. This book provides them with an overview of the developments worldwide in a high-level form. While parts of the book go into technological details, we provide an overview to every chapter to allow the reader to understand the main concepts being discussed.

Organization of the Book

In Chapter 1 we share some of our experience with the work on this topic, illustrate the history, and provide you with our reasons for writing this book.

To our surprise, it was quite difficult to group the work on emergency services in a coherent fashion. We nevertheless gave it a try and separated the work on location (see Chapter 2) from emergency services architectures (see Chapter 3). It turns out that the work on certain location protocols and formats is very much driven by the architectural spirit the designers had in mind at the start of their work. Over time, others joined the work, and the desire to support the same protocol for many different deployment models then influenced further work, and extensions then offered different architectural models.

Roughly at the same time as the standardization work took place, various national organizations started brainstorming about transitioning toward the standardized next-generation emergency services architecture, taking their country-specific deployment constraints into account. We present four of these projects in Chapter 4. While each of these projects has their own history and their own challenges, they illustrate the complexity of the work to accomplish the goal of global emergency services interoperability.

In Chapter 5 we cover a topic that often surfaces in discussions, namely security. Today's emergency services system is already subject to a number of attacks (in the form of false calls, hoax calls, and "swatting"), consuming valuable resources of call-takers and first responders. In some cases, lives are put at risk. Many worry that the problems we see on the Internet with malware, botnets, and distributed denial-of-service attacks will leave the future emergency services system even more vulnerable. In this chapter, we have tried to untangle various security threats and discuss ways to mitigate them.

IP-based communication architectures offer more flexibility and extensibility than the legacy communication architecture, since it is based on core Internet design principles, such as the end-to-end principle. This also provides better support for communication services for persons with disabilities in the form of multimedia communication. In Chapter 6 we exclusively focus on this topic.

Offering emergency services support for citizens and individuals is expensive, and in many cases key stakeholders do not have enough economic incentives to offer support for it. For this reason, regulatory agencies pay attention to this topic to steer the development. In Chapter 7 we illustrate the regulatory developments in the United States and in Europe. We expect to see many changes in this area as more and more users switch to IP-based services and legacy technology ceases to exist.

In Chapter 8 we illustrate three government-funded projects that support the work on IP-based emergency services and highlight their goals and observations.

In Chapter 9 we present a list of organizations that many of our readers may not be familiar with. With short descriptions of what these organizations are working on and what their goals are, we hope to widen your understanding of the bigger picture.

We offer our conclusions and an outlook in Chapter 10.

Scope

It may surprise you that emergency services is a large field. This book covers only a small part of the landscape, since it focuses mainly on the communication chain from an emergency caller dialing 9-1-1, 1-1-2, or any other emergency number that is supported in a certain region, to the call-takers, who receive your call at a Public Safety Answering Point (PSAP) and dispatch emergency personnel (also called first responders).

In this book we do not cover the following:

- The communication from the PSAP toward the first responders (which in most parts of the world still use dedicated radio technology, such as Terrestrial Trunked Radio (TETRA))
- The communication from public safety authorities toward citizens and individuals before, during, and after emergency situations. This body of work is often called "early warning," and there is a large community working on standards and operational matters. Interestingly, the community working on early warning is largely distinct from those working on the classical 9-1-1, 1-1-2,... emergency service communication
- A detailed description of the legacy telecommunication system and its emergency services architecture
- Details about the underlying communication infrastructure. As such, this book does not replace the need to read books about the IP Multimedia Subsystem (IMS), Voice-over-IP system, Real-Time Web Communication, or about the eXtensible Messaging and Presence Protocol (XMPP)
- Any form of social media for usage in disaster response and emergency management

If you would like to give us your feedback, please send us an email (hgs@cs.columbia.edu or hannes.tschofenig@gmx.net) or visit our webpage (http://ip-emergency.net).

Hannes Tschofenig
Nokia Siemens Networks

Henning Schulzrinne
Columbia University

Acknowledgments

The work on this book took a long time, much longer than we had initially anticipated. While working on advancing the emergency services and while writing the book, we talked to many experts in the field to ask for their input and their feedback. We both have been very active in the IETF, NENA, and EENA. While many discussions take place in online mailing lists and phone conference calls, there have also been many workshops where we have met other experts face-to-face. Owing to our involvement in various discussions, we have also been in contact with various emergency services organizations all over the world, software developers, communication service providers, Internet Service Providers, and various other stakeholders, who contribute to the success of the emergency services infrastructure.

It is difficult, if not impossible, to list a few persons in this section since so many have shaped our thinking in the work on emergency services and have contributed to all the topics included in this book. If you look at the individual chapters and follow the references, you will notice that many individuals have worked on the specifications and each document has a story; some documents have taken years to complete and have received a lot of attention. On top of the work on the specifications, there is an even larger group who develop code, and integrate various protocol implementations into complete products, and finally there is the wider community that deploys and uses those products, ranging from voice application client software to a PSAP call-center installation.

We would particularly like to thank our contributors for providing various chapters in this book. They have obviously done a significant part of the work to turn this book into a reality.

We would also like to thank our contact persons at John Wiley & Sons for their patience and for their guidance. In particular, we would like to thank Anna Smart, Susan Barclay, and Tiina Wigley.

A big thanks also goes to Jean Mahoney for reviewing and correcting several chapters in this book. Additionally, we would like to thank Spencer Dawkins, Andrew Newton, and Jean-Jacques Sahel for their review comments.

Finally, in advance, we thank those readers whom we expect to contact us with questions and suggestions for a future edition of this book, since the emergency services topic is by no means a completed area of work.

The editors and contributors welcome comments, questions or suggestions. Feedback can be sent to the editors via email (hgs@cs.columbia.edu or hannes.tschofenig@gmx.net) or by posting your remarks on our webpage (http://ip-emergency.net).

Acronyms

3G	3rd Generation (Mobile Systems)
3GPP	3rd Generation Partnership Program
3GPP2	3rd Generation Partnership Project 2
AAA	Authentication, Authorization and Accounting
AC	Access Class
AFLT	Advanced Forward Link Trilateration
A-GNSS	Assisted GNSS
A-GPS	Assisted GPS
ALI	Automatic Location Identification
AN	Access Network
ANI	Automatic Number Identification
AP	Access Point
APCO	Association of Public Safety Communications Officials
ASN	Access Service Network
ASN-GW	ASN Gateway
ASP	Application Service Provider
ASWG	Address Standard Working Group
ATIS	Alliance for Telecommunications Industry Solutions
ATIS-ESIF	Alliance for Telecommunications Industry Solutions – Emergency Services Interconnection Forum
B2BUA	Back-to-Back User Agent
BCF	Border Control Function
BEREC	Body of European Regulators for Electronic Communications
BS	Base Station
BTS	Base Transceiver Station
CA	Certificate Authority
CAD	Computer-Aided Dispatch
CAP	Common Alerting Protocol
CDMA	Code Division Multiple Access
CGALIES	Coordination Group on Access to Location Information by Emergency Services
CID	Call Information Database
CID	Cell ID
cid	Content Indirection
CLDXF	Civic Location Data Exchange Format

CLF	Connectivity Session Location and Repository Function
CN	Core Network
COMREG	Commission for Communications Regulation
CPE	Customer Premises Equipment
CRL	Certificate Revocation List
CRN	Contingency Routing Number
CRTC	Canadian Radio-television and Telecommunications Commission
CS	Circuit Switched
CSCF	Call Session Control Function
CSG	Closed Subscriber Group
CSN	Connectivity Service Network
CSP	Communication Service Provider
DHCP	Dynamic Host Configuration Protocol
DNS	Domain Name Server (or Service or System)
DoS	Denial of Service
DSAC	Domain Specific Access Control
DSL	Digital Subscriber Line
E9-1-1	Enhanced 9-1-1
EAP	Extensible Authentication Protocol
E-CID	Enhanced CID
ECRF	Emergency Call Routing Function
ECRIT	Emergency Context Resolution with Internet Technologies
E-CSCF	Emergency Call Session Control Function
EENA	European Emergency Number Association
EHA	Emergency Handling Authorities
EMM	EPS Mobility Management
eNodeB	Evolved Node B
E-OTD	Enhanced Observed Time Difference
EPLMN	Equivalent PLMN
EPS	Evolved Packet System
EPSG/OGP	European Petroleum Survey Group/International Association of Oil and Gas Producers
ERDB	Emergency Service Zone Routing Data Base
ERT	Emergency Route Tuple
ESGW	Emergency Services Gateway
ESGWRI	Emergency Services Gateway Route Identifier
ESINet	Emergency Services IP Network
E-SLP	Emergency SLP
ESM	EPS Session Management
E-SMLC	Evolved Serving Mobile Location Center
ESN	Emergency Service Number, Electronic Serial Number, Emergency Service Network
ESNet	Emergency Services Network
ESQK	Emergency Services Query Key
ESRD	Emergency Services Routing Digits
ESRK	Emergency Services Routing Key

ESRP	Emergency Services Routing Proxy
ESZ	Emergency Services Zone (Same as ESN)
ETSI	European Telecommunications Standards Institute (ETSI)
E-UTRAN	Evolved Universal Terrestrial Radio Access Network
FCC	Federal Communications Commission
FGDC	US Federal Geographic Data Committee
FICORA	Finnish Communication Regulatory Authority
FLAP	Flexible Location server to AMF Protocol
Geopriv	Geolocation and Privacy
GeoRSS	Geodetic Really Simple Syndication
GERAN	GSM EDGE Radio Access Network
GML	Geographic Markup Language
GMLC	Gateway Mobile Location Center
GMM	GPRS Mobility Management
GMSC	Gateway MSC
GNSS	Global Navigation Satellite System
GPRS	General Packet Radio Service
GPS	Global Positioning System
GSM	Global Standard for Mobile Communication
GUTI	Globally Unique Temporary Identity
HA	Home Agent
HELD	HTTP Enabled Location Delivery
H-GMLC	Home GMLC
HLR	Home Location Register
HPLMN	Home PLMN
HSA	Home Subscription Agent
H-SLP	Home SLP
H-SPC	Home SPC
HSS	Home Subscriber Server
HTTP	Hypertext Transfer Protocol
IAIC	IP to ISP Address Converter
IANA	Internet Assigned Numbers Authority
IAP	Internet Access Provider
I-CSCF	Interrogating Call Session Control Function
IEEE	Institute of Electrical and Electronics Engineers
IETF	Internet Engineering Task Force
ILP	Internal Location Protocol
IM	Instant Messaging
IMEI	International Mobile Equipment Identity
IMS	IP Multimedia Subsystem
IMSI	International Mobile Subscriber Identity
IP	Internet Protocol
IPsec	Internet Protocol Security
ISP	Internet Service Provider
ITU	International Telecommunications Union
LA	Location Agent

LBS	Location Based Services
LbyR	Location by Reference
LbyV	Location by Value
LC	Location Controller
LCP	Location Configuration Protocol
LCS	Location Service
LIF	Location Interoperability Forum
LIMS-IWF	Location IMS – Interworking Function
LIS	Location Information Server
LMU	Location Measurement Unit
LNG	Legacy Network Gateway
LO	Location Object
LoST	Location to Service Translation
LPP	LTE Positioning Protocol
LR	Location Requester, Location Recipient
LRF	Location Retrieval Function
LRO	Last Routing Option
LS	Location Server
LTD	Long-Term Definition
LTE	Long-Term Evolution
LVF	Location Validation Function
MAC	Media Access Control
MCC	Mobile Country Code
MDN	Mobile Directory Number
ME	Mobile Equipment
MIN	Mobile Identification Number
MIP	Mobile IP
MLP	Mobile Location Protocol
MLS	Mobile Location Service
MME	Mobility Management Entity
MNC	Mobile Network Code
MO	Mobile Originating (call)
MS	Mobile Station
MSAG	Master Street Address Guide
MSC	Mobile Switching Center
MSISDN	Mobile Station International ISDN Number
MSRP	Message Session Relay Protocol
NAD83	North American Datum 1983
NAI	Network Access Identifier
NAP	Network Access Provider
NAPTR	Naming Authority Pointer
NAS	Network Attachment Subsystem
NAT	Network Address Translation
NAVD88	North American Vertical Datum of 1988
ND&S	Network Discovery and Selection
NENA	National Emergency Number Association

NG112	Next Generation 112
NG9-1-1	Next Generation 9-1-1
NGO	Non-Governmental Organization
NICC	Network Interoperability Consultative Committee
NRA	National Regulatory Authority
NRM	Network Reference Model
NSP	Network Service Provider
NWG	Network Working Group
OFCOM	Office of Communications
OGC	Open Geospatial Consortium
OMA	Open Mobile Alliance
O-TDOA	Observed Time Difference of Arrival
PAI	P-Asserted Identity
PCC	Policy and Charging Control
PCP	Privacy Checking Protocol
PCRF	Policy Control and Charging Rules Function
P-CSCF	Proxy Call State Control Function
Phone	Either ME or UE
PIDF	Presence Information Data Format
PIDF-LO	Presence Information Data Format – Location Object
PKI	Public-Key Infrastructure
PLMN	Public Land Mobile Network
PMD	Pseudonym Mediation Device functionality
PMIP	Proxy Mobile IP
PPAC	Paging Permission Access Control
PPR	Privacy Profile Register
PS	Packet Switched
PSAP	Public Safety Answering Point
PSK-TLS	Pre-Shared Key TLS
PSTN	Public Switched Telephone Network
QoP	Quality of Position
QoS	Quality of Service
RADIUS	Remote Access Dial-In User Service
RAU	Routing Area Update
RDF	Routing Determination Function
RDS	Root Discovery Server
RFC	Request For Comment
R-GMLC	Requesting GMLC
RNC	Radio Network Controller
RRC	Radio Resource Control
RRLP	Radio Resource LCS Protocol
RTCP	Real Time Control Protocol
RTP	Real Time Transport Protocol
RTSP	Real Time Streaming Protocol
RTT	Real Time Text
SAI	Service Area Identity

SAS	Standalone SMLC
SBC	Session Border Control
S-CSCF	Serving CSCF
SDES	Session Description Protocol Security Descriptions
SDO	Standards Development Organization
SDP	Session Description Protocol
SET	SUPL Enabled Terminal
SGSN	Serving GPRS Support Node
SIM	Subscriber Identity Module
SIP	Session Initiation Protocol
SLA	SUPL Location Agent
SLC	SUPL Location Center
SLP	SUPL Location Platform
SMS	Short Message Service
SPC	SUPL Positioning Center
SRDB	Selective Routing Database
SRTP	Secure Real Time Transport Protocol
SRVCC	Single Radio Voice Call Continuity
SS	Subscriber Station
SSAC	Service Specific Access Control
STA	Station (an entity with 802.11 interface)
SUPL	Secure User Plane Location
TAU	Tracking Area Update
TCP	Transmission Control Protocol
TDD	Telecommunications Device for the Deaf and Hard-of-Hearing
TLS	Transport Layer Security
TMSI	Temporary Mobile Subscriber Identity
TTY	Teletypewriter (a.k.a. TDD)
UA	User Agent
UAC	User Agent Client
UAS	User Agent Server
UDP	User Datagram Protocol
UE	User Equipment
ULP	Userplane Location Protocol
UMB	Ultra Mobile Broadband
U-NAPTR	Straightforward URI-Enabled NAPTR
URI	Uniform Resource Identifier
URL	Uniform Resource Locator
URN	Uniform Resource Name
USI	Universal Services Interface
USIM	UMTS Subscriber Identity Module
U-TDOA	Uplink-Time Difference of Arrival
UTRAN	Universal Terrestrial Radio Access Network
VCC	Voice Call Continuity
VDB	Validation Database
V-GMLC	Visited GMLC

VLR	Visitor Location Register
VLS	Visited Location Server
VMSC	Visited MSC
VoIP	Voice over IP
VPC	VoIP Positioning Center
VPLMN	Visited PLMN
VPN	Virtual Private Network
VSP	VoIP Service Provider
WCDMA	Wideband Code Division Multiple Access
WebRTC	Web Real-Time Communication
WiFi	WLAN based on IEEE802.11
WiMAX	Worldwide Interoperability for Microwave Access
WLP	WiMAX Location Protocol
XML	eXtensible Markup Language
XMPP	eXtensible Messaging and Presence Protocol

1

Introduction

In an August 2011 interview, Federal Communications Commission (FCC) Chairman Julius Genachowski said that "It's hard to imagine that airlines can send text messages if your flight is delayed, but you can't send a text message to 911 in an emergency." He continued "The unfortunate truth is that the capability of our emergency-response communications has not kept pace with commercial innovation, has not kept pace with what ordinary people now do every day with communications devices" [1].

Vice Presidents of the European Commission Neelie Kroes and Siim Kallas have decided to work together to ensure every European can access a 112 smart-phone application, in their own language. This announcement was made in February 2012 [2] on the European 112 day when surveys revealed that "74% of Europeans don't know what emergency number to call when traveling in the European Union."

In May 2012 Verizon Wireless announced that it will offer text-to-911 service nationwide to the public as the first US wireless carrier [3].

A lot has changed since the first standardization work on IP-based emergency services began more than 10 years ago. Today, the understanding of Internet applications and their potential has improved significantly. This book will help you to understand how we reached the point where we are today and will also show you a complicated ecosystem that still requires lots of deployment and education work.

We start this journey into the emergency services field with a short history, told from the viewpoint of one of the editors of this book (Hannes Tschofenig). We then provide a short introduction to the relevant part of the Internet Protocol that makes this work technically demanding, and conclude with an overview of the core building blocks for emergency services that will be discussed in the rest of the book.

1.1 History

In November 2004 Jon Peterson and I (HT) attended the 61st IETF Meeting in Washington, DC, to discuss the formation of a new Working Group in the IETF, in a Birds of a Feather ("BoF") meeting [4]. The name of the proposed group was "Emergency Context Resolution

Internet Protocol-Based Emergency Services, First Edition. Hannes Tschofenig and Henning Schulzrinne.

```
Jon: Is there interest in the room to work
on the problem outlined in the proposed
Charter?

hum ... seems broad interest!

Nobody opposed to create a Working Group.
```

Figure 1.1 IETF 61 ECRIT BOF: Decision to form a Working Group.

with Internet Technologies (ECRIT)." Such BOF meetings require a fair amount of preparation to have a credible proposal for the audience to evaluate. However, the group does not expect a worked-out solution, but rather requires a critical mass of people to work on a topic and a good starting point. The BOF chairs, namely Jon and I, had worked on a Charter text proposal based on the technical material that was created by Brian Rosen and Henning Schulzrinne the year before this BOF. Speakers were also recruited to talk about requirements, the state of the art, and the design considerations. Jon Peterson was the area director of the Transport Area, and the ECRIT BOF fell under his supervision. As such, I could not have had a better co-chair for my first BOF, since Jon was experienced and the community knew him.

In any case, the audience was convinced that the IETF should start their work on IP-based emergency services standardization and the meeting minutes [5] reflect the decision, as shown in Figure 1.1. The Tao of IETF [4] describes the process of humming in the IETF as follows: "Sometimes consensus is determined by 'humming' – if you agree with a proposal, you hum when prompted by the chair. Most 'hum' questions come in two parts: you hum to the first part if you agree with the proposal, or you hum to the second part if you disagree with the proposal. Newcomers find it quite peculiar, but it works. It is up to the chair to decide when the Working Group has reached rough consensus."

Not every BOF receives such support; sometimes there is a fierce fight if many oppose the formation of a new Working Group or argue to radically change the Charter.

Support from the community does not instantly lead to the formation of a new Working Group. Instead, the Internet Engineering Steering Group (IESG) reviews the outcome of the BOF [4]. However, this is uncontroversial and quickly finalized. At the 62nd IETF Meeting in Minneapolis, early March, the ECRIT Working Group met for the first time. The IESG had selected Marc Linsner and me to co-chair the group. I had not worked with Marc prior to that time, but we quickly got along very well. This was the starting point for a fair amount of work that followed over the subsequent years in the IETF.

Late 2005 and early 2006, after a number of ECRIT Working Group meetings, Marc and I started wondering why there was suddenly such a huge amount of activity in the standardization on emergency services. We heard about various efforts either starting or already ongoing in other standards developing organizations.

Early July 2006, the 3GPP CT1 and the IETF ECRIT Working Group organized a joint meeting on emergency calls. This meeting was organized by Hannu Hietalahti (3GPP TSG CT chairman at that time), Marc Linsner and me. The discussions were fruitful and we reached a better understanding of the architectural approaches the two organizations had in mind, even though it was clear that our views were far apart.

Although we did not have the complete picture of the entire standardization landscape, we wondered whether it would make sense to arrange a workshop to meet all those people working in other organizations. The goal of such a workshop would be to reach a better understanding of what everyone else was working on. We hoped that the sharing of information would avoid overlapping and conflicting standards efforts. We were convinced that emergency services would have to interoperate on a global scale in order to avoid problems for those calling for help. Of course, it was not clear whether other organizations would be interested in discussing their work, and various aspects (such as policies regarding Intellectual Property Rights) could easily become meeting show-stoppers. We also knew it would be time-consuming to find our counterparts from other organizations, and to prepare the workshop.

We nevertheless gave it a try and started with the preparation of the first standards development organizations (SDOs) Emergency Services Coordination Workshop [6], which took place in New York hosted by Henning Schulzrinne at Columbia University on 5th and 6th October 2006. The workshop was organized by Marc Linsner (IETF ECRIT), Hannu Hietalahti and Atle Monrad (3GPP TSG CT), Henning Schulzrinne, and me. The agenda was simply a list of speakers from various organizations talking about their standardization efforts.

Our initial workshop was a big success: we managed to get many of the stakeholders to attend the meeting and to talk about their work. We suddenly had a much better idea of what everyone else was working on. It was also a lively workshop: we were interrupted by a fire alarm twice in the workshop building, and it was clear that the participants had very different views about where the standardization efforts should be going and what global interoperability meant for them. Although the architectural disconnections should not have been surprising, I was surprised again and again during the meeting. As one of the meeting organizers, I had to pause the meeting for some time because some of the participants had gotten into a huge argument, and I was worried that this would derail the entire workshop.

Despite the difficulties, we were convinced that further workshops were needed to improve the cooperation between the different standards bodies. After our second workshop in Washington, DC, we organized workshops in Europe and even went to Kuala Lumpur, Malaysia, in November 2009.

As we progressed our cooperation with other organizations, and regularly met with them, we realized that just hosting workshops was not enough. Marc, Henning, and I were active in different Working Groups in the National Emergency Number Association (NENA), where Brian Rosen led the work on NENA i3 – NENA's vision for the long-term technical architecture. NENA had a large number of members from the emergency services community, and their feedback on and input to the standards process were crucial. Hence, it was obvious to many of the IETF ECRIT Working Group participants that they should also participate in the NENA work process, which was heavily driven by weekly or bi-weekly conference calls.

During our workshop in Brussels, we contacted the European counterpart of NENA, namely the European Emergency Number Association (EENA). EENA had a very different history than NENA and was initially focused only on lobbying for citizen rights. NENA, on the other hand, had been weaker on the lobbying side but was much stronger on the technical and operational side.

The emergency services communities,[1] which we also invited to our workshops, needed venues for discussions as well, and our highly technical discussions were not necessarily all that they were looking for. Many of them had questions about operating and funding

[1] In this context, an emergency services community refers to the public authorities in a specific country or part of a country who are responsible for maintaining the emergency services infrastructure, and their vendors and service providers who contribute the equipment and communications infrastructure.

new emergency services technology. We added tutorial tracks to our workshops to bring newcomers up to speed, but, for many, the tutorials were still too technical. We knew that in the long run we would not be able to organize our workshop series for this extended audience of standards professionals and the emergency services community. And that was not even the entire stakeholder community. We had contacted state and federal regulators, disability groups, Internet Service Providers, researchers, emergency services organizations, advisory bodies, technology providers, over-the-top application server providers as well as telecommunication operators.

Consequently, we started to work more closely with EENA and NENA. For example, in April 2011, we organized our emergency services coordination workshop with the EENA emergency services workshop in Budapest: we provided the technical track and EENA staff organized everything else, including the venue.

In February 2009, I received the "Outstanding Vision for 112" award from EENA for my work on IP-based emergency services. Later I was asked to co-chair EENA's Next Generation 112 Technical Committee. Since I was familiar with NENA's work style and had the technical background, I worked with Gary Machado and the rest of the EENA Advisory Board on changing the shape of EENA.

The work in ECRIT took much, much longer than we expected: we added new Charter items to address new requirements and new technology; lots of work was also done in the IETF Geopriv Working Group on location and location protocols. When I was elected to the Internet Architecture Board (IAB) in 2010, I had to step down from my Working Group co-chair role to ensure that I had enough time for my IAB duties. Marc continued as a co-chair and was joined by Richard Barnes, who also co-chaired the IETF Geopriv Working Group with Alissa Cooper. In 2012 Richard decided to step down and Roger Marshall, who is a very active NENA member, took over his position.

At the time of writing, there are still various specifications in progress. Deployments are picking up, although in a way that we had never expected. We always thought that the innovation in communication protocols would also push emergency services forward, but instead application service providers were very reluctant to move into the emergency services space, largely because of liability and regulation fears. Instead, new developments in the communication space (e.g., instant messaging, Voice over IP, social networks) failed to lead to improvements in the ability to call for help. This reinforces the observation by FCC Chairman Julius Genachowski stated at the beginning of this chapter. On the side of emergency services authorities, however, we saw many changes. Authorities learned that Internet Protocols and the new off-the-shelf products led to lower capital investment and lower operational expenses. By focusing on future-proof technology, many of the investments were in IP-based systems rather than into legacy technology. IP also provided them with new flexibility that allowed many countries to reduce the number of Public Safety Answering Points (PSAPs),[2] leading to a more efficient emergency services organization.

Over all these years, I have met many people working on emergency services and tried to understand their views to better see the big picture of IP-based emergency services. Working with Henning Schulzrinne over the years, we both thought it would be valuable for others to

[2] RFC 5012 [7] defines a Public Safety Answering Point (PSAP) as "a facility where emergency calls are received under the responsibility of a public authority." In simplified terms, a PSAP is a call center with call-takers waiting for incoming emergency calls from persons in need of help.

see that same big picture without having to spend 10 years or more in this field. We have asked those whom we have met in our workshops and in our standardization work to contribute their views on the emergency services to this book.

1.2 Overview

IP-based emergency services is best understood as an extension of existing communication architectures that are used for everyday communication between friends, coworkers, and businesses. Consequently, there are two parts that have evolved independently from each other, namely the communication architectures used by all of us (e.g., various forms of instant messaging tools, VoIP clients, social networks, voice integration into gaming platforms), and the emergency service networks with the call-takers as the other communication partners.

The drivers behind change for these two ecosystems are different. Day-to-day communication platforms are evolving to meet the needs of the users. The Internet has allowed companies and individuals to easily provide new services. Installing and operating a new Session Initiation Protocol (SIP) VoIP server or an Extensible Messaging and Presence Protocol (XMPP) server requires neither a huge time investment for a technically skilled person nor a huge financial investment. In most cases, all that is needed is a hosting provider that allows the software of the preferred communication software to be installed and run. Companies offering their services to millions of users on the Internet of course need a different level of sophistication and resource investment, but, as history has shown, even small startups can deploy exciting services that are accepted and used on a daily basis by many end customers. The speed of innovation is astonishing if we only look at the types of services offered via browser-based or downloadable applications, such as smart-phone applications. While there is lot of easily recognizable development happening in the communication section, these mechanisms have not been designed with emergency services in mind. In 2012, it is not possible in most parts of the world[3] to use instant messaging to start a real-time communication with call-takers working at a PSAP, to establish video calls, to transmit pictures, or to transmit highly accurate location information available in so many of today's smart phones.

There are many reasons for this development. Many communication service providers see emergency services support as a burden. This impression was partially created by the regulatory community, which is fragmented throughout the world, and the desire to aim for the perfect solution. For those who are mandated to provide emergency services support, there is a liability obligation that comes with it that is often very fuzzy. What service provider would want to spend the time and resources to monitor regulatory developments throughout the world (which are unfortunately not synchronized even within regions), and to invest in systems whose designs only suit a subset of the stakeholders? Without the strict obligation to offer emergency services functionality in our communication software, many emergency services authorities do not see the need to accept this new form of communication. Consequently, one would get the impression that no progress had been made by any of the emergency services authorities in upgrading their emergency services network and their PSAPs. This is interestingly enough

[3] Although it would not be too far fetched to say that it is not possible at all to communicate with PSAPs using modern communication technology. There are a few exceptions where emergency services authorities showed leadership and are trying to accept Internet multimedia communication, for example as part of pilot projects.

not correct either. In many countries, we can observe a shift to IP-based emergency services network deployments mainly due to four reasons:

1. lower costs due to the commodity nature and increased competition,
2. lower operational costs compared to the legacy (circuit-switched) infrastructure,
3. future-proofing since the rest of the entire industry is moving to IP as well, and
4. more functionality and better extensibility.

From a functionality point of view, this allows an emergency services network to connect PSAPs with each other and to re-route calls. Re-routing can happen for various reasons, including load balancing, better utilizing call-taker skills (e.g., language skills), re-routing incorrectly routed calls, conference bridging capabilities, and so on. Furthermore, multimedia data from service providers as well as from end devices can be received and passed on to other parties to improve the situational awareness.

In some countries, the transition to an all-IP-based emergency services network was also used to re-think the current emergency services organization. As a result, the existing processes and organizational structures were simplified, typically leading to a reduction in the number of PSAPs. This has been a trend in Europe for the last few years.

Figure 1.2 shows the vision of an IP-enabled emergency services architecture graphically. The left side of the picture shows end devices attached to different access networks and communication service providers. In the Internet many applications are provided by companies

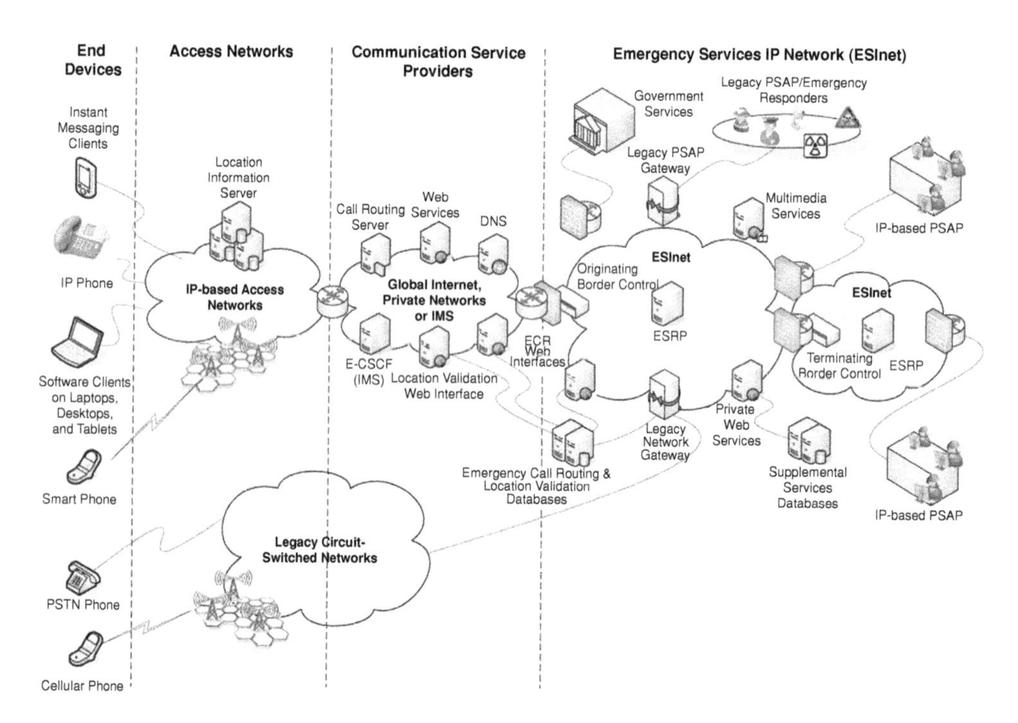

Figure 1.2 Communication architecture overview. (Courtesy of Guy Caron, ENP.)

independent from the access network operator, and therefore the separation between the access network and the communication service provider is indicated explicitly, whereas in the circuit-switched telephony world the application functionality was provided by the access network operator. The right side of Figure 1.2 depicts the IP-based emergency services network. This book covers technical specifications and developments that concern both sides. For communication architectures we focus on those systems that utilize Session Initiation Protocol (SIP), since most emergency services standardization efforts have focused on SIP.

As described in earlier paragraphs, various communication architectures exist today and Figure 1.2 just refers to them as originating networks. Three core features must be provided by such an communication architecture in order to interwork smoothly with an IP-based emergency services network:

1. ability to identify an emergency call;
2. ability to communicate location and/or a location reference; and
3. ability to convey multimedia content.

We will describe these three core building blocks in more detail in section 1.3. Some of the communication architectures in use today provide support for this functionality, such as the SIP-based IP Multimedia Subsystem (IMS) architecture, and the SIP-based VoIP architecture. At the time of writing, there are other communication architectures that do not yet support emergency services, such as the Real-Time Communication in WEB-browsers [8] and the Extensible Messaging and Presence Protocol (XMPP) [9]. There are also many non-standardized and proprietary communication architectures, such as Skype, and many smart-phone applications that do not support emergency services.

You may wonder why there is not just one communication architecture used by everyone. There are probably many reasons, but the history and background of the people who did the work often had a huge influence on the direction of the standardization work. There are also different business models that motivate the work in different standards developing organizations. We will describe one such differentiator in terms of the chosen design assumptions, which will also explain the developments described in other sections of this book.

The Internet architecture follows a layered design, which is a common design pattern for communication systems in general, and allows the replacement of different components (such as a new radio technology, or new applications) while only impacting the neighboring layers. In student textbooks, the Internet architecture has the link layer, Internet layer, transport layer, and the applications layer. In the real world, the layering is much more complex, and it is sometimes hard to assign specific protocols to specific layers (since protocols can be used in a very flexible way and tunneled inside other protocols). The responsibility of providing implementations and deployments of certain layers is also distributed to different parties. In many deployments, the provider of physical connectivity (e.g., a wire) and the provider offering connectivity to the Internet are different companies. Furthermore, those companies offering Internet connectivity are very often different from those offering application layer services. This separation of functionality is a consequence of the end-to-end principle rather than the layering alone. The end-to-end principle states that end-to-end functions can best be realized by end-to-end protocols [10]. While the end-to-end principle still leaves room for discussion and interpretation, the fact is that many application deployments today happen independently of those who provide Internet access. Needless to say, that those who provide

Internet access would like to provide applications as well, or benefit from the deployment of the applications in some way, very much like they did in the past with the Plain Old Telephone Service (POTS). Consequently, a design that assumes that the Internet access provider also offers application services (as is done with the IMS architecture) is different from a design that assumes a separation between the two parties. Such fundamental design assumptions lead to different communication architectures and consequently also to different designs for the emergency services system (even though it is possible to reuse the same building blocks).

The differences between XMPP-based, Skype-based, and over-the-top SIP-based VoIP deployments are, on the other hand, less dramatic. All three assume independence from the Internet access provider, but they are different in the protocol choice. SIP is an IETF standard and is today widely used for voice traffic, Skype software clients use a proprietary protocol that only relies on SIP for interconnecting with other non-Skype-based systems, and XMPP is also an IETF protocol that has found widespread usage for instant messaging.

The term "smart-phone app(lication)" just refers to an implementation of some protocol. It does not indicate whether a standardized or a proprietary protocol has been implemented. As such, a smart-phone app may be an implementation of any of the protocols above, even an SIP-based VoIP client.

1.3 Building Blocks

1.3.1 Recognizing Emergency Calls

In the early days of Public Switched Telephone Network (PSTN)-based emergency calling, callers would dial a local number for the fire or police department. It was recognized in the 1960s that trying to find this number in an emergency caused unacceptable delays. Thus, most countries have been introducing single nationwide emergency numbers, such as 9-1-1 in North America and 1-1-2 in all European Union countries. This became even more important as mobile devices started to supplant landline phones. As can be seen from the introduction of 1-1-2 in Europe, this education effort takes many years, and the old emergency numbers are therefore still in use today as well (in addition to the European-wide 1-1-2 number).

In many countries, different types of emergency services, such as police or mountain rescue, are still identified by separate numbers. Unfortunately, there are more than 60 different emergency numbers in use worldwide, many of which also have non-emergency uses in other countries, so that simply storing the list of numbers in all devices is not feasible. Furthermore, hotels, university campuses, and larger enterprises often use dial prefixes, so that an emergency caller may have to dial 0-1-1-2 to reach the fire department.

With the introduction of smart phones, new user interface designs emerged as well. For this reason, some devices may use dedicated emergency calling buttons or similar user interface elements to initiate an emergency call. Such mechanisms need to be carefully designed so that they are not accidentally triggered, for example, when the device is in a pocket.

Instead of conveying the actual dial string in a protocol message once the user has entered it, a symbolic representation is used instead. This allows unambiguous emergency call identification and automatic treatment of calls. The mechanism used for this emergency call marking uses the Uniform Resource Names (URNs) defined in RFC 5031 [11], such as urn:service.sos.

1.3.2 Obtaining and Conveying Location Information

Location information is needed by emergency services for three reasons: routing the call to the right PSAP, dispatching first responders (e.g., policemen), and determining the emergency service dial strings that are supported in a specific area. It is clear that the location has to be automatic for the first and third purposes, but experience has shown that automated, highly accurate location information is vital to dispatching as well, rather than relying on the caller to report his or her location to the call-taker. This increases accuracy and avoids dispatch delays when the caller is unable to provide location information due to language barriers, lack of familiarity with his or her surroundings, stress, physical or mental impairment. For this reason, automatic location retrieval for emergency calls is a mandatory requirement in nearly all countries in the world.

Location information for emergency purposes comes in two representations: geospatial (also called geodetic), that is, longitude and latitude; and civic, that is, street addresses similar to postal addresses. Particularly for indoor location, vertical information (floors) is also very useful. Civic locations are most useful for fixed Internet access, including wireless hotspots, and are often preferable for specifying indoor locations; while geodetic location is frequently used for cell phones. However, with the advent of femto- and pico-cells, civic location is both possible and probably preferable since accurate geodetic information can be very hard to acquire indoors.

The requirements for location accuracy differ between routing and dispatch. For call routing, city- or even county-level accuracy is often sufficient, depending on how large the PSAP service areas are. First responders, however, benefit greatly when they can pinpoint the caller to a particular building or, better yet, apartment or office for indoor locations, and an outdoor area of at most a few hundred meters outdoors. This avoids having to search multiple buildings, for example, for medical emergencies.

Various protocol mechanisms have been developed to obtain location information from Location Servers operated by different entities, if the calling device does not yet have location information available. Chapter 2 discusses various location protocols that have been developed to allow the calling device, the PSAP, or some other proxy on behalf of them to request location information.

1.3.3 Routing Emergency Calls

Once an emergency call is recognized, the call needs to be routed to the appropriate PSAP. Each PSAP is only responsible for a limited geographic region, its service region. In addition to the geographical region, different PSAPs often only provide their service for specific services, such as police, fire, ambulance, and so on. There is a wide range of different deployment models used throughout the world, and a description of the different PSAP models for Europe can be found elsewhere [12, 13].

The number of PSAPs serving a country varies quite a bit: Sweden, for example, has 18 PSAPs; and the United States has approximately 6200 (at the time of writing). Therefore, there is roughly one PSAP per 500,000 inhabitants in Sweden, and one per 50,000 in the United States. As all-IP infrastructure is rolled out, smaller PSAPs may be consolidated into regional PSAPs.

Emergency calls are primarily routed based on the emergency caller's location information. Routing may also take place in multiple stages, taking into account factors such as the number of available call-takers, the load situation of a PSAP, or the language capabilities of the call-takers. We expect that dynamic routing will be the predominant mechanism in the near future. In order to perform the initial location-based routing step closer to the PSAP, information about the service boundaries of PSAPs need to be known to end devices or to VoIP Service Provider (VSPs). This information has somehow to be exchanged and shared. Today, Excel sheets are used [14] to exchange phone numbers associated with PSAP service boundaries. A call-taker then has to manually analyze the situation, look up the appropriate phone number, and re-route the call. This may seem a reasonable initial step for the small number of transnational emergency calls that are misrouted in today's emergency services system, but it is not a future-proof approach for a next-generation IP-based emergency services system. The inability of emergency centers to quickly re-route calls due to an overload situation in a mass incident has already been shown in various incidents today.

2

Location: Formats, Encoding and Protocols

Location is one of the most important components in an emergency services system. First, it is needed for emergency call routing, since in most cases it is desirable to reach the PSAP that is closest to the emergency caller. Second, the PSAP needs precise location to dispatch first responders to the emergency scene.

There are two types of location formats: civic location and geospatial (sometimes also called geodetic) location information. Civic location describes location in a postal address-like form. The following address, for example, is a valid civic address for dispatching first responders to the Nokia Siemens Networks office: "Linnoitustie 6, 02760, Espoo, Finland, Building A, Floor 3." Geospatial information, however, expresses location in the form of latitude, longitude, and optionally altitude. The complexity of the represented geospatial location format varies depending on the described location shapes, which are often the result of the process for calculating location (a process usually referred as "positioning").

Typically, one would assume that civic and geospatial information is defined in such a way that it can be encoded for use with different underlying protocols without any impact on the semantics. Unfortunately, this is not quite how the standardization community approached it. Two types of encodings became popular: one encoding uses the Extensible Markup Language (XML), which can typically be found in protocols that use HTTP or SIP; and the other, a binary encoding, is used by size-constrained protocols (such as the Dynamic Host Configuration Protocol (DHCP)), or link layer protocols.

The binary encoding of location information has been reused by other groups and standards organizations. In addition to DHCP, the format has been reused for RADIUS and Diameter [15], by IEEE 802.11 (as described in section 3.4), and by the Link Layer Discovery Protocol – Media Endpoint Discovery (LLDP-MED).[1]

As described later in this chapter, there are subtle differences between location information conveyed in different protocols.

[1] Note that LLDP-MED is not described in this book since it is considered unsuitable for emergency services purposes. It only provides location of access points rather than end hosts.

Internet Protocol-Based Emergency Services, First Edition. Hannes Tschofenig and Henning Schulzrinne.

Table 2.1 Examples of civic address labels

Label	Description	Example
Country	The country is identified by the two-letter ISO 3166 code	United States
A1	National subdivisions (state, region, province, prefecture)	New York
A2	County, parish, gun (Japan), district (India)	King's County
A3	City, township, shi (Japan)	New York
A4	City division, borough, city district, ward, chou (Japan)	Manhattan
PC	Postal code	10027-0401
PLC	Place type*	School

*Since it often causes confusion, it is worthwhile to mention that the values that go into the "Place type" (PLZ element) are defined in a separate location types registry created by RFC 4589 [23]. The registry can be found in ref. [24].

To make it even more complicated, an emergency services architecture typically requires a sequence of interactions to take place, with different entities requesting location information. From a protocol point of view, it matters what information is available for a location query and who transmits it. Consequently, it is sometimes challenging to completely separate protocols from specific architectural considerations, at least in the way Standards Developing Organizations (SDOs) have approached these topics.

In this chapter we focus on the basic building blocks of location: civic and geospatial location formats, their encoding, and some of the most important protocols.

Civic Location Information

There is only one civic location information format that has been standardized for use with emergency services, although the specification is split into two parts defining different encodings (with the same expressiveness from a functionality point of view): RFC 4776 [16] defines the civic address tokens for use with DHCP (and also describes the encoding in a binary format), and RFC 4119 [17] defines the XML-based encoding. Note that there are other location formats standardized, for example, XEP-0080 [18], a specification for use with XMPP, or the W3C Geolocation API [19], a JavaScript-based API mostly used by Web applications, but it is not clear at the time of writing whether these other architectures with their location formats are indeed suitable for emergency services purposes.[2]

Table 2.1 shows examples of standardized civic address labels. After the initial specification work had been completed, it was noticed that extensions were needed, and those are defined in RFCs 5139 [21]. More recently, further specification work [22] has been started to define additional civic address extensions, such as mile posts and pole numbers, and discusses how future extensions can be dealt with, in a backwards-compatible manner.

Even while working on the civic address tokens for RFC 4119 and RFC 4776, it was clear that profiles of the civic address formats will be needed to reflect the usage of country-specific civic address naming conventions and to define rules on how to populate the different fields in a

[2] In ref. [20] Bernard Aboba and Martin Thomson investigate how emergency services functionality can be integrated in the ongoing work on Web Real-Time Communication (WebRTC).

well-defined way. RFC 5774 [25] provides guidance on how to create civic location profiles and contains address considerations for Austria as an example. In section 2.1 Ed Wells, an active NENA member, shares his experience on how to create such a profile in the case of the NENA Civic Location Data Exchange Format (CLDXF). We believe that his write-up will provide valuable insights for others who would like to go through a similar process in other countries.

Geodetic Location Information

The standardization community took two different directions for geodetic location information:

- *Binary Encoding.* RFC 3825 [26] defines a binary encoding of geodetic location information mainly for use with the Dynamic Host Configuration Protocol (DHCP), which was later revised by RFC 6225 [27].[3] A description of the binary location encoding in DHCP is provided in section 2.2 by Martin Thomson and Bernard Aboba. Martin and Bernard are active participants in the IETF Geopriv Working Group and co-authors of RFC 6225.
- *XML-Based Encoding.* The other direction in the work on geospatial location information was based on reusing the Geography Markup Language (GML). Unlike the binary format described in the previous paragraph, GML uses an XML encoding and offers more flexibility for expressing geospatial location information. This flexibility allowed a number of location shapes to be supported. Carl Reed from the Open Geospatial Consortium (OGC) describes the developments around GML in section 2.3. The recommendations given in RFC 5491 [28] cover the usage of GML for different location shapes when GML is used in a Presence Information Data Format – Location Object (PIDF-LO) and thereby updates RFC 4119, the initial PIDF-LO specification.

Location Protocols

So far, we have only discussed the location formats and their encoding. Often, the entity that knows the location (often called the Location Information Server) and the entity that wants to obtain location (called the Location Recipient) are not necessarily co-located in the same physical device.[4] Hence, there is a need for a protocol to request and retrieve location information. Depending on the purpose of the protocol interaction, different terminology is used, as explained below.

- *Location Configuration.* RFC 6280 [29] offers the following definition:[5] "A Location Configuration Protocol (LCP) is one mechanism that can be used by a Device to discover its own location from a Location Information Server (LIS). LCPs provide functions in the way they obtain, transport, and deliver location requests and responses between an LIS and

[3] The binary encoding has since been introduced into other protocols as well, as mentioned earlier. However, for use in emergency services the core specification only is described in this book.

[4] Specifying consistent terminology suitable for various different deployment scenarios to describe the full location life-cycle has always been difficult but an attempt was made with RFC 6280 [29]. This RFC also addresses privacy and security extensively.

[5] A problem statement and requirements document about Location Configuration Protocols was published with RFC 5687 [30].

a Device such that the LIS can trust that the location requests and responses handled via the LCP are in fact from/to the Target." The IETF Geopriv Working Group had published two types of LCPs, an LCP based on DHCP and another one based on HTTP. The DHCP-based solution descriptions are specified within refs. [16, 27] and the HTTP version is described in RFC 5985 [31]. James Winterbottom, one of the lead contributors of RFC 5985, explains the HTTP-based LCP in section 2.4. The DHCP-based LCP is described in section 2.2.
- *Location Distribution.* RFC 6280 [29] also defines the term "Location Distribution" and refers to the interaction of a Location Server that distributes location objects to Location Recipients. Examples of these protocols are SIP Location Conveyance [32], the SIP event notification mechanism [33] (potentially in combination with location filters [34]), or HELD with the third-party location identity extension [35]. For emergency service call setup procedures, the location distribution mechanism is a core component and will therefore be covered in Chapter 3 on emergency services architectures.

The location protocol-specific work by 3GPP and OMA also standardizes positioning related protocol extensions specific for their radio technologies in addition to location configuration and location distribution protocols. Jan Kåll, a 3GPP standardization professional, describes the location protocol work done by the 3GPP in section 2.6, and Khiem Tran, who is active in the OMA, explores prior OMA location protocol work in section 2.5.

2.1 Applying the PIDF-LO civicAddress Type to US Addresses

Ed Wells
Technical Editor [6]

This section[7] describes the NENA Next Generation 9-1-1 US Civic Location Data Exchange Format (CLDXF) Standard, which was created by NENA to support the exchange of US civic location data about 9-1-1 calls, and to do so in a format consistent with the international civicAddress type defined within the IETF's Presence Information Data Format – Location Object (PIDF-LO) standard. The section is presented in three parts:

1. *Introduction: The Context and Purpose of PIDF-LO and CLDXF.* A summary of the context and purpose of CLDXF and its relation both to PIDF-LO and a third address data standard, the US Thoroughfare, Landmark, and Postal Address Data Standard, which was developed concurrently and endorsed by the US Federal Geographic Data Committee (FGDC) [36].
2. *CLDXF Elements.* A description of the scope of each group of CLDXF address elements; the PIDF-LO elements used within it; the US context in which the elements were applied;

[6] On behalf of: CLDXF Work Group (WG), Data Structures Subcommittee, Core Services Committee, National Emergency Number Association (NENA).
[7] This section was finished in September 2012, before NENA finalized the CLDXF standard. It is based on the second public-review draft of the CLDXF. Because this section is intended to present the work of the CLDXF WG, it quotes and paraphrases extensively from the draft CLDXF.

key details about the individual elements as defined and used within the CLDXF; and the key issues resolved in aligning them with their corresponding PIDF-LO elements.
3. *Conclusion.* A conclusion summarizing the issues encountered and resolved in applying the PIDF-LO civicAddress elements to US addresses, the overall suitability of the PIDF-LO civicAddress type as a basic framework for exchanging US address data, and the usefulness of profiles as a formal mechanism for aligning standards.

2.1.1 Introduction: The Context and Purpose of PIDF-LO and CLDXF

2.1.1.1 IETF, PIDF-LO, and the Civic Address (CA) Data Structure

The Internet Engineering Task Force (IETF) Geopriv Working Group creates Internet standards for both communicating location and protecting the privacy of location information. Because the Internet operates worldwide, the Geopriv standards must accommodate location information worldwide. Finding no existing location data standards that could be used worldwide for a wide variety of applications (including emergency calling), the Geopriv Working Group undertook to create a general XML-based location data structure.

The Geopriv work builds on a more general XML data structure used to provide "presence" information – information about a person's current state. The general Presence Information Data Format (PIDF) was originally defined by IETF RFC 3863 [37]. RFC 4119 [17] extended RFC 3863 to define the location of a "presentity" as a Presence Information Data Format – Location Object (PIDF-LO). It defined two Location Objects (PIDF-LO): geographic coordinates (PIDF-LO Geo), and civic addresses (PIDF-LO CA). RFC 5139 [21] updates RFC 4119 and developed the civic address work further to define a civicAddress type. Currently, the Geopriv Working Group is finalizing their work on additional civic address extensions (see ref. [22]). The fields in the civicAddress type are intended to handle all civic addresses worldwide.

2.1.1.2 NENA and the NG9-1-1 CLDXF

The National Emergency Number Association (NENA) is a professional organization solely focused on 9-1-1 policy, technology, operations, and education issues. Among many key activities, NENA works to facilitate the creation of an IP-based Next Generation 9-1-1 (NG9-1-1) system, including the establishment of NG9-1-1 standards. NENA has based its Next Generation 9-1-1 emergency calling architecture [38, 39] on the IETF emergency services architecture [40].

To provide a means for Public Safety Answering Points (PSAPs) to exchange US 9-1-1 call location information, the NENA Core Services Committee/Data Structures Subcommittee/CLDXF Work Group (WG) has created the Next Generation 9-1-1 (NG9-1-1) Civic Location Data Exchange Format (CLDXF) Standard. The CLDXF is one component of the larger suite of NG9-1-1 standards.

The first public-review draft of the CLDXF was posted for public comment in May 2011. All comments were adjudicated and a second public-review draft was posted in September 2012. Adoption is expected thereafter.

2.1.1.3 CLDXF as a Profile of IETF PIDF-LO

To support information exchange across national boundaries, the CLDXF civic location address is generated in the IETF PIDF-LO civicAddress type. As such, the CLDXF standard conforms to the framework set by the PIDF-LO civic address standard.

Formally, the CLDXF is the US profile of the IETF PIDF-LO civic address standard. The CLDXF restricts the PIDF-LO by omitting certain elements not found in US addresses, and by narrowing the PIDF-LO element definitions to make them United States-specific. In addition, the CLDXF extends the PIDF-LO by inclusion of an additional, United States-specific, element.

2.1.1.4 CLDXF Purpose

The NENA NG9-1-1 CLDXF standard supports the exchange of US civic location address information about 9-1-1 calls, both within the United States and internationally. The standard covers civic location addresses within the United States, including its outlying territories and possessions. The CLDXF standard defines the detailed data elements needed for address data exchange.

As a data exchange standard, the NENA NG9-1-1 CLDXF is not intended to support civic location address data management. It is assumed that address information will be transmitted call by call, as part of the call record, and that any local address data repository would be external to the call information. Therefore the standard does not provide for an address identifier, address metadata, or address data quality tests.

2.1.1.5 CLDXF and the US Federal Geographic Data Committee (FGDC) Address Data Standard

Within the United States, concurrently with the CLDXF WG's development of the CLDXF, the Address Standard Working Group (ASWG) developed the US Thoroughfare, Landmark, and Postal Address Data Standard for the FGDC [36]. The ASWG was formed by the Urban and Regional Information Systems Association (URISA) in 2005, and it worked under the authority of the US Census Bureau and the FGDC. (An early draft of the FGDC standard provided material for the IETF Geopriv Working Group in creating the initial version of the PIDF-LO civicAddress type.) The FGDC endorsed the standard in February 2011. The FGDC has posted a PDF version of the standard for free public use at [41].[8]

The FGDC standard is intended to support address data management and exchange to meet the needs of local address authorities, emergency service providers, postal authorities, and regional and national address aggregators. The FGDC standard defines address data content, attributes, and address-specific metadata; address classes; address data quality tests; and an XML Schema Definition (XSD) for address data exchange. The CLDXF WG and the ASWG have worked closely together to align the CLDXF and the FGDC standard as closely as possible within the constraints of their respective purposes. The two Working Groups have

[8] A searchable version, funded by an FGDC grant, is available from http://www.spatialfocus.com/twiki/bin/view/SpatialFocus/SearchableFGDCAddressStandard. It will be moved to the FGDC website soon.

prepared a profile, to be issued with each standard upon their adoption, that details the precise relationship between the two standards.

2.1.1.6 CLDXF Content

The CLDXF standard breaks civic addresses into simple components, each constructed as a simple XML element. For each element, the CLDXF standard gives: the element name; corresponding PIDF-LO element name; definition; definition source; examples; data type; domain of values (if any); whether the element is mandatory, conditional or optional; the minimum and maximum occurrences permitted; and, finally, explanatory notes. The elements are presented in six groups:

1. Country, State, and Local Place Names Elements
2. Road Name Elements
3. Address Number Elements
4. Landmark Name Element
5. Subaddress Elements
6. Place-Type Element (Address Attribute)

The next part of this section describes the elements in each group. It sets forth the scope of each group and the PIDF-LO elements used within it; the US context in which the elements were applied; key details about the individual elements as defined and used within the CLDXF; and the key issues resolved in aligning them with their corresponding PIDF-LO elements.

2.1.2 CLDXF Elements

2.1.2.1 Country, State, and Local Place Name Elements

Scope
Country, state, and local place name elements specify the general vicinity within which an address is located.

PIDF-LO Elements Used in CLDXF for Country, State, County, and Local Place Name Elements
CLDXF uses PIDF-LO country for country names, A1 for state names, A2 for county names, A3 for municipality names, A4 for unincorporated community names, A5 for unincorporated neighborhood names, PCN for postal community names, and PC for ZIP/Postal Codes. CLDXF does not use PIDF-LO's A6 element.

US Context for Country, State, County, and Local Place Name Elements
Place names denote areas within which individual street addresses and landmarks are found. Place names give the general location of an address, and are often needed to distinguish between identical road name and address number combinations in a given area. Although place names may seem as simple as "city–state–postcode," within the United States place names often turn out to be surprisingly ill-defined and confusing.

Confusion occurs because, within the United States, place names arise from three different processes: legislative, postal, and unofficial. Each process is independent of the others, and all three are useful and sometimes necessary to locate an address. To provide for all three processes, the CLDXF includes eight place name elements:

- Legislative – Country, State, County, Incorporated Municipality
- Postal – Postal Community, Postal Code
- Unofficial – Unincorporated Community, Neighborhood Community

The Country, State, County, Incorporated Municipality elements (the four elements for legislated place names) must be given in every address record. The Postal Community and Postal Code are optional, but strongly recommended if they have been assigned to the address by the US Postal Service.

Unofficial place names are optional, but should be given as needed. Incorporated municipalities, unincorporated areas, and postal delivery areas can be quite extensive, and often contain duplicate road names and address ranges. In these cases, unofficial place names may be useful or necessary in specifying more precisely where an address is located, and in differentiating between similar addresses in the same general area.

Legislative places are created by law. Because they create taxing and police jurisdictions, they are well-documented and precisely mapped. Every address in the United States lies within a state, county, and (within a county) either an incorporated municipality or the unincorporated portion of the county. The four elements for legislative place names are as follows.

1. Country (e.g., "US") – The two-letter abbreviation for the country name, as given in the ISO 3166-1 standard [42].
2. State (e.g., "WA") – The two-letter abbreviation for the state (or territory, or federal district) name, as given in USPS Publication 28, Appendix B [43].
3. County (e.g., "King") – The name of the county (or equivalent thereof – terms and forms of government vary greatly among the states and territories), as given in the official list maintained by the US Census Bureau.
4. Incorporated Municipality (e.g., "Seattle") – The name of a city, town, or other incorporated local government (if any). For addresses in unincorporated areas, enter "Unincorporated" for this element.

Postal places are denoted by the Postal Community and Postal Code elements. Within the United States, the US Postal Service (USPS) has sole authority to recognize Postal Community names and assign their corresponding ZIP Codes. The USPS assigns and changes them as required for efficient mail delivery, without regard to county or incorporated city boundaries. Thus postal places should be assumed to include a different set of addresses than legislative places, even if the postal and legislative place names are the same. In addition, the following should be noted.

1. The USPS City State File is the authoritative register of US postal place names and their associated ZIP Codes. The USPS City State File often recognizes multiple place names for a given ZIP Code. In all cases, one name is preferred (the "actual city" name in

USPS terminology) and the other(s) are acceptable ("other acceptable cities" in USPS terminology). Both the preferred and acceptable names may be used within CLDXF.
2. Strictly speaking, Postal Communities are not areas, but collections of delivery points; the USPS has never defined ZIP Code "boundaries."
3. Either the five-digit ZIP Code or the nine-digit ZIP+4 Code may be given for the Postal Code element.

Unofficial place names include the hamlets, neighborhoods, subdivisions, shopping districts, crossroads, and other locales within large municipalities and unincorporated areas. These place names come into existence through informal usage and recognition. The places have no precise definition in US place name geography, they have no general powers of government, and they are not controlled or registered by any authority. Their definition and boundary are often imprecise. Nevertheless, unofficial place names can be useful and necessary in locating an address, because incorporated municipalities, unincorporated areas, and postal delivery areas can be quite extensive and often contain duplicate road names and address ranges. Unofficial place names should be included whenever they might be helpful in getting first responders to the right place faster. The CLDXF therefore includes two elements for unofficial place names:

1. Unincorporated Community (e.g., "Harlem," in New York City) – for areas on the scale of a community, ward, borough, village, hamlet, and so on, or larger; and
2. Neighborhood Community (e.g., "Cypress Meadows," a subdivision near Tampa) – for areas smaller than an Unincorporated Community, on the scale of a neighborhood or small shopping district.

Country, State, and Local Place Name Element Descriptions

- *Country.* Countries are shown by their ISO 3166-1 two-letter abbreviation.
- *State.* The United States is divided administratively into 50 states, the federal District of Columbia (DC), and five territories. They are represented by their two-letter postal abbreviations (USPS Publication 28 [43], also found in ISO 3166-2).
- *County.* All US states and territories are divided completely into counties or county equivalents. Thus every US address is in a county or county equivalent. Every CLDXF record must include the name of the county or county equivalent where the address is located. (County "equivalents" are recognized because of the differing jurisdictional structures and terminologies established by the states, territories, and DC.) The authoritative reference for counties and county equivalents and their names is the ANSI INCITS 31:2009 standard. It is maintained by the US Census Bureau [44].
- *Incorporated Municipality.* Within the United States, an incorporated municipality is a local government created under state law with general local police and taxation powers. Most but not all incorporated municipalities lie within one county. In most states, incorporated municipalities need not comprise the whole of a county. Areas outside of any incorporated municipality are called unincorporated areas.

  The CLDXF requires that a municipality name be included with every address record. If the address is in an unincorporated area, then "Unincorporated" should be entered for this item. There is no authoritative register of US incorporated municipalities.

The legal terms denoting incorporated municipalities vary from state to state (e.g., city, town, borough, village, township). The differences are not important in constructing CLDXF records.

- *Postal Community.* The US Postal Service (USPS) uses place names to identify post offices. The Postal Community name frequently differs from the Incorporated Municipality name and Unincorporated Community name associated with an address.

  The USPS often recognizes multiple names for a given post office and its postal code. Any of the names is acceptable in a CLDXF record. The USPS City State file shows what names are acceptable as Postal Community names.
- *Postal Code.* To expedite mail delivery, the USPS assigns a five-digit ZIP Code to each USPS post office or delivery station. The ZIP Code can be extended by a hyphen and four additional digits that identify a specific delivery address range (ZIP+4 Code). The USPS does not provide mail service to all addresses, and has not assigned ZIP Codes to all addresses. Therefore, the Postal Code is optional, but strongly recommended if the USPS has assigned a ZIP Code to the address. Either the five-digit ZIP Code or the nine-digit ZIP+4 Code is accepted.
- *Unincorporated Community.* Unincorporated communities include small hamlets, neighborhoods, subdivisions, and other portions of unincorporated areas; and neighborhoods and other unofficial divisions of incorporated municipalities. Because both unincorporated areas and municipalities can be quite extensive, Unincorporated Community names can be useful in locating an address. The Unincorporated Community name may be used as needed to indicate where within a county or an incorporated municipality an address is located.

  Unincorporated communities comprise a wide variety of settlements within the United States. Because they are unofficial, their definition and boundary are often imprecise. If an unincorporated community does not have a precise, mapped boundary, some description or rule should be given for determining whether an address is in the community.

  The differences between Unincorporated Community, Neighborhood Community, Landmark Name, and Building are not always clear. Some general distinctions are given in section 2.1.2.4.
- *Postal Code.* The Neighborhood Community, like the Unincorporated Community element, denotes an area defined outside of legislative or postal administrative processes. The Neighborhood Community and Unincorporated Community differ in the scale of the place they name. A Neighborhood Community has the scale of a subdivision or small commercial area. An Unincorporated Community has the scale of a community, ward, borough, village, hamlet, and so on, or larger.

  A CLDXF record may include an Unincorporated Community name, a Neighborhood Community name, both, or neither. These elements should be used to indicate as clearly as possible where addresses are located, especially when duplicate road names and address ranges occur within the area.

  The differences between Unincorporated Community, Neighborhood Community, Landmark Name, and Building are not always clear. Some general distinctions are given in section 2.1.2.4.

Place Name Issues

Substantial discussion was required to determine how to apply PIDF-LO's hierarchical place name elements (Country and A1 through A5, plus PCN and PC) to the non-hierarchical, overlapping, decentralized place names, both official and unofficial, used in US addresses.

The FGDC standard, by contrast, provides a non-hierarchical place name structure: an address may include any number of Place Names, which together constitute a Complete Place Name, with place name attributes indicating the order in which they should be listed and the type of place each separate name represents.

2.1.2.2 Road Name Elements

Scope
Road name elements give the parsed components of the road name within an address.

PIDF-LO Elements Used in US Street Names
1. The CLDXF uses the following PIDF-LO road name elements: PRM (Street Pre-Modifier), PRD (Leading Street Direction), RD (Street Name), STS (Street Type Suffix), POD (Trailing Street Direction), and POM (Street Post-Modifier).
2. The CLDXF also uses a civic address extension element, STP (Street Type Prefix).
3. The CLDXF uses one United States-specific extension element that is not found in PIDF-LO, but which is required to represent certain US road names: Street Type Prefix Separator.
4. The CLDXF does not use the following PIDF-LO elements because they are not found in US road names: RDSEC (Road Section), RDBR (Road Branch), and RDSUBBR (Road Sub-Branch).

US Context for Street Name Elements
The CLDXF follows PIDF-LO in requiring that complete road names be parsed into their component simple elements. The CLDXF defines and describes eight simple elements needed for parsing any complete road name in the United States. Each word or phrase of a road name fits into one of the elements. When the elements are reassembled in order, the complete road name is reconstructed. If each element is correct, then the complete road name will also be correct. The eight elements are as follows:

1. Street Pre-Modifier (e.g., "Alternate" in Alternate Route 8; "The" in The Oaks Drive)
2. Leading Street Direction (e.g., "North" in North Fairfax Drive)
3. Street Type Prefix (e.g., "Avenue" in Avenue A; "County Route" in County Route 88)
4. Street Type Prefix Separator (e.g., "of the" in Avenue of the Americas)
5. Street Name (e.g., "Fairfax" in North Fairfax Avenue)
6. Street Type Suffix (e.g., "Avenue" in North Fairfax Avenue)
7. Trailing Street Direction (e.g., "East" in Seventh Street East)
8. Street Post-Modifier (e.g., "Extended" in East End Avenue Extended)

Basic Parsing Rules for US Road Names. Parsing is the process of resolving a road name into its component simple elements. There are two basic rules for parsing US road names:

1. The eight elements, if they appear in a complete road name, always appear in the following order: Street Pre-Modifier, Leading Street Direction, Street Type Prefix, Street Type Prefix Separator, Street Name, Street Type Suffix, Trailing Street Direction, Street Post-Modifier.
2. A Street Name element must be given before any of the other seven elements can be given. All of the other elements are optional. (No road name is likely to include all eight elements.)

Street Pre-Modifier and Street Post-Modifier. A word or phrase that precedes or follows the Street Name but is separated from it by a Leading or Trailing Street Direction, Street Type Prefix or Suffix, or both. In addition, words such as "The" or "Old" may, at the discretion of the local addressing authority, be classed as Street Pre-Modifiers, if they are placed outside the Street Name element so that the Street Name element can be used in creating a sorted (alphabetical or alphanumeric) list of road names. A Street Pre-Modifier precedes the Street Name; a Street Post-Modifier follows it. Street Pre-Modifiers and Street Post-Modifiers are unusual in US road names.

Road Name Issues

During the drafting process, the PIDF-LO civicAddress type was found not to include three elements needed for systematic parsing of road names: Street Pre-Modifier, Street Type Prefix, and Street Post-Modifier. Street Pre-Modifier and Street Post-Modifier were added by RFC 5139. A work in progress IETF draft [22] adds the Street Type Prefix as a civic extension to PIDF-LO.

A fourth required element, Street Type Prefix Separator, has not been added to PIDF-LO. It is included in CLDXF as a United States-specific extension of the PIDF-LO civicAddress type. Ref. [22] provides a means for declaring local extensions within PIDF-LO.

2.1.2.3 Address Number Elements

Scope

Address number elements show where along a road a feature is found.

PIDF-LO Elements Used in US Address Numbers

CLDXF uses the following PIDF-LO elements: HNP (Address Number Prefix), HNO (Address Number), HNS (Address Number Suffix), and MP (Milepost).

US Context for Address Number Elements

Address numbers indicate, by sequence and parity, where along a thoroughfare the numbered feature is found. Within the United States, address numbers are typically integer numbers assigned sequentially along a thoroughfare, with even numbers on one side and odd numbers on the other side, according to one of three rules:

1. by distance (e.g., 1000 numbers per mile, or one per 10.56 feet of road frontage on either side);
2. by blocks (e.g., starting a new 100 range at each cross-street, where the cross-streets are closely and regularly spaced as in a city grid);
3. sequentially, assigning numbers consecutively to features along the road (this system is little used because of the problems that arise when features are later divided and intermediate numbers are needed).

In the CLDXF, address numbers are broken into three component elements: Address Number Prefix, Address Number, and Address Number Suffix. A fourth element, Milepost, may be given in place of or in addition to the address number.

Milepost signs are typically posted along federal, state, and county roads, such as interstate highways, state routes, and county roads. Within the Unites States, mileposts are rarely used as address numbers, except in areas of Puerto Rico, where they are commonly used as assigned addresses ("Km 2.7, Carretera 175").

However, mileposts are often used by road maintenance agencies to locate road infrastructure such as interchanges, bridges, and emergency call boxes. They are also often used by emergency agencies to give the location of incidents along roads with milepost signs, such as crash sites and brush fires. They are sometimes used by the public to give the locations of roadside businesses, farm entrances, inspection stations, roadside rest stops, railroad crossings, park and campground entrances, recreational vehicle parks, and truck stops and other features along limited-access highways and in areas where address numbers have not been assigned. When used, they are functionally and semantically similar to address numbers, so they are grouped with the address number elements within the CLDXF.

Address Number Element Descriptions

Address Numbers and Mileposts are optional in the CLDXF, because there are roads in the United States where neither has been created.

Address Number. The integer portion of the address number. The Address Number must be an integer (no fractions, no decimals, no leading zeros, no text or punctuation) to support sequential sorting, tests for even/odd party, and in/out of range tests. An Address Number is required before an Address Number Prefix or Suffix can be created.

Address Number Prefix. Alphanumeric text prepended to the Address Number. In a few parts of the United States, alphanumeric extensions precede the Address Number (e.g., "A" in A100; "5-" in 5-143; "194-0" in 194-03; "N89W167" in N89W16758; "0" in 0121). In these cases, the letters, numbers, hyphens, and leading zeros that precede the Address Number integer all comprise the Address Number Prefix. Address Number Prefixes are known in only five areas of the United States, although they may be found in other areas:

1. In Puerto Rico, letters are often placed before the Address Number element (e.g., A107, A109, etc.). The letter ("A") is an Address Number Prefix.
2. In parts of New York City, and parts of Hawaii, address numbers are hyphenated (e.g., 4-09, 4-11, etc.). The hyphen, everything preceding the hyphen, and a leading zero if it immediately follows the hyphen ("4-," "4-0"), comprise an Address Number Prefix.
3. In certain municipalities in Wisconsin and northern Illinois, a map grid reference is placed before the Address Number (e.g., N6W32 1603). The map grid reference ("N6W32") is an Address Number Prefix.
4. In one area of Portland, Oregon, the Address Numbers have leading zeros (e.g., 0121, which indicates a different location than 121). The leading zero ("0") is an Address Number Prefix.

Address Number Suffix. Fractions or letters sometimes appended to Address Numbers. Many jurisdictions include a few address numbers with alphanumeric extensions (e.g., 123 A; 123 1/2). The extensions ("A," "1/2") are Address Number Suffixes. The Address Number Suffix includes the hyphen, if present.

Milepost. A distance traveled along a route such as a road or highway, typically indicated by a milepost sign (e.g., "Milepost 1303" in Milepost 1303, Alaska Highway). The word ("Milepost" or its equivalent) and the number (1303, in this example) are combined in a single text element. A Milepost may or may not correspond to an assigned address number. A single address record may have a Milepost element, an Address Number element, or both.

Address Number Issues

During the drafting process, the PIDF-LO civicAddress type was found not to include two required address elements: Address Number Prefix and Milepost. Ref. [22] adds them as civic extensions to PIDF-LO.

Substantial discussion was required to determine whether and how to incorporate milepost measures into CLDXF. The CLDXF WG chose to use the PIDF-LO Milepost element.

The Milepost element combines the "milepost" word with the numeric distance value into a text element. Without numeric values, sequence and distance calculations between Mileposts become unnecessarily cumbersome. An alternative, used in the FGDC standard, is to treat the "milepost" word as an Address Number Prefix, and the milepost integer as an Address Number. (Decimal values, including the decimal point itself, can be placed in the Address Number Suffix.)

2.1.2.4 Landmark Name Element

Scope

A landmark name is the name by which a prominent feature is publicly known (e.g., the Empire State Building).

Applying PIDF-LO to US Landmark Names

The CLDXF uses the PIDF-LO LMK element to represent landmark names.

US Context for Landmark Name Element

A landmark name specifies a location by naming it. Within CLDXF the name is given in the Landmark Name element. A Landmark Name is optional unless no Street Name is given, in which case a Landmark Name is required.

The name by itself does not relate the landmark to any street system or coordinate reference system and therefore provides no information about where to find the feature. Landmarks are therefore often associated with street addresses, and address records often include a both a Landmark Name and a street address. A Landmark Name can be associated with a street address simply by including the Landmark Name in the record with appropriate road name and address number elements.

Within the CLDXF, a Landmark Name may be given either in place of, or in addition to, a street address. Thus both of the following addresses are acceptable within CLDXF:

1. Statue of Liberty National Monument, Liberty Island, New York, NY 10004 (includes a Landmark Name but no street address)
2. The White House, 1600 Pennsylvania Avenue NW, Washington, DC 20500 (includes both a Landmark Name and a street address)

The differences between the Landmark Name, Unincorporated Community, Neighborhood Community, and Building elements are not always clear and distinct. The following should be considered as general principles:

1. A landmark is under a single use or ownership or control, while neighborhoods and communities are not. Thus a neighborhood or a community generally includes numerous separate addresses, while a landmark, even if it covers an extensive area, might be considered to be a single "master address" (often containing multiple subordinate landmarks).
2. If a building is identified by a name that is unique within the community, the name should be placed in the Landmark Name element (e.g., "Empire State Building," "Ohio State University, Derby Hall").
3. If a building is identified by a number or letter that distinguishes it from others at the same address, the identifier should be placed in the Building element (e.g., "Terminal 3" in John F. Kennedy International Airport, Terminal 3; "Building A" in 456 Oak Street, Building A, Apartment 206).

These general principles apply to most cases and are useful as general distinctions, but exceptions and marginal cases are easily found.

Landmark names are given to both natural and man-made features. In general, names of natural landmarks are not used in addresses and are therefore excluded from the scope of CLDXF.

Landmark Name Element Description

One element, Landmark Name, has been defined to hold landmark names. It is a free-form text field. Landmark names often denote extensive areas, which may contain smaller named landmarks (e.g., individual buildings within a college campus). In such cases one landmark name may function as a single "master address" containing multiple subordinate names. If there are multiple landmark names within this data element, they must be separated with a pipe "|" symbol. In such cases, order the names from largest geographic area to smallest.

Landmark Name Issues

The pipe symbol may cause problems if a CLDXF output file is read into a pipe-delimited file format. An alternative, used in the FGDC standard but not in CLDXF, would be to allow multiple landmark names in this element, separated by tags. RFC 5139 [21] permits multiple names within the Landmark Name element, and this change will likely be after public review of the current CLDXF draft.

2.1.2.5 Subaddress Elements

Scope

Subaddress elements identify separate portions within an addressed feature, such as floors, apartments, suites, rooms, and seats.

PIDF-LO Elements Used in US Subaddresses

The CLDXF uses the following PIDF-LO elements to represent subaddresses: BLD, FLR, UNIT, ROOM, SEAT, and LOC.

US Context for Subaddress Elements

Subaddresses occur within a wide variety of residential and commercial buildings, from single basement apartments to multi-structure office parks, as well as countless specialized structures and complexes such as airports, piers, warehouses, manufacturing plants, hospitals, parking garages, stadiums, and military bases. The CLDXF follows the PIDF-LO civicAddress type in providing a structured set of six elements to hold subaddress information: Building, Floor, Unit, Room, Seat, and Additional Location Information.

The differences between the Building, Landmark Name, Neighborhood Community, and Unincorporated Community elements are not always clear. Some general distinctions are given in section 2.1.2.4.

Subaddress Element Descriptions

A subaddress is a separate, identifiable portion of an addressed feature, the whole of which is identified by a landmark name or a road name, usually in conjunction with an address number.

CLDXF defines the following subaddress elements, which may be used in any combination as needed:

1. Building – one among a group of buildings that have the same address number and road name.
2. Floor – a floor, story, or level within a building.
3. Unit – a group or suite of rooms within a building that are under common ownership or tenancy, typically having a common primary entrance.
4. Room – a single room within a building.
5. Seat – a place where a person might sit within a building (e.g., a cubicle in a large room).
6. Additional Location Information (AddLoc) – a part of a building or subaddress that is not a Building, Floor, Unit, Room, or Seat (e.g., a building wing or corridor).

Subaddress Issues

The Building–Floor–Unit–Room–Seat–AddLoc hierarchy poses some difficulties for implementation. Not all site and building subaddress components fit easily into this set of six elements, and the elements can be difficult to distinguish in practice. Specifically, the following difficulties arise:

1. Not all interior spaces are structured into a Building–Floor–Unit–Room–Seat hierarchy (e.g., airport terminals, hospital wings, transit terminal bays and platforms, factory production lines and stations).
2. The types can be difficult to distinguish in practice. A large hotel may include towers, meeting and lobby areas, basement parking levels and ramps, utility subbasements, and elevator shafts for multiple towers. Is it one building or many? Do the six steps up from the lobby to the bar create a new "floor? The hotel rooms may include a bathroom with a closeable door. That would make it a "unit," not a room, even if they are called "rooms"

within the hotel. The difficulties are likely to be greatest in the largest, most complex, buildings, where subaddresses matter the most.

An alternative construct would discard the six existing elements in favor of one new one, an "interior" element comprising:

1. a "type" value (e.g., room, suite, concourse, etc.);
2. an "identifier" value (e.g., Room "306," "Presidential" Suite, Concourse "A"); and
3. an attribute indicating whether the identifier follows or precedes the type value.

Any word or phrase (or none at all) could be used as a type; multiple type-identifier pairs could be included if needed (e.g., "Concourse A, Gate 32"), and there would be no requirement that anything be classified as a Building, Floor, Unit, Room, Seat, and so on. This construct is used in the FGDC standard and it was proposed (but not adopted) for PIDF-LO in 2010 [45].

2.1.2.6 Place-Type Element (Address Attribute)

Scope

The Place-Type element identifies the type of feature at an address (e.g., residence, business, hospital, etc.). The Place-Type element is not part of the address. It is an attribute of the address.

PIDF-LO Element Used for US Place-Type Attribute

The CLDXF uses the PIDF-LO PLC element to represent Place-Type values. Allowable Place-Type values are restricted to terms given in the IETF Location Types Registry. The Location Types Registry is intended to provide a standard set of categories and terms for "describing the types of places a human or end system might be found" (IETF RFC 4589 [23]).

US Context for Place-Type Attribute

Identifying the type of feature present at a particular address is often valuable to emergency responders and therefore Place-Type is included in CLDXF as an optional element.

Place-Type Attribute Description

Place-Type is a text element. Allowable CLDXF values are restricted to those found in the IETF Location Types registry.

Often, one address contains multiple types of places (e.g., "restaurant" and "bar"), or is used to identify several types of features (e.g., parcel, building, building entrance, utility meter, utility pole, incident location, etc.) that occur at the same location. Per RFC 4589 [23], if multiple Place-Type values apply to an address, then the Place-Type element should be included multiple times within the record, containing a different value each time.

Place-Type Issues

The Location Types Registry is intended for use with other types of locations in addition to addresses, such as mobile computing environments, so it includes Place-Type terms that do not pertain to address features, such as "airplane," "bicycle," "automobile," "truck," "bus," "watercraft," and "underway" (a vehicle in motion).

The registry provides (as of September 2012) 44 terms, with informal definitions and examples. However, the terms are neither exclusive nor exhaustive (i.e., the terms overlap with each other to an undefined extent, and they do not cover all possible types). No data model or principles of classification are given, so users have little guidance on how to handle ambiguous cases, or when a new term is needed.

The NENA CLDXF Working Group considered creating an exhaustive, exclusive, systematic, formally defined set of Place-Type categories, but concluded that no such classification system could be created. The relevant categories depend on the purpose of the classifier. For example, fire fighters may classify places according to the structural characteristics of a building (low/high-rise; wood-frame, steel-frame, brick, etc.), while police may classify places according to the nature of the activities occurring at an address (residential versus office; bar versus restaurant; big-box retail versus convenience store; etc.), and emergency medics, telephone service providers, other utilities, address administrators, city planners, social service providers, and others all may have other concerns and points of view. Clearly, no single system can meet all of these purposes when each purpose implies a different logical basis for defining categories. A set of categories that suits one purpose will be ambiguous or incomplete when used for a different purpose.

As a result, the Location Types Registry as currently defined will probably prove impossible to develop into a single, complete, systematic, all-purpose taxonomy. Place-Type is a useful attribute, but the registry will have to be revised and improved over time if it is to serve its intended purpose. One possible avenue of development would be to define a set of purposes, and then define specialized location-type registries for each purpose.

Currently, the Place-Type element serves as a place-holder and marker of a problem to be solved. Users are encouraged (i) to make use of the attribute as needed, and (ii) to add locally defined values to the international registry, to provide material for further development.

2.1.3 Conclusion

The CLDXF WG has applied the PIDF-LO framework and the civicAddress type to the task of providing a structure for the exchange of civic location information about 9-1-1 calls. The CLDXF WG encountered and resolved several issues in applying the PIDF-LO civicAddress type to US addresses:

1. The PIDF-LO place name elements A1 through A6 are defined as a hierarchy. US place names are non-hierarchical. After substantial discussion, the CLDXF WG was able to define the A1 through A5 elements, along with the Postal Community element, so as to accommodate US place name practices.
2. The PIDF-LO civicAddress type as defined in RFC 5139 [21] is missing four elements needed to represent certain components of US road names and address numbers: Street Type Prefix, Street Type Prefix Separator, Address Number Prefix, and Milepost. Ref. [22] adds three of the elements as civic extensions to PIDF-LO. It would also provide a means for declaring the fourth (Street Type Prefix Separator) as a United States-specific extension within CLDXF.

3. The proposed PIDF-LO Milepost element is structured as text, with no numeric value to support sequence and distance calculations. It has been included in the CLDXF as defined in ref. [22].
4. The PIDF-LO subaddress element hierarchy (Building–Floor–Unit–Room–Seat–AddLoc) does not fit all buildings or sites, and the different elements can be difficult to distinguish in practice. The CLDXF includes the elements as defined in PIDF-LO, with user notes that state the limitations.
5. The Location Types Registry does not now provide a systematic taxonomy of Place-Type values. The CLDXF WG considered proposing such a taxonomy, but concluded that CLDXF serves too many diverse purposes to be served by one single taxonomy. User notes state the limitations and encourage further development of the Location Types Registry.

The CLDXF WG work to date shows that PIDF-LO civicAddress type (with the extensions defined in ref. [22]) provides a basic framework that can be adapted to meet US NG9-1-1 address data needs. It provides for all of the elements needed for complete exchange of information about any US address found in a 9-1-1 call record.

The CLDXF was created as a formal profile of PIDF-LO, and it incorporates a second profile (Appendix A) that relates CLDXF in complete detail to the US FGDC US Thoroughfare, Landmark, and Postal Address Data Standard [36]. In this context, profiles have proven to be a flexible and powerful mechanism for aligning different address data standards while reconciling the differences imposed by the constraints of their different business purposes. As such, profiles have provided a means by which NENA was able to work in close coordination with IETF and ASWG so that each standard strengthens the others, and the needs of the entire address data community are better served.

The CLDXF WG looks forward to the application of the PIDF-LO civicAddress type within other countries. As other countries develop their own profiles, the resulting experience should strengthen the PIDF-LO standard as a whole.

2.2 DHCP as a Location Configuration Protocol (LCP)

Martin Thomson and Bernard Aboba
Skype

RFC 3825 [26] defines a binary encoding of geodetic location information for usage with the Dynamic Host Configuration Protocol (DHCP). The Geopriv WG [46] worked on a revision to RFC 3825, which has been published as RFC 6225 [27]. The revision provides for configuration of geospatial location (latitude, longitude, and altitude) for both DHCPv4 [47] and DHCPv6 [48] clients. The focus of this text is on the binary encoding of geodetic location information. The XML-based encoding of geodetic location information, which offers much richer expressiveness and a larger number of location shapes, can be found in section 2.3.

Note that civic address configuration for DHCP clients is defined in a separate specification, namely RFC 4776 [16]. As described in Chapter 2 there is a common Civic Address (CA) label types registry established with RFC 4119 [17], which is used in distributing location information in DHCP, in IEEE link layer protocols, in RADIUS and Diameter, as well as in XML-based containers (like a PIDF-LO). The Civic Address (CA) label structure is described in Chapter 2 and the challenges of civic address information is explained in section 2.1.

How does DHCP-based location configuration work? Typically DHCP-based location configuration is used to configure clients on wired networks where the location can be derived from a wiremap by the DHCP server, using the Circuit-ID Relay Agent Information Option (RAIO) defined (as Sub-Option 1) in RFC 3046 [49]. To provide location, the DHCP server correlates the Circuit-ID with the geographic location where the identified circuit terminates (such as the location of the wall jack).

The DHCP-based geospatial location format found in RFC 3825 is also supported for configuration of IEEE 802.11 stations, as provided in IEEE 802.11k-2008 [50]. In IEEE 802.11 [51], the Access Point answers location queries sent by the Station, providing the geospatial location of either the station or the Access Point, depending on the query.

2.2.1 *What's New in RFC 6225?*

RFC 6225 defines a DHCPv6 option format as well as a new GeoLoc option 144. The DHCPv4 option defined in RFC 3825 (GeoConf option 123) is also included. The DHCPv4 GeoLoc option defined in RFC 6225 supports the specification of location uncertainty; the GeoConf option originally defined in RFC 3825 instead uses resolution. What is the difference between the resolution concept introduced in RFC 3825, and the uncertainty concept introduced in RFC 6225?

Resolution refers to the accuracy of a reported location, as expressed by the number of valid bits in each of the Latitude, Longitude, and Altitude fields.

Uncertainty is a quantification of errors. Any method for determining location is subject to some sources of error; uncertainty describes the amount of error that is present. Uncertainty might be the coverage area of a wireless transmitter, the extent of a building or a single room.

Uncertainty is usually represented as an area within which the target is located. Each of the three axes can be assigned an uncertainty value. In effect, this describes a rectangular prism, which may be used as a coarse representation of a more complex shape that fits within it. When representing locations from sources that can quantify uncertainty, the goal is to find the smallest possible rectangular prism that this format can describe. This is achieved by taking the minimum and maximum values on each axis and ensuring that the final encoding covers these points. This increases the region of uncertainty, but ensures that the region that is described encompasses the target location.

A point within the region of uncertainty is selected to be the encoded point; the centroid of the region is often an appropriate choice. The value for uncertainty is taken as the distance from the selected point to the furthest extreme of the region of uncertainty on that axis. Uncertainty applies to each axis independently. This is demonstrated in Figure 2.1, which shows a two-dimensional polygon that is projected on each axis. The ranges marked with "U" is the uncertainty. The vertical arrows delimit the vertical uncertainty range, and the horizontal arrows delimit the horizontal uncertainty range.

2.2.2 *DHCPv4 and DHCPv6 Option Formats*

Since the original DHCPv4 GeoConf option 123 defined in RFC 3825 relied upon the concept of resolution, but did not support uncertainty, RFC 6225 creates a new Geoloc option 144 that uses uncertainty. Within the DHCPv6 option, only uncertainty can be expressed.

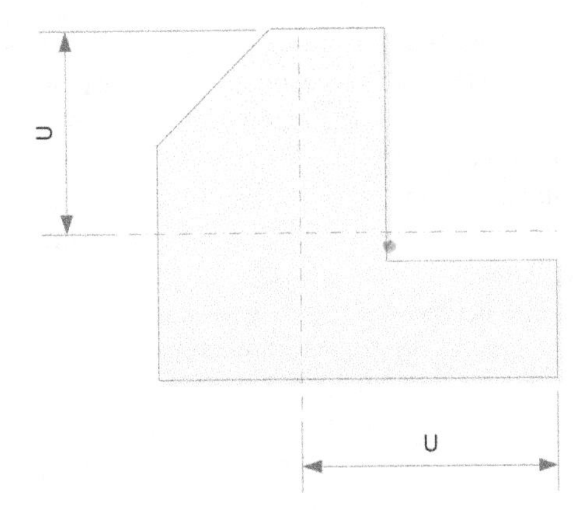

Figure 2.1 Uncertainty. (Copyright © 2011 IETF Trust and the persons identified as the document authors. All rights reserved. http://tools.ietf.org/html/rfc6225.)

The GeoConf option format includes a resolution parameter for each of the dimensions of location. Since this resolution parameter need not apply to all dimensions equally, a resolution value is included for each of the three location elements. The DHCPv6 option does not support a resolution parameter. The DHCPv4 GeoLoc option and the DHCPv6 option format utilize an uncertainty parameter.

The DHCPv6 option format, as defined in Section 2.1 of RFC 6225, is shown in Figure 2.2. The fields have the following meaning:

- *Code (16 bits).* The code for the DHCPv6 option (63).
- *OptLen (16 bits).* This option has a fixed length, so that the value of this octet will always be 16.

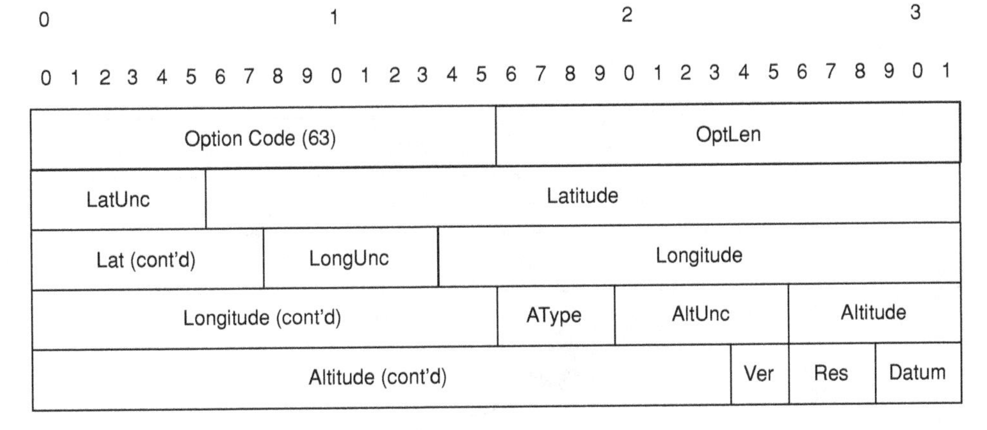

Figure 2.2 DHCPv6 GeoLoc option format. (Copyright © 2011 IETF Trust and the persons identified as the document authors. All rights reserved. http://tools.ietf.org/html/rfc6225.)

- *LatUnc (6 bits).* When the Ver field = 1, this field represents latitude uncertainty. The contents of this field is undefined for other values of the Ver field.
- *Latitude (34 bits).* A fixed-point value consisting of 9 bits of integer and 25 bits of fraction.
- *LongUnc (6 bits).* When the Ver field = 1, this field represents longitude uncertainty. The contents of this field is undefined for other values of the Ver field.
- *Longitude (34 bits).* A fixed-point value consisting of 9 bits of integer and 25 bits of fraction.
- *AType (4 bits).* Altitude Type.
- *AltUnc (6 bits).* When the Ver field = 1, this field represents altitude uncertainty. The contents of this field is undefined for other values of the Ver field.
- *Altitude (30 bits).* A value defined by the AType field.
- *Ver (2 bits).* The Ver field is 2 bits, providing for four potential versions. Only version 1 is defined for the DHCPv6 option. The Ver field is always located at the same offset from the beginning of the option, regardless of the version in use.
- *Res (3 bits).* The Res field is reserved. These bits have been used by IEEE 802.11y [52], but are not defined within the DHCPv6 option.
- *Datum (3 bits).* The Map Datum used for the coordinates given in this Option.

The DHCPv4 GeoLoc option format from Section 2.2.2 of RFC 6225 is shown in Figure 2.3. We do not show the DHCPv4 GeoConf option, which can be found in Section 2 of RFC 3825 and Section 2.2.1 of RFC 6225. The fields have the following meaning:

- *Code (8 bits).* The code for the DHCPv4 GeoLoc option (144).
- *Length (8 bits).* The length of the DHCPv4 GeoLoc option, in octets. For Version 1, the option length is 16.
- *LatUnc (6 bits).* For Version 1, this field represents latitude uncertainty. The field is undefined for other Version values.
- *Latitude (34 bits).* A fixed-point value consisting of 9 bits of signed integer and 25 bits of fraction.

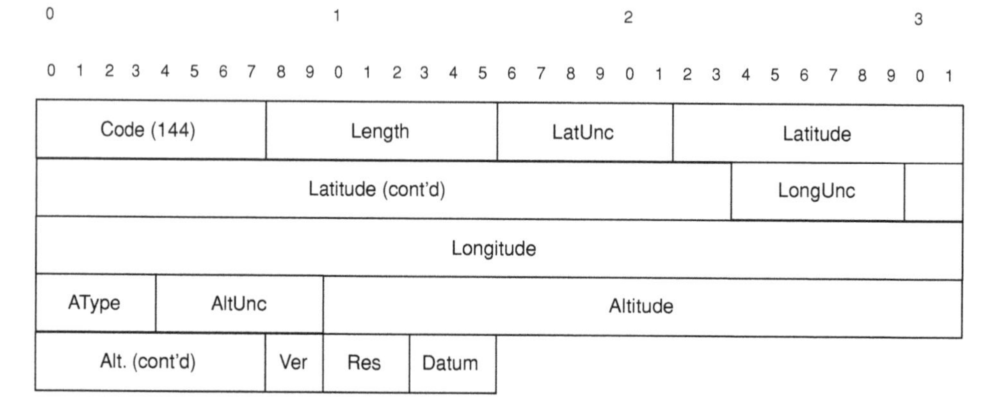

Figure 2.3 DHCPv4 GeoLoc option format. (Copyright © 2011 IETF Trust and the persons identified as the document authors. All rights reserved.)

- *LongUnc (6 bits).* For Version 1 this field represents longitude uncertainty. The field is undefined for other Version values.
- *Longitude (34 bits).* A fixed-point value consisting of 9 bits of signed integer and 25 bits of fraction.
- *AType (4 bits).* Altitude Type.
- *AltUnc (6 bits).* For Version 1, this field represents altitude uncertainty. The field is undefined for other Version values.
- *Altitude (30 bits).* The interpretation of this value depends on the selected AType value.
- *Ver (2 bits).* The Ver field is 2 bits, providing for four potential versions. For the GeoLoc option, only Version 1 is defined in RFC 6225. The Ver field is always located at the same offset from the beginning of the option, regardless of the version in use.
- *Res (3 bits).* The Res field is reserved. These bits have been used by IEEE 802.11y [52], but are not defined for the DHCPv4 GeoLoc option.
- *Datum (3 bits).* The Map Datum used for the coordinates given in this option.

2.2.3 *Option Support*

2.2.3.1 Client Support

DHCPv4 clients implementing RFC 6225 must support receiving the DHCPv4 GeoLoc Option 144 (version 1), and may support receiving the DHCPv4 GeoConf Option 123 originally defined in RFC 3825.

DHCPv4 clients request the DHCPv4 server to send GeoConf Option 123, GeoLoc Option 144, or both via inclusion of the Parameter Request List option. As noted in Section 9.8 of RFC 2132:

> This option is used by a DHCP client to request values for specified configuration parameters. The list of requested parameters is specified as *n* octets, where each octet is a valid DHCP option code as defined in this document.
>
> The client MAY list the options in order of preference. The DHCP server is not required to return the options in the requested order, but MUST try to insert the requested options in the order requested by the client.

When DHCPv4 and DHCPv6 clients implementing RFC 6225 do not understand a datum value, they must assume a World Geodetic System 1984 (WGS 84) datum (European Petroleum Survey Group (EPSG) 4326 or 4979, depending on whether there is an altitude value present) and proceed accordingly. Assuming that a less accurate location value is better than none, this ensures that some (perhaps less accurate) location is available to the client.

2.2.3.2 Server Option Selection

A DHCPv4 server implementing RFC 6225 must support sending GeoLoc Option 144 version 1 and should support sending GeoConf Option 123 in responses.

A DHCPv4 server that provides location information should honor the Parameter Request List included by the DHCPv4 client in order to decide whether to send GeoConf Option 123, GeoLoc Option 144, or both in the Response.

2.2.4 *Latitude and Longitude Fields*

The latitude and longitude values are encoded as 34-bit, two's-complement, fixed-point values with 9 integer bits and 25 fractional bits. The exact meaning of these values is determined by the datum; the description in this section applies to the datums defined in this document. This document uses the same definition for all datums it specifies.

When encoding, latitude and longitude values are rounded to the nearest 34-bit binary representation. This imprecision is considered acceptable for the purposes to which this form is intended to be applied and is ignored when decoding.

Positive latitudes are north of the equator, and negative latitudes are south of the equator. Positive longitudes are east of the Prime Meridian, and negative (two's-complement) longitudes are west of the Prime Meridian.

Within the coordinate reference systems defined in this document (datum values 1–3), longitude values outside the range of −180 to 180 decimal degrees or latitude values outside the range of −90 to 90 degrees must be considered invalid. Server implementations should prevent the entry of invalid values within the selected coordinate reference system. Location consumers must ignore invalid location coordinates and should log errors related to invalid location.

In the DHCPv6 GeoLoc Option 63 and the DHCPv4 GeoLoc Option 144, the Latitude and Longitude Uncertainty fields (LatUnc and LongUnc) quantify the amount of uncertainty in each of the latitude and longitude values, respectively. A value of 0 is reserved to indicate that the uncertainty is unknown; values greater than 34 are reserved.

A point within the region of uncertainty is selected to be the encoded point; the centroid of the region is often an appropriate choice. The value for uncertainty is taken as the distance from the selected point to the furthest extreme of the region of uncertainty on that axis. This is demonstrated in Figure 2.1.

2.2.5 *Altitude*

The altitude is expressed as a 30-bit, fixed-point, two's-complement integer with 22 integer bits and 8 fractional bits. How the Altitude value is interpreted depends on the type of altitude and the selected datum. Three Altitude Types are defined in this document: unknown (0), meters (1), and floors (2).

2.2.5.1 No Known Altitude (AT = 0)

In some cases, the altitude of the location might not be provided. An Altitude Type of 0 indicates that the altitude is not given to the client. In this case, the Altitude and Altitude Uncertainty fields can contain any value and must be ignored.

2.2.5.2 Altitude in Meters (AT = 1)

If the Altitude Type has a value of 1, the Altitude is measured in meters. The altitude is measured in relation to the zero set by the vertical datum.

2.2.5.3 Altitude in Floors (AT = 2)

A value of 2 for Altitude Type indicates that the Altitude value is measured in floors. This value is relevant only in relation to a building; the value is relative to the ground level of the building. In this definition, numbering starts at ground level, which is floor 0 regardless of local convention.

Non-integer values can be used to represent intermediate or sub-floors, such as mezzanine levels. For instance, a mezzanine between floors 4 and 5 could be represented as 4.1.

2.2.6 Datum

The Datum field determines how coordinates are organized and related to the real world. Three datums are defined in this document, based on the definitions in ref. [53]:

1. *WGS84 (Latitude, Longitude, Altitude).* The World Geodesic System 1984 [54] coordinate reference system. This datum is identified by the European Petroleum Survey Group (EPSG) and International Association of Oil and Gas Producers (OGP) with the code 4979, or by the URN "urn:ogc:def:crs:EPSG::4979." Without Altitude, this datum is identified by the EPSG/OGP code 4326 and the URN "urn:ogc:def:crs:EPSG::4326."
2. *NAD83 (Latitude, Longitude) + NAVD88.* This datum uses a combination of the North American Datum 1983 (NAD83) for horizontal (Latitude and Longitude) values, plus the North American Vertical Datum of 1988 (NAVD88) vertical datum. This datum is used for referencing location on land (not near tidal water) within North America. NAD83 is identified by the EPSG/OGP code of 4269, or the URN "urn:ogc:def:crs:EPSG::4269." NAVD88 is identified by the EPSG/OGP code of 5703, or the URN "urn:ogc:def:crs:EPSG::5703."
3. *NAD83 (Latitude, Longitude) + MLLW.* This datum uses a combination of the North American Datum 1983 (NAD83) for horizontal (Latitude and Longitude) values, plus the Mean Lower Low Water (MLLW) vertical datum. This datum is used for referencing location on or near tidal water within North America. NAD83 is identified by the EPSG/OGP code of 4269, or the URN "urn:ogc:def:crs:EPSG::4269." MLLW does not have a specific code or URN.

All hosts must support the WGS84 datum (Datum 1).

2.3 Geography Markup Language (GML)

Carl Reed
Open Geospatial Consortium (OGC)

2.3.1 Introduction

Expressing and communicating location elements in an unambiguous manner are a critical component of any mobile IP emergency service application. This section describes the uses of the OGC Geography Markup Language (GML) and related supporting standards work for encoding and transporting location elements within the emergency services context. Although the description in this section focuses on the usage of GML for encoding geodetic location information for use with citizen-to-authority applications, GML has also been successfully applied to authority-to-authority and authority-to-citizen communication. We begin with a short overview of the OGC followed by a description of GML and supporting standards or best practices, such as coordinate reference systems (CRSs). We conclude with examples of how GML is used in existing standards, such as PIDF-LO, CityGML, and GeoRSS.

2.3.2 Overview of the OGC

Founded in 1994, the Open Geospatial Consortium (OGC) is a global industry consortium with a vision to "Achieve the full societal, economic and scientific benefits of integrating location resources into commercial and institutional processes worldwide." Inherent in this vision is the requirement for geospatial standards and strategies to be an integral part of business processes.

The OGC consists of more than 465 members – geospatial technology software vendors, systems integrators, government agencies, and universities – participating in a consensus process to develop, test, and document publicly available geospatial interface standards and encodings for use in information and communications industries. Open interfaces and protocols defined by OpenGIS Standards are designed to support interoperable solutions that "geo-enable" the Web, wireless and location-based services, and mainstream IT, and to empower technology developers to make complex spatial information and services accessible and useful to all kinds of applications. As such, the mission of the OGC is to serve as a global forum for the development, promotion, and harmonization of open and freely available geospatial standards. Therefore, the OGC has a major commitment to collaborate with other standards development organizations that have requirements for using location-based content, such as ISO [55], OASIS [56], the IETF [57], NENA [58], and OMA [59].

2.3.3 The OGC Geography Markup Language (GML)

One of the key encoding standards developed and maintained by the OGC Membership is the Geography Markup Language (GML). GML is an XML grammar defined to express geographical features. GML serves as a modeling language for geographic systems as well as an open interchange format for geographic transactions on the Internet. Note that the concept of feature in GML is a very general one and includes not only conventional "vector" or discrete objects, but also coverages[9] and some elements of sensor data.[10] The ability to integrate all forms of geographic information is key to the utility of GML.

OGC work on the GML standard began in 1998. GML was first formally approved as an OGC standard in 2001. GML became an ISO standard in 2007. GML 3.2.1[11] is the most current revision of the joint OGC-ISO standard and Version 3.3 – extended schemas and encoding rules – has been published early 2012.

[9] A coverage is a subtype of feature that has a coverage function with a spatio-temporal domain and a value set range of homogeneous one- to *n*-dimensional tuples. A coverage may represent one feature or a collection of features – to model and make visible spatial relationships between, and the spatial distribution of, Earth phenomena.

[10] For an integrated approach to defining, tasking, modeling, and accessing sensors in a sensor network, the OGC recommends using the OGC Sensor Web Enablement framework [60].

[11] However, most existing GML applications are based on GML 3.1.1 and even 2.1.

2.3.3.1 GML Development and Model

The original GML model was described using the World Wide Web Consortium's Resource Description Framework (RDF). Subsequently, the OGC introduced XML schemas into GML's structure to help connect the various existing geographic databases, whose relational structure XML schemas more easily define. The 3.x.x version of GML is a major evolution GML development. For GML 3.0, the information model was harmonized with a number of ISO documents. Therefore, the current version of GML specifies XML encodings, conformant with ISO 19118, and several of the conceptual classes defined in the ISO 19100 series of International Standards and the OpenGIS Abstract Specification. These conceptual models include those defined in:

- ISO/TS 19103 – Conceptual schema language (units of measure, basic types);
- ISO 19107 – Spatial schema (geometry and topology objects);
- ISO 19108 – Temporal schema (temporal geometry and topology objects, temporal reference systems);
- ISO 19109 – Rules for application schemas (features);
- ISO 19111 – Spatial referencing by coordinates (coordinate reference systems);
- ISO 19123 – Schema for coverage geometry and functions.

GML contains a rich set of primitives that are used to build application-specific schemas or application languages. These primitives include the following:

- Feature
- Geometry
- Coordinate Reference System
- Topology
- Time
- Dynamic feature
- Coverage (including geographic images)
- Unit of measure
- Directions
- Observations
- Map presentation styling rules

Understanding the use of "Feature," "Geometry," and "CRS" is critical in the use and development of any GML-based encoding.

2.3.3.2 GML Features

A feature[12] is an abstraction of real-world phenomena (ISO 19101); it is a geographic feature if it is associated with a location relative to the Earth. So a digital representation of the real world may be thought of as a set of features. The state of a feature is defined by a set of properties, where each property may be thought of as a {name, type, value} triple.

[12] From the GML 3.2.1 standard document.

The number of properties a feature may have, together with their names and types, is determined by its type definition. Geographic features with geometry are those with properties that may be geometry-valued. A feature collection is a collection of features that may itself be regarded as a feature; as a consequence, a feature collection has a feature type and thus may have distinct properties of its own, in addition to the features it contains.

Following ISO 19109, the feature types of an application or application domain is usually captured in an application schema. A GML application schema is specified in XML Schema.

2.3.3.3 GML Geometry

Geometry provides the means for the quantitative description, by means of coordinates and mathematical functions, of the spatial characteristics of features, including dimension, position, size, shape, and orientation. The mathematical functions used for describing the geometry of an object depend on the type of coordinate reference system used to define the spatial position. Geometry is the only aspect of geographic information that changes when the information is transformed from one geodetic reference system or coordinate system to another.

In a traditional Geographic Information System (GIS), some well-known geometry types are points, lines, and polygons.

2.3.3.4 Relationship Between GML Feature and GML Geometry

GML defines features distinct from geometry objects. A feature is an application object that represents a physical entity, for example, a building, a river, or a person. A feature may or may not have geometric aspects. A geometry object defines a location or region instead of a physical entity, and hence is different from a feature. The distinction between features and geometry objects in GML contrasts with models used in other GISs that make no such distinction. That is, although some other GISs define features and geometry objects interchangeably as items on a map, GML maintains them as separate entity types.

In GML, a feature can have various geometry properties that describe geometric aspects or characteristics of the feature (e.g., the feature's Point or Extent properties). GML also provides the ability for features to share a geometry property with one another by using a remote property reference on the shared geometry property. Remote properties are a general feature of GML borrowed from RDF.

2.3.3.5 Coordinate Reference Systems

Geographic information contains spatial references that relate the features represented in the data to positions in the real world. Coordinates are unambiguous only when the coordinate reference system (CRS) [61, 62] to which those coordinates are related has been fully defined. A coordinate reference system is a coordinate system that is related to the real world by a datum. An abstract model for expressing coordinate reference systems may be found in OGC Abstract Specification Topic 2 (a.k.a. ISO 19111): Spatial referencing by coordinates.

Without knowing the CRS for a given set of geometry, there is absolutely no way for an application (or a user) to know or display the geometry on a map. Therefore, defining and communicating CRS information as part of any geospatial content or encoding/payload standard is critical. The expression of the CRS must further be done in a consistent, well-known manner. The normative reference used the OGC for all CRS definitions is the EPSG registry. The Geodesy Subcommittee of OGP maintains the EPSG Geodetic Parameter Dataset.

The most commonly used CRS in many applications and location payloads is WGS84. However, simply stating WGS84 is meaningless in terms of any standards work. This is because WGS84 defines a standard coordinate frame for the Earth, a standard spheroidal reference surface (the datum or reference ellipsoid) for raw altitude data, and a gravitational equipotential surface (the geoid) that defines the nominal sea level. This is why in examples or definitions used in standards that are based on GML, the normative text often states much more detail, such as in the GML 3.1.1 PIDF-LO Shape Application Schema for use by the Internet Engineering Task Force [63]:

> These are identified using the European Petroleum Survey Group (EPSG) Geodetic Parameter Dataset, as formalized by the following Open Geospatial Consortium (OGC) URNs:
>
> - 3D: WGS84 (latitude, longitude, altitude), as identified by the URN urn:ogc:def:crs:EPSG::4979. This is a three-dimensional CRS.
> - 2D: WGS84 (latitude, longitude), as identified by the URN urn:ogc:def:crs:EPSG::4326. This is a two-dimensional CRS.
>
> The most recent version of the EPSG Geodetic Parameter Dataset should be used. The CRS shall be specified using the above un-versioned URN notation only; implementations do not need to support user-defined CRSs.
>
> Implementations must specify the CRS using the srsName attribute on the outermost geometry element. The CRS must not be re-specified or changed for any sub-elements. The srsDimension attribute should be omitted, since the number of dimensions in these CRSs is known.

The following is a GML snippet showing the use of a CRS:

```
<entry>
    ...
    <georss:where>
       <gml:Point
             srsName="urn:ogc:def:crs:EPSG:6.6:4979"
             srsDimension="3">
             <gml:pos>42.3453 -156.2342 45</gml:pos>
       </gml:Point>
    </georss:where>
</entry>
```

2.3.3.6 GML Profiles and Application Schemas

GML profiles are restricted subsets of GML. These profiles are intended to simplify usage of GML in a specific application domain and to facilitate rapid adoption of the standard. The following profiles, as defined by the GML specification, have been published or proposed for public use:

- a Point Profile [64] for applications with point geometric data but without the need for the full GML grammar;

- a GML Simple Features profile [65] supporting vector feature requests and transactions, for example, with an OGC Web Feature Service Interface instance [66];
- a GML profile for GMJP2 (GML in JPEG 2000);
- a GML profile for RSS.

Note that Profiles are distinct from GML application schemas. Profiles are part of GML namespace and defined as restricted subsets of GML.

Application schemas are XML vocabularies defined using GML and which live in an application-defined target namespace. Application schemas can be built on specific GML profiles or use the full GML schema set. In either case, the can be extended to accommodate the requirements of a particular information community.

PIDF-LO, GeoRSS GML, and CityGML as discussed below are application schemas.

2.3.3.7 Developing Successful GML Application Schemas

GML is a very rich grammar for modeling and encoding geospatial content. Simply reading the standard leads to a common belief that GML is overly complex and heavyweight. This is not an accurate assessment. As with programming, the first step in a successful use of GML is to understand and document the use cases and requirements for sharing geospatial content for your application, organization or information community. The next step is to achieve community consensus on the information model for the content you wish to share. Without an information model, there is no way that you can effectively understand or successfully use GML! From Wikipedia:

> An information model in software engineering is a representation of concepts, relationships, constraints, rules, and operations to specify data semantics for a chosen domain of discourse. It can provide sharable, stable, and organized structure of information requirements for the domain context.

There are many operational examples of GML deployments based on first defining the information model. Many examples of GML application schemas based on information models are available [67]. The truly hard work is in developing the information model for the respective application domain. Encoding the model as GML follows once the model has been defined and documented (hopefully using the Unified Modeling Language (UML)).

2.3.3.8 Presence Information Data Format – Location Object (PIDF-LO)

From an emergency response perspective, location is a critical element of information that must be captured and communicated. This is one reason why the collaboration of standards work between the IETF, NENA, the OGC, and other standards organizations is so important. One example of this collaboration is RFC 4119 [17]. "This document extends the XML-based Presence Information Data Format (PIDF) to allow the encapsulation of location information within a presence document." Two forms of location object are specified: civic and geodetic. The civic form is used for geographic identifiers such as addresses and landmarks and geodetic is for coordinate-based location objects. The geodetic location object "GML" 3.1.1 shall be this mandatory format (a MUST implement for all PIDF implementations supporting the "Geopriv" element). More specifically, RFC 4119 references a GML application schema titled "GML 3.1.1 PIDF-LO Shape Application Schema for use by the Internet Engineering Task Force (IETF)" [63]. Historically, in 2004, RFC 4119 just referenced GML, which led a few of individuals, namely James Winterbottom, Martin Thomson, and Hannes Tschofenig, to write an IETF GML application schema document. Based on discussions within the IETF Geopriv Working Group, the decision was that the Geopriv group (and the IETF) is not the right place to develop the application schema and that the OGC would be the correct standards forum because of the expertise that is available in the OGC membership. The text from the draft was then turned into the OGC Best Practice document and the original IETF draft was hugely simplified and mostly replaced by references to the respective OGC document. Since then, RFC 4119 has been updated. First, there is the update via RFC 5139 [21] with respect to the civic location information. Second, there is the update of the geodetic location information via RFC 5491 [28], and that document normatively references the OGC PIDF-LO application schema.

This document was submitted to the OGC in 2007 and was approved to be an OGC Best Practice. The next step will be to approve the document as a normative OGC standard.

The document defines a GML Application Schema (AS) for encoding specific geometric shapes as required in PIDF-LO, specifically for such location service and mobile emergency service requirements for ellipses, prisms, and spheres. Specifically, the Application Schema defines the following geometry types for use in PIDF-LO:

- Point = gml:Point element is used when there is no known uncertainty.
- Polygon = gml:Polygon element with four or more coordinates defining the boundary.
- Polygon Upward Normal = the upward normal of a polygon determines the orientation of the surface. The upward normal of a polygon is important for the definition of the Prism.
- Circle = the circular area is used for coordinates in two-dimensional CRSs to describe uncertainty about a point. The definition is based on the one-dimensional geometry in GML, gml:CircleByCenterPoint.
- Ellipse = an elliptical area describes an ellipse in two-dimensional space. The ellipse is described by a center point, the length of its semi-major and semi-minor axes, and the orientation of the semi-major axis.
- Arc Band = an arc band is a section of a circular area that is constrained to an arc between two radii. The arc band shape is most useful when radio timing technologies are used to determine location, based on a single wireless transmitter.
- Sphere = the sphere is a volume that provides the same information as a circle in three dimensions. The sphere shall be specified using a three-dimensional CRS.
- Ellipsoid = the ellipsoid is a volume that is based on the ellipse shape, with the addition of a third dimension. A single value, vertical uncertainty, is added to the ellipse to form an ellipsoid.
- Prism = the prism is a volume that is commonly used to represent a shape that has a constant cross-section along one axis. For the purposes of PIDF-LO, a prism is most useful when representing a building, or single floor of a building.

The location object (LO) and the associated GML 3.1.1 Shape Application Schema are also referenced by a number of other Internet standards, such as the following.

- LoST: A Location to Service Translation Protocol [68] – satisfies the requirements [7] for mapping protocols. LoST provides a number of operations, centered around mapping locations and service URNs to service URLs and associated information.
- HTTP Enabled Location Delivery (HELD) [31] – defines an XML-based protocol that enables the retrieval of Location Information from a Location Generator.

Finally, the geometry types used to specify the LO are consistent with other payload and encoding standards based on GML and ISO 19107 spatial schema.

2.3.3.9 GeoRSS

In addition to the ability to carry a consistent location object as an information component of the mobile Internet, there is a need to carry location content as an element of feeds, including consumer-based alerting applications. A 2005 activity labeled GeoRSS resulted in a simple, lightweight encoding for location-enabling RSS feeds. This encoding has been used in numerous alerting and warning applications, especially in Europe and Asia.

Originally proposed in 2005,[13] GeoRSS has become a *de facto* standard for encoding location as part of a Web feed. GeoRSS is simple proposal for geo-enabling, or tagging, "really simple syndication" (RSS) feeds with location information. GeoRSS proposes a standardized way in which location is encoded with enough simplicity and descriptive power to satisfy most needs to describe the location of Web content. GeoRSS may not work for every use, but it should serve as an easy-to-use geotagging encoding that is brief and simple with useful defaults but extensible and upwardly compatible with more sophisticated encoding standards such as the PIDF-LO and CityGML application schemas.

Currently, there are two GeoRSS[14] serializations:[15] GeoRSS GML and GeoRSS Simple. GeoRSS GML is a formal GML Profile, and supports a greater range of features than GeoRSS Simple, notably coordinate reference systems other than WGS84 latitude and longitude. GeoRSS is designed for use with Atom 1.0, RSS 2.0, and RSS 1.0, although it can be used just as easily in non-RSS XML encodings. GeoRSS Simple has greater brevity, but also has limited extensibility.

GeoRSS is based on an information model that is documented in UML (Figure 2.4). As discussed above, one of the key aspects of correctly and effectively using GML is first to define an information or content model. After the requirements definition, developing a GeoRSS information model was the first step in the development of GeoRSS.

Figure 2.4 is a depiction of the GeoRSS information model. This is a very simple information model and shows that there are four basic geometry types supported (point, line, box, and polygon), with some basic additional properties, such as Elevation and Radius.

[13] The original GeoRSS version was developed by Allan Doyle, Ron Lake, Josh Lieberman, Carl Reed, and Raj Singh.

[14] GeoRSS originally served informally as an extension to the W3C geo (point) vocabulary.

[15] From Wikipedia: Serialization is the process of saving an object onto a storage medium (such as a file, or a memory buffer) or to transmit it across a network connection link such as a socket either as a series of bytes or in some human-readable format such as XML.

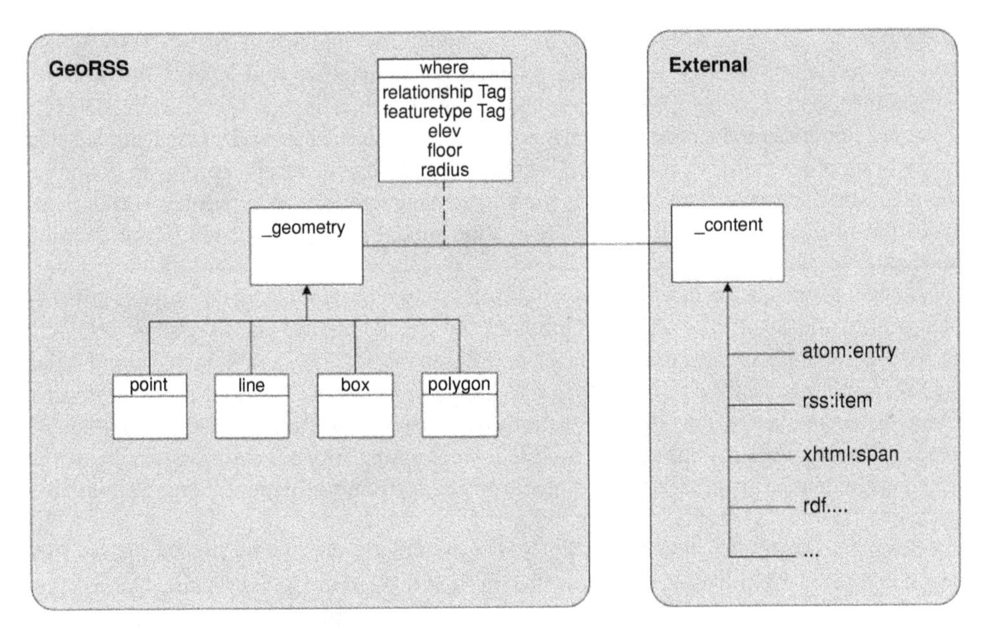

Figure 2.4 GeoRSS information model. (Copyright © 2006 Open Geospatial Consortium, Inc. All Rights Reserved. http://www.opengeospatial.org/ogc/document, Figure re-used from OGC Document 06-050r3, http://www.opengeospatial.org/pressroom/pressreleases/580.)

These same types are also supported in the PIDF-LO encoding, and there are such similarities between many of the information models that are then encoded using GML (such as CityGML, see below).

GeoRSS is now broadly implemented and supported by numerous consumer applications, such as Google Earth, MS Virtual Earth, Flickr, and Mapquest. There are also a variety of GeoRSS usages in the alerting and warning communities, such as the USGS earthquake and Asian tsunami alerting applications. A really interesting portal is "The Global Disaster Alert and Coordination System" (GDACS) that provides near real-time alerts about natural disasters around the world and tools to facilitate response coordination, including media monitoring, map catalogs and a Virtual On-Site Operations Coordination Center [69].

GeoRSS can also be easily used in alerting applications, such as traffic, weather, and so forth.

2.3.3.10 CityGML

A critical requirement for the emergency services community is to have access to building information, such as the building footprint, entrances and exits, location of fire fighting assets, and location of hazardous materials. Ultimately, first responders need building floor plans. The problem is that this type of information is typically available from multiple sources and provided in a variety of formats. The solution is to have a community-defined model and encoding for sharing building or urban model information. City Geography Markup Language (CityGML) is one solution.

CityGML is an innovative concept for the modeling and exchange of 3D city and landscape models that is quickly being adopted on an international level. CityGML is a common information model for the representation of 3D urban objects. The model defines the classes and relations for the most relevant topographic objects in cities and regional models with respect to their geometrical, topological, semantic, and appearance properties. Included are generalization hierarchies between thematic classes, aggregations, relations between objects, and spatial properties. The CityGML information model is well grounded in the Building Information Model (BIM).

In contrast to other 3D vector formats, CityGML is based on a rich, general-purpose information model in addition to geometry and graphics content that allows one to employ virtual 3D city models for sophisticated analysis tasks in different application domains like simulations, urban data mining, facility management, and thematic inquiries. Targeted application areas include urban and landscape planning; architectural design; tourist and leisure activities; 3D cadastres; environmental simulations; mobile telecommunications; emergency services, disaster management; homeland security; vehicle and pedestrian navigation; training simulators; augmented reality; and mobile robotics.

Beginning in 2002, the model for CityGML was initially defined by the members of the Special Interest Group 3D (SIG 3D) of the initiative Geodata Infrastructure North-Rhine Westphalia (GDI NRW) in Germany. The SIG 3D is an open group consisting of more than 70 companies, municipalities, and research institutions working on the development and commercial exploitation of interoperable 3D models and geo-visualization.

The information model was then encoded as a GML application schema and submitted to the OGC for consideration as an international standard. CityGML version 1.0 was approved in August 2008. Version 2.0 was approved late 2011 and includes a number of additions, such as underground infrastructure, enhancements (such as better definition for texture maps), and additional elements of the information model (such as for floor plans).

The following are the design features for CityGML [70]:

- Geospatial information model (ontology) for urban landscapes based on the ISO 191xx family
- GML3 representation of 3D geometries, based on the ISO 19107 spatial schema model
- Representation of object surface characteristics (textures, materials)
- Taxonomies and aggregations
  - Digital Terrain Models as a combination of (including nested) triangulated irregular networks (TINs), regular rasters, break and skeleton lines, mass points
  - Sites (currently buildings; bridges and tunnels in the future)
  - Vegetation (areas, volumes, and solitary objects with vegetation classification)
  - Water bodies (volumes, surfaces)
  - Transportation facilities (both graph structures and 3D surface data)
  - City furniture
  - Generic city objects and attributes
  - User-definable (recursive) grouping
- Multiscale model with five well-defined consecutive Levels of Detail (LOD)
  - LOD 0 – regional, landscape
  - LOD 1 – city, region
  - LOD 2 – city districts, projects

  - LOD 3 – architectural models (outside), landmarks
  - LOD 4 – architectural models (interior)
- Multiple representations in different LODs simultaneously; generalization relations between objects in different LODs
- Optional topological connections between feature (sub)geometries
- Application Domain Extensions (ADE): specific "hooks" in the CityGML schema allow one to define application-specific extensions, for example, for noise pollution simulation, or to augment CityGML by properties of the new National Building Information Model Standard (NBIMS) in the Unites States.

Perhaps of even greater importance to emergency services and related citizen-to-authority applications as required by first responders is the ability to share indoor content, especially for routing and location of related emergency response assets. Thus, there has been considerable recent research and development for using CityGML for encoding indoor information, such as floor plans.

2.3.4 Conclusion

GML is a powerful encoding language for geospatial content. As more communities utilize GML encodings for sharing geospatial content, interoperability will be significantly increased. The ability to share geospatial content efficiently and effectively enhances community collaboration and applications that require timely geospatial content, such as emergency services, while at the same time reducing costs and risk. GML is being enhanced. A new version of GML, namely version 3.3, was released early 2012. The main changes between the versions 3.2.1 and 3.3 are:

- support for localizable strings;
- support for additional date representations from ISO 8601;
- support for an additional, compact encoding of commonly used geometry types;
- support for a new encoding of Triangulated Irregular Networks (TINs) using the compact geometry encoding and supporting more robust TIN elements consistent with the TIN type from SQL (ISO/IEC 13249-3);
- support for Linear Referencing (OGC Abstract Specification Topic 19, ISO 19148);
- support for referenceable grids (CV_ReferenceableGrid from ISO 19123); and
- extensions to the XML encoding rule including clarifications on the encoding of code lists and code-list-valued properties.

All extensions are made in separate XML namespaces to support a more modular use of components from the GML schema.

2.4 A Taxonomy of the IETF HELD Protocol

James Winterbottom
CommScope

2.4.1 The LIS and HELD

The Location Information Server (LIS) is the standard Internet service for providing location information about target devices in a local access network. The LIS provides a location service analogous to a DHCP server providing general host configuration information, or a DNS server providing host name to IP address resolution.

Internet services generally have two main components: a discovery component that is used to identify a service to devices in the access network; and a using component that enables devices to access the service. Access to an Internet service is performed in a common way regardless of the underlying access technology. A DHCP server, for example, is accessed in the same way regardless of whether the underlying physical access network is DSL, WiFi, cable or fibre. This is also true for the Internet location service: a device is able to request location information from an LIS in exactly the same way regardless of the nature of the physical access network. A protocol used by a device to acquire its own location from an LIS is called a Location Configuration Protocol (LCP). The IETF recognizes two primary LCPs: DHCP, which operates at the link layer; and the HTTP Enabled Location Delivery (HELD) protocol, which operates at the application layer.

An LCP is used by a device to request location information from an LIS in a consistent way. It is the role of the LIS, however, to abstract how location is determined from how location is acquired. To do this, the LIS needs to understand the physical access network and be able to put a context around parameters and measurements obtained from the network so that the location of devices attached to the network can be determined.

A LIS can provide location information in two forms, a location value and a location reference, and these concepts are shown in Figure 2.5. A location value, or literal location, is a direct representation of where a device is at a given point in time, and may be represented as a civic street address and/or a geodetic location expressed using latitudes, longitudes, and distances. A location reference is generally expressed as a URI, and can be accessed by a third party to obtain the location of the device. Constraints on who can access a location URI and what information is provided when the URI is accessed is described later in this section.

The remainder of this section will concentrate on how an LIS can be located in an access network, and how the HELD protocol is used as an LCP in different network topologies to communicate with the LIS. It will also examine how third parties can request the location of target devices inside the access network serviced by the LIS, and how this service is being used in various architectures.

2.4.2 LIS Discovery

Discovery is an important part of many Internet services and is equally important to the Internet location service. It is desirable to have a single service discovery mechanism, but the reality of IP networks and the Internet mean that this is not always possible. The IETF is defining two discovery techniques that enable LIS discovery through the majority of broadband and LAN technologies. However, some network deployments, particularly those consisting of virtual private networks (VPN), can pose problems in certain circumstances.

The primary LIS discovery mechanism, defined in RFC 5986 [71], comprises three main steps:

1. acquiring the access network domain name using the defined DHCP option;

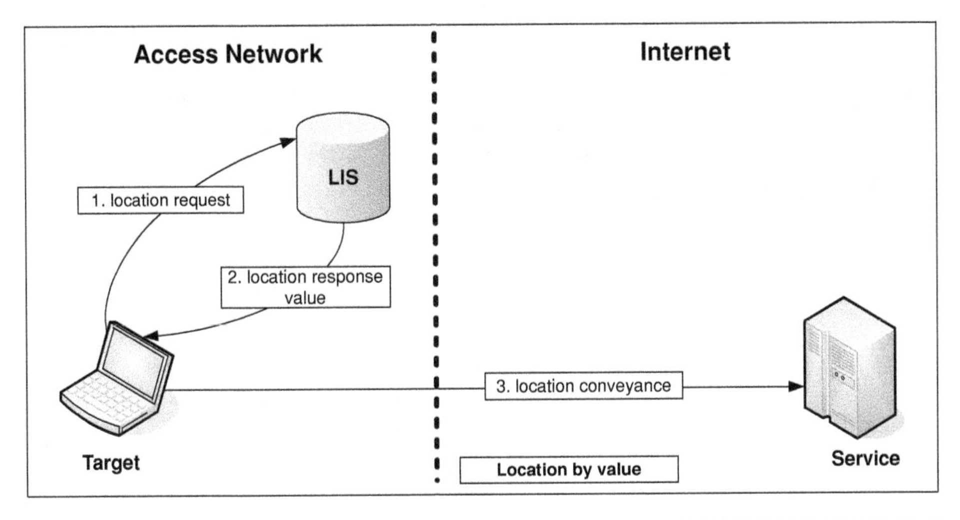

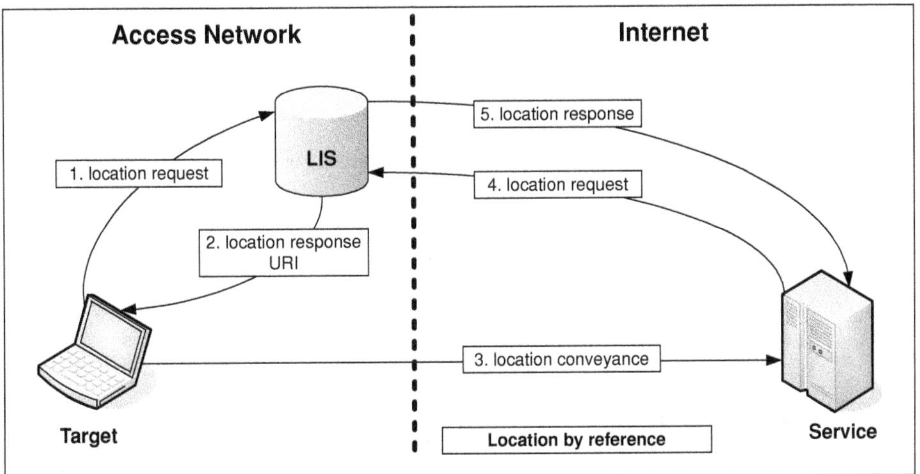

Figure 2.5 LIS location types.

2. applying DNS U-NAPTR [72] to resolve the domain name and HELD service tag to discover the LIS URI; and
3. verifying that the LIS discovered is able to provide location information to the Target, which is done by sending a location request to the LIS.

Modern hosts often have multiple network interfaces. Where this is the case, the steps above are performed on each interface in turn until an LIS that is able to provide the location of the Target is successfully discovered. A LIS signals its inability to locate the Target by responding to a location request with a notLocatable error code. A discovery client that receives this error code makes no further requests to that LIS URI and proceeds to the next interface in its list. If all interfaces in the list have been tried and no LIS is discovered that can locate the device,

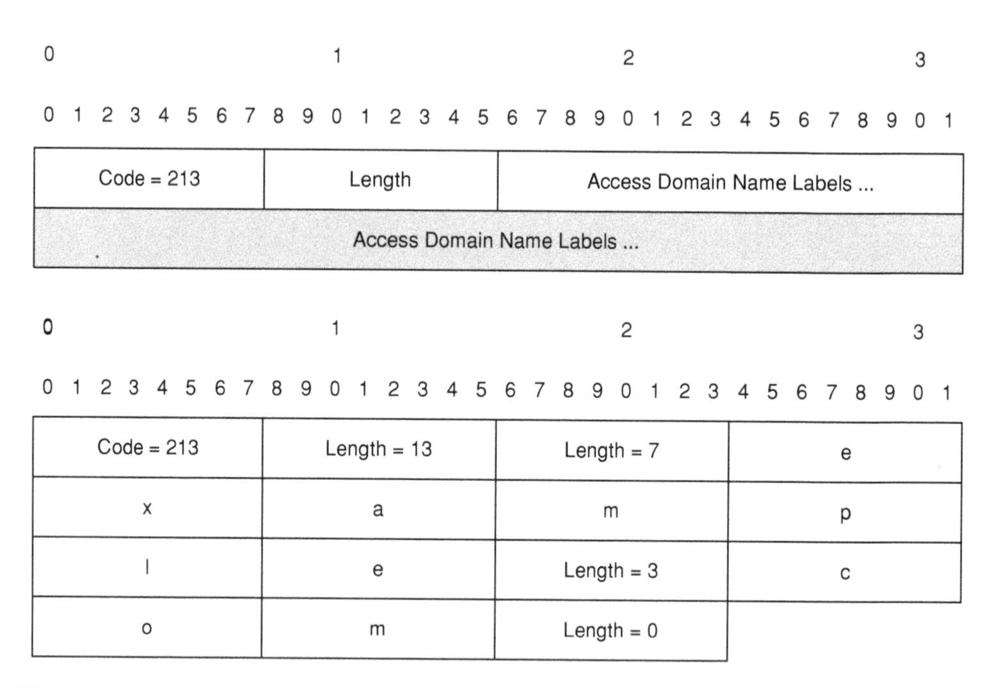

Figure 2.6 DHCPv4 access domain name encoding for "example.com." (From RFC 5986. Copyright © 2009. IETF Trust and the persons identified as authors of the code. All rights reserved. http://datatracker.ietf.org/doc/rfc5986/.)

then pre-configured domain names may be tried in place of the domain name discovered using the DHCP domain name option.

2.4.2.1 Access Network Domain Name DHCP Options

The LIS discovery specification defines options for both DHCPv4 and a DHCPv6 to provide the access domain name to the DHCP client. The DHCP options are containers that convey the access domain name, which is encoded in accordance with Section 3.1 of RFC 1035 [73].

The DHCP versions 4 and 6 encoding schemes are shown in Figures 2.6 and 2.7, respectively.

2.4.2.2 DNS U-NAPTR

DNS U-NAPTR [72] takes an Application Service tag, Application Protocol tag, and domain name, and tries to resolve this into a service URI. The service URI gives the client the identity of a host that can understand the requested protocol and can service their application request for the access domain in which they reside. The LIS discovery specification defines the Application Service tag, LIS, and the Application Protocol tag, HELD, so that a HELD-based LIS can be discovered using U-NAPTR.

The example in Figure 2.8 demonstrates the use of the Application Service and Protocol tags. In the example, NAPTR is used iteratively to delegate responsibility for the HELD LIS service from areas 51 and 7 to a converged service "converged.example.com."

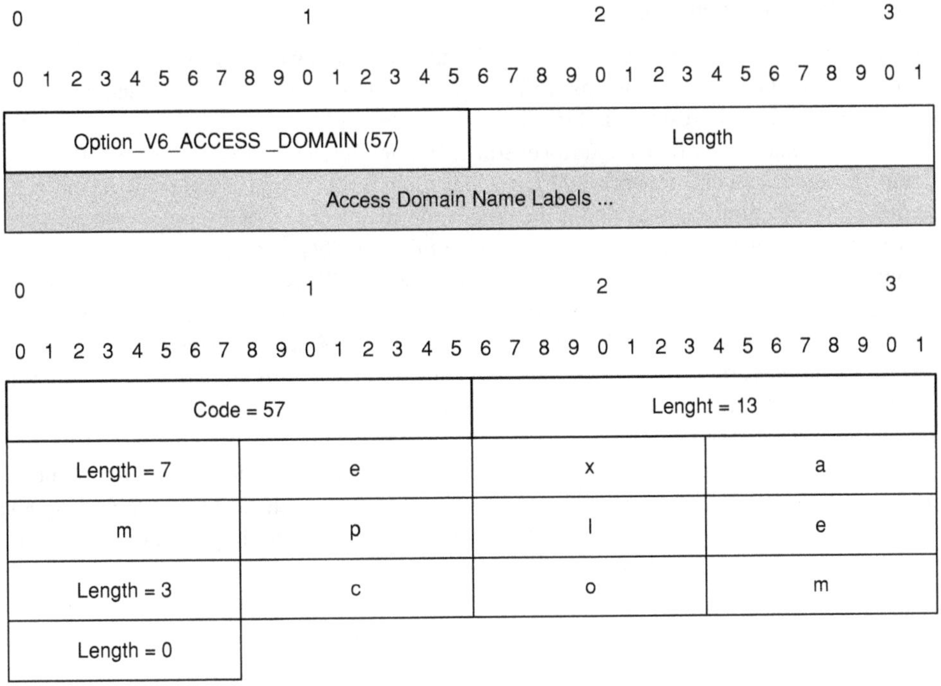

0 1 2 3

0 1 2 3 4 5 6 7 8 9 0 1 2 3 4 5 6 7 8 9 0 1 2 3 4 5 6 7 8 9 0 1

Option_V6_ACCESS _DOMAIN (57)	Length
Access Domain Name Labels ...	

0 1 2 3

0 1 2 3 4 5 6 7 8 9 0 1 2 3 4 5 6 7 8 9 0 1 2 3 4 5 6 7 8 9 0 1

Code = 57		Lenght = 13	
Length = 7	e	x	a
m	p	l	e
Length = 3	c	o	m
Length = 0			

Figure 2.7 DHCPv6 Access domain name encoding for "example.com." (From RFC 5986. Copyright © 2009. IETF Trust and the persons identified as authors of the code. All rights reserved. http://datatracker.ietf.org/doc/rfc5986/.)

```
area51.example.net.
        ;;          order pref flags
        IN NAPTR 100    10    ""   "LIS:HELD" (             ; service
            ""                                              ; regex
            converged.example.com.                          ; replacement
            )
        area7.example.net.
        ;;          order pref flags
        IN NAPTR 100    10    ""   "LIS:HELD" (             ; service
            ""                                              ; regex
            converged.example.com.                          ; replacement
            )
        converged.example.com.
        ;;          order pref flags
        IN NAPTR 100    10    "u"   "LIS:HELD" (            ; service
            "!*.!https://lis.example.org:2292/?c=ex!"       ; regex
            .                                               ; replacement
            )
```

Figure 2.8 DNS LIS U-NAPTR example.

2.4.2.3 Residential Gateways

The discovery steps described in RFC 5986 [71] are fine for newer devices, or devices where the owner of the DHCP server is able to control the values of the DHCP options, but this is generally not the case in residential broadband deployments.

Residential gateways often include several different functions, such as a DHCP server, modem, Ethernet switch, network address translations (NAT), and a router. It is the NAT function of a residential gateway that is the biggest impediment to LIS discovery as it provides the insulation between the private LAN and the public broadband service.

Many residential broadband networks do not use DHCP as the configuration protocol from the access provider (ISP) to the residential gateway. This renders the DHCP option for determining the access network domain name largely unworkable. Manual configuration of network operating parameters into residential gateways can lead to problems if a residential gateway is moved. This is of particular concern with parameters associated with location information and location servers.

There are millions of residential gateways deployed in broadband networks throughout the world, and a discovery solution is required that can operate without necessitating upgrade or replacement of these existing residential gateways. LIS discovery in residential broadband networks is described in ref. [74], and its principle purpose is to aid devices in determining the correct LIS to use when they are behind a NAT or firewall, such as a residential gateway.

The basic network topology for which ref. [74] addresses LIS discovery is illustrated in Figure 2.9. This discovery technique assumes that the location of a residential gateway is sufficiently granular so as to represent the location of devices attached to it. For most houses and small offices, this assumption is fine. The discovery technique also assumes that the LIS at an ISP is able to provide location information for the residence in which the access is terminated. A variety of solutions for different networks exist to allow the LIS to locate the residential gateway, many of which are described in refs. [75] and [76].

The basic mechanism described in ref. [74] is simple: discover the device IP address; use reverse DNS to discover the domain the device is operating in; then, use the standard LIS discovery procedure described earlier; repeat as necessary.

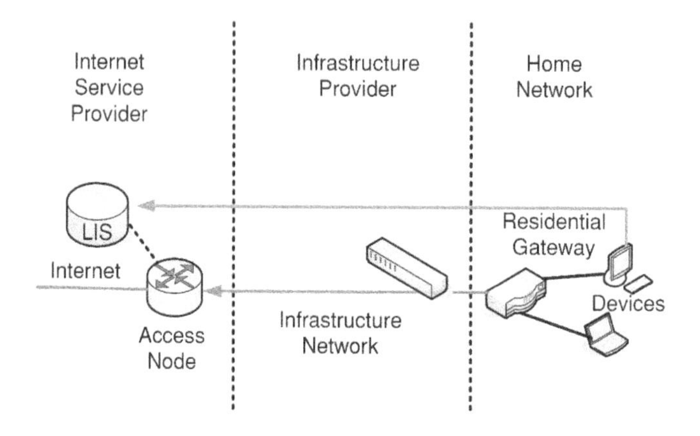

Figure 2.9 Residential broadband network topology.

The devil, of course, is in the detail, and this relates to IP address determination. In particular, if the device is behind a NAT in a residential gateway, it is the public IP address of the NAT that is required. Ref. [74] describes a number of options, with the more commonly discussed option using Session Traversal Utilities for NAT (STUN), which is specified in RFC 5389 [77]. When using STUN, the device has a pre-configured STUN server to which it sends a "Binding Request" message. The device then uses the "XOR-MAPPED-ADDRESS" parameter in the response message from the STUN server to determine its outward-facing IP address.

Having discovered its publicly facing IP address, the device needs to discover the associated domain name to which the IP address belongs. It does this by performing a reverse DNS lookup, either to the ".in-addr-arpa." tree for an IPv4 address, or to the ".ip6.arpa." tree for an IPv6 address. For example, if the device's external IP address was 10.2.0.52, then the query contains 52.0.2.10.IN-ADDR.ARPA.

Each of the components of the IP address becomes a label in the searched domain. The device starts with a complete domain with all labels from the IP address as shown in the example above. If the domain returned from this query does not subsequently yield a NAPTR record with the desired "LIS:HELD" service tag, then the initial query is repeated but with a shortened domain. The domain is shortened by removing the leftmost label. A subsequent domain query based on the example above would be 0.2.10.IN-ADDR.ARPA. This domain shortening can be performed twice, if need be, equating to searches for the /24 and /16 network prefixes.

The outcome of this process is the identity of the LIS capable of providing location information about a device with a specific IP address. This approach allows not only a device behind a NAT to discover an LIS, but for a third-party node to discover the LIS servicing a particular IP address. This latter characteristic has benefits for interim next-generation emergency architectures where location is not conveyed directly by the calling device.

2.4.3 *Basic HELD*

The previous sections provided details on how a device discovers the LIS serving its access network. This section describes the basic HELD semantics. The HELD protocol consists of query and response messages encoded as XML documents. The HELD specification allows for these messages to be transported over any protocol that meets a basic set of requirements, but specifically it defines a binding for HTTP, and this is the assumed transport for the remainder of this chapter.

HELD comprises three message types:

- "locationRequest"
- "locationResponse"
- "error"

where the locationRequest is the way in which a device requests location from the LIS, the locationResponse returns the location information to the device, while an error message indicates to the requesting device why location information cannot be provided.

```
POST /location HTTP/1.1
Host: lis.example.com:43951
Content-Type: application/held+xml;charset=utf-8
Content-Length: 87

<?xml version="1.0"?>
<locationRequest xmlns="urn:ietf:params:xml:ns:geopriv:held"/>
```

Figure 2.10 HTTP location request. (From RFC 5986. Copyright © 2009. IETF Trust and the persons identified as authors of the code. All rights reserved. http://datatracker.ietf.org/doc/rfc5986/.)

2.4.3.1 Location Requests

Location is requested from an LIS by sending it a locationRequest message. When HELD is being used as an LCP, the source IP address of the incoming request is used as the means to identify the subject of the location request. That is, the Target.

The URI of an LIS is an http: or https: URI. The locationRequest message is an XML document included in the body of an HTTP POST to the LIS URI. To make sure that the LIS knows that message is a HELD message, the device sets the content type to "application/held+xml." A basic location request looks similar to the example in Figure 2.10.

Location Type

In the previous example, the requestor is asking for any location information that the LIS may have. HELD supports qualifying the location request by including a "locationType" parameter (Figure 2.11). The locationType parameter is used to tell the LIS what kind of location information the requestor would like. The value of locationType is a list, and it may consist of one or more of the following:

- any – the LIS should provide whatever information it can (this is the default);
- geodetic – the LIS should provide location in geodetic form;
- civic – the LIS should provide location as a street address;
- locationURI – the LIS should provide a set of location URIs, which when dereferenced will yield the location of the Target (this is how HELD supports location by reference).

Using the locationType qualifier in a request allows the requestor to indicate what form they would prefer their location information to take, but it does not mandate what the LIS must

```
<locationRequest xmlns="urn:ietf:params:xml:ns:geopriv:held">
    <locationType>
        geodetic
        civic
    </locationType>
</locationRequest>
```

Figure 2.11 Location request with locationType.

```
<locationRequest xmlns="urn:ietf:params:xml:ns:geopriv:held">
    <locationType exact="true">
       geodetic
       locationURI
    </locationType>
</locationRequest>
```

Figure 2.12 Location request using "exact" attribute.

provide. For example, if the request contains a location type preference of geodetic but the LIS is only able to provide civic location information, then the LIS will respond with a civic address. Similarly, if the requestor requests multiple location types in the same request and the LIS is only able to satisfy some of them, then only the types that the LIS can provide will be provided in the response.

In some cases the requestor wants only what they ask for, and if the LIS cannot provide it then they want an error. HELD supports this behaviour through the use of the exact attribute (Figure 2.12), which is applied to the locationType parameter. By default, the exact attribute has a value of false, and so may be omitted from the request. However, when it is set to true, the LIS must return all of the requested location types or else return an error. Partial responses are not supported.

Response Time

Different types of networks offer different location determination techniques, and some types of wireless networks can support multiple determination methods. Experience has shown that, generally, the more accurate the location desired, the longer it takes to determine it. Applications, however, have a finite time in which location is useful to them, and many applications can use a location even if it is not as accurate as is it could be. These characteristics of location determination for applications are addressed in HELD by the responseTime attribute (Figure 2.13) and it applies to the whole location request.

The response time attribute is expressed in milliseconds and provides the LIS with a means of arbitrating between alternative location determination methods when more than one exists, allowing the LIS the opportunity to return the most accurate location it can within the allotted time. The response time is only an indicator of how much time the LIS has to find a location

```
<locationRequest xmlns="urn:ietf:params:xml:ns:geopriv:held"
      responseTime="8000">
   <locationType exact="true">
      geodetic
     locationURI
   </locationType>
</locationRequest>
```

Figure 2.13 Location request using "responseTime" attribute.

```
<locationRequest xmlns="urn:ietf:params:xml:ns:geopriv:held"
      responseTime="emergencyRouting">
   <locationType exact="true">
      geodetic
   </locationType>
</locationRequest>
```

Figure 2.14 Location request using “responseTime="emergencyRouting"” attribute.

and respond to the requestor. A LIS must to try to accommodate a request within this time, but it may take longer if there are no determination methods that can provide a result within the requested response time. In this case, whether the LIS takes longer and responds with a location or returns an error is an implementation decision and may vary depending on the magnitude of the time difference.

If no response time is specified in the location request, then the LIS provides the most accurate location that it can.

There are two special values for response time that are configured locally on the LIS by the LIS provider. These are the emergencyRouting and emergencyDispatch values. When the LIS receives a location request with a response time value set to emergencyRouting or emergencyDispatch (Figure 2.14), it substitutes the corresponding value set by the operator, which was likely provided to it by the local emergency service providers. These values ensure that the location provided by the LIS will most likely meet certain standards within an acceptable period of time. It is important to note that these are time values, not an indicator of the intended use of the location information by the location recipient.

2.4.3.2 Location Responses

The LIS responds to a successful locationRequest message with a locationResponse message. The location response message may contain a PIDF-LO [17] and/or a location URI set. When HTTP is used to convey HELD messages, the locationResponse is always conveyed in the body of a 200 OK message (Figure 2.15).

Returning a PIDF-LO

As described elsewhere in this book, the PIDF-LO is a container object capable of carrying both civic and geodetic location information. Location objects returned in HELD must comply with the formatting rules specified in RFC 5491 [28]. This ensures that other entities receiving the location information can interpret it in the way in which the LIS intended.

The entity parameter is mandatory in the present document, but in many cases an LIS servicing a HELD request will not know the actual identity of a user. Furthermore, privacy constraints prohibit the inclusion of device identity information without explicit permission from the user, and since providing this permission is out of scope for HELD, the LIS must assume that it has not been given. To address this problem, the LIS generates an unlinked pseudonym and converts it into a URI, which is then used as the entity value in the PIDF document.

```
HTTP/1.1 200 OK
    Server: Example LIS
    Date: Sat, 20 Feb 2010 03:42:29 GMT
    Expires: Sat, 20 Feb 2010 03:42:29 GMT
    Cache-control: private
    Content-Type: application/held+xml;charset=utf-8
    Content-Length: 913

    <?xml version="1.0"?>
     <locationResponse xmlns="urn:ietf:params:xml:ns:geopriv:held">
      <presence xmlns="urn:ietf:params:xml:ns:pidf"
      entity="pres:3650n45g6j8@lis.example.com">
       <tuple id="b650sf789nd">
        <status>
         <geopriv xmlns="urn:ietf:params:xml:ns:pidf:geopriv10">
          <location-info>
           <Point xmlns="http://www.opengis.net/gml"
           srsName="urn:ogc:def:crs:EPSG::4326">
            <pos>-34.407 150.88001</pos>
           </Point>
          </location-info>
          <usage-rules
           xmlns:gbp="urn:ietf:params:xml:ns:pidf:geopriv10:basicPolicy">
           <retransmission-allowed>false</retransmission-allowed>
           <gbp:retention-expiry>2010-02-17T03:42:28+00:00
           </gbp:retention-expiry>
          </usage-rules>
          <method>Wiremap</method>
         </geopriv>
        </status>
        <timestamp>2010-02-16T03:42:28+00:00</timestamp>
       </tuple>
      </presence>
     </locationResponse>
```

Figure 2.15 Location response. (Modified from RFC 5985. Copyright © 2010 IETF Trust and the persons identified as the document authors. All rights reserved.)

Location information is included in a tuple element. The device and person elements are not used, as both reveal identity information about the user, and no means to obtain explicit user permission for this is available.

The usage rules that comply with the default values specified in RFC 4119 are used. These consist of a retention-expiry value of 24 hours from the time of the response and retransmission-allowed value is set to false. No other usage rules are specified.

The value for the method element should reflect the mechanism used to determine the location of the Target device, and valid values for this element are defined in an IANA registry [78].

The timestamp element should always be sent, and should be set to the time that the location was determined, or, in the case of a database lookup, the time the data was extracted from

```
<locationResponse xmlns="urn:ietf:params:xml:ns:geopriv:held">
 <locationUriSet expires="2012-02-01T13:00:00.0Z">
  <locationURI>https://lis.example.com:2020/22175xoceeod309fdmr</locationURI>
  <locationURI>sip:8192+22175xoceeod309fdmr@lis.example.com</locationURI>
 </locationUriSet>
</locationResponse>
```

Figure 2.16 Location URI set in location response.

the database. Timestamps should generally be expressed in GMT, but may include specific timezone offsets from GMT.

Returning a Location URI Set

Location URIs are returned in the location response in a locationUriSet element (Figure 2.16). The location URI can contain any number of location URIs, each in its own element. In general, an LIS provides a location URI for each URI scheme it supports for location dereferencing, and each URI scheme is only provided once.

The URI set contains an absolute expiry time after which the URIs are no longer valid. Dereferencing an expired location URI results in a protocol error; for HTTP this is a "404 Not Found" error.

2.4.3.3 HELD Errors

HELD errors are returned when the LIS is unable to process a request for some reason. Like location response messages, when HTTP is the transport protocol, HELD error messages are returned in the body of a 200 OK message. The basic structure of the error message is simple: it consists of a mandatory code parameter that indicates what the error is, and an optional message element that may contain a textual description of the problem for human consumption (Figure 2.17).

```
HTTP/1.1 200 OK
Server: Example LIS
Date: Sat, 20 Feb 2010 03:42:29 GMT
Expires: Tue, 20 Feb 2010 03:49:20 GMT
Cache-control: private
Content-Type: application/held+xml;charset=utf-8
Content-Length: 182

<?xml version="1.0"?>
<error xmlns="urn:ietf:params:xml:ns:geopriv:held"
   code="locationUnknown">
   <message xml:lang="en">Unable to determine location</message>
</error>
```

Figure 2.17 HELD error example. (Modified from RFC 5985. Copyright © 2010 IETF Trust and the persons identified as the document authors. All rights reserved.)

Table 2.2 HELD error codes

Code	Description
cannotProvideLiType	The LIS was unable to provide the exact type of location information requested. This code is only returned when the exact attribute is set to true
generalLisError	An unspecified error occurred on the LIS. The requestor should try again after a short delay
locationUnknown	The LIS is unable to determine the location of the device at this time. The requestor should delay their request for a short time before retrying
notLocatable	The LIS is unable to locate the device because the device is outside the accessed network serviced by the LIS. The requestor must not try to request location for this device again
requestError	The location request was badly formed
timeout	The time specified in the responseTime parameter has expired and the LIS was unable to determine the location of the device
unsupportedMessage	The message received by the LIS was valid XML but not understood
xmlError	The request contained an XML error

HELD error codes are defined in an IANA registry [79]. This allows new error codes to be added to HELD as extensions are added. A description of each HELD error code is included in Table 2.2.

2.4.4 HELD Target Identities and Third-Party Requests

The basic HELD protocol, which has been focused on to this point, is intended to be used by a device requesting its own location, that is, it is a location configuration protocol (LCP). A basic assumption in HELD is that the device's IP address is the best identifier to use as a root key for determining the location of the device; but this is not always the case, so a device needs to have a way of providing the LIS with an identifier other than (or as well as) its IP address when requesting its location. This need gave rise to the HELD identity extensions specification [35].

One of the advantages of using an application protocol, such as HELD, is that it is adaptable. The initial intent of the HELD identity extensions was to provide the ability for a device to use more than IP address to identify itself to the LIS. However, it quickly became apparent that this same mechanism could be used by a trusted third party to request the location of a device from the LIS, and this has subsequently become the primary focus of the HELD identity extensions document.

Location information is pivotal to the correct operation of next-generation emergency services, yet almost none of the VoIP clients deployed at the time of writing are location-capable or location-aware. This means that they are unable to provide location information with the emergency call, or to do route determination prior to making call. The HELD identity extensions allow a third-party routing node to request this information from the LIS

```
<device xmlns="urn:ietf:params:xml:ns:geopriv:held:id">
        </identity-1>
        </identity-2>
        ...
        </identity-n>
</device>
```

Figure 2.18 Device identity extensions structure.

on behalf of the calling device, enabling general emergency call support without requiring modifications to legacy VoIP clients. As a result of this capability, the HELD identity extensions are finding uses in a range of transitional next-generation emergency architectures and specifications.

2.4.4.1 Identity Extensions Structure and Types

The HELD identity extensions define an XML container that may contain any number of identifiers (Figure 2.18). The schema included in the specification defines a rich set of identifiers for a wide range of access network technologies, but includes extension points so that additional or even proprietary identifiers can be added if need be.

Device identifiers may be discrete, for example, an Ethernet MAC address, or may be combined together to form a unique identifier, for example, an IP address and TCP port might identify a device behind a NAT (Figure 2.19).

The identity extensions specification provides a lot of details on each identity type that it defines. For completeness these types are provided in Table 2.3. However, for examples on each identity type, the reader is referred to the HELD identity extensions specification [35].

The use of telephone numbers and the like is often discussed when the topic of device identifiers is raised in telephony circles. The HELD identify extensions address these in a number of ways.

General directory numbers, PBX extensions, and the like are specified using a tel URI in the uri element (Figure 2.20). Tel URIs and how to specify them are described in RFC 3966 [80].

The identity extensions specification explicitly defines types for cellular mobile device identifiers. This is because these types have explicit meanings and there is no easy way to express which type is which using a tel URI without the introduction of additional parameters into the URI scheme and their associated complexity.

```
<device xmlns="urn:ietf:params:xml:ns:geopriv:held:id">
        <ip v="4">192.168.1.100</ip>
        <tcpport>9192</tcpport>
</device>
```

Figure 2.19 Complex device identity.

Table 2.3 HELD identity extensions

Identifier	Description
ip	The IP address of the target to locate. This identifier has an additional v version attribute that is used to specify if the address is an IPv4 or an IPv6 address
mac	The MAC address of the target to locate
tcpport	The TCP port number attributed to the device. Usually used in combination with an IP address
udpport	The UDP port number attributed to the device. Usually used in combination with an IP address
nai	The network address identifier (NAI) of the device or user. This type of identifier is used a lot in PPP and RADIUS-based networks such as DSL, WiFi, and WiMAX
uri	This may be something like an SIP URI, a presence URI, or even a tel URI
hostname	The fully qualified hostname (FQDN) of the device
msisdn	Mobile Station International ISDN Number (MSISDN). This is a cellular telephony identifier represented as an E.164 number between 6 and 15 digits long
imsi	International Mobile Subscriber Identity (IMSI). This is a cellular telephony identifier represented as an E.164 number between 6 and 15 digits long
imei	International Mobile Equipment Identifier (IMEI). This is an identifier used in cellular networks and represents a unique serial number for a device. It can be up to 15 digits long
min	Mobile Identification Number (MIN). This is a unique number assigned to CDMA devices
mdn	Mobile Directory Number (MDN). This is an E.164 number used to identify CDMA subscribers
duid	DHCP Unique Identifier. This is often referred to as the Client identifier in DHCP circles and is expressed in option 61 in DHCPv4, and option 1 in DHCPv6

2.4.4.2 Including Identity Extensions in HELD Messages

The HELD messages are extensible and easily support the inclusion of additional data. The device element is easily added to the HELD locationRequest (Figure 2.21).

If the LIS does not understand identity extensions, then, under the rules of XML, it must ignore all elements that it does not understand, in which case the source IP address of the

```
<device xmlns="urn:ietf:params:xml:ns:geopriv:held:id">
        <uri>tel:800-555-1234;phone-context=+1</uri>
</device>
```

Figure 2.20 A tel URI example.

```
<locationRequest xmlns="urn:ietf:params:xml:ns:geopriv:held"
      responseTime="5000">
   <locationType exact="true">
      geodetic
   </locationType>

   <device xmlns="urn:ietf:params:xml:ns:geopriv:held:id">
      <ip v="4">192.168.1.100</ip>
   </device>

</locationRequest>
```

Figure 2.21 Location request with identity extension.

requesting entity is used as the target identifier. For third-party queries, this almost always results in an error of notLocatable being returned to the requesting entity.

There are cases where the LIS requires explicit identity extensions in order to be able to determine the location of a device. Often the requesting device has this information but does not know that the LIS requires it, so it is not provided in an initial location request. This scenario is catered for in the identity extensions specification by the introduction of a new error code and element that are included in the HELD error message.

The badIdentifier error code indicates that the requestor was not authorized to use a certain identifier, or that the required identifier was not present in the request. In the latter case, the LIS should provide a list of identifiers that are acceptable for identifying the target; these are included in the requiredIdentifiers element (Figure 2.22). This element is a list of identifiers that are qualified by the XML namespace in which they are defined. By taking this approach, the element automatically caters for any new identifiers that may be defined in the future.

2.4.5 HELD Measurements

In addition to providing alternative identifiers to its IP address when making a location request, a device might also be able to measure certain characteristics of its surrounding environment.

```
<error xmlns="urn:ietf:params:xml:ns:geopriv:held"
               code="badIdentifier">
  <message xml:lang="en">MAC address required</message>
  <requiredIdentifiers xmlns:di="urn:ietf:params:xml:ns:geopriv:held:id">
     di:mac
  </requiredIdentifiers>
</error>
```

Figure 2.22 Required identifiers example.

```
<lm:measurements xmlns:lm="urn:ietf:params:xml:ns:geopriv:lm"
                        time="2010-04-29T14:33:58"
                        expires="2010-04-29T17:33:58">

   <!-- Explicit measurements go here -->

</lm:measurements>
```

Figure 2.23 HELD measurement container.

These measurements may include a variety of things, including radio-frequency information about the serving and nearby access networks. HELD does not require this information to be provided to the LIS in order for the location of a device to be determined. However, if the device is able to provide this information, and the LIS is able to verify it to some degree, then such measurements may provide enhancements to accuracy or reduce the amount of time required to determine the location of the device.

The HELD measurements specification [81] is broken up into three main parts:

- a container, in which measurements are transported (Figure 2.23);
- a base-types XML schema that defines types from which measurements can be constructed; and
- measurement schemas, which define measurements for specific access network types.

As with other HELD extensions, the measurement extensions are flexible, and it is easy to add new measurement types to support new access network technologies as they become available. It is anticipated that most new measurement types can be constructed from the defined base types.

The time attribute indicates when the measurements were actually captured. This attribute is not mandatory to use, and indeed in relatively static environments may provide little benefit. In dynamic environments, such as those involving wireless access, it is recommended that this attribute always be used so that measurement freshness can be estimated.

The expires attribute indicates the time after which the measurements must no longer be used. Like the time attribute, this attribute is not mandatory but it is recommended that it be included where measurements are likely to go stale in relatively short periods of time.

The measurements specification defines measurement schemas for a range of access technologies and supports not only measurements that may be acquired by a target device, but also measurements that may be passed from one LIS to another LIS. On the surface, this type of function may seem a little peculiar. However, in access networks where the Internet service is provided by an operator other than the infrastructure provider, wholesale DSL services, for example, a single LIS may not have all of the information or relationships necessary to determine the location of the device, so measurements must be passed from one to the other. Examples of where this is necessary are described in [75] and [76].

Table 2.4 HELD measurement specification access measurement types

Access technology	Description	
LLDP	Link Layer Discovery Protocol as described in 802.1AB	
DHCP Relay Information	DHCP relay information is described in detail in RFC 3046 [49]	
802.11 Wireless LAN (WiFi)	RF signal measurements for 802.11 range or technologies. This is quite extensive and represents both measurements that the device can take as well as measurements taken at the AP	
Cellular wireless measurements	Schema is provided for GSM, UMTS, LTE, and CDMA cell measurements. Specific radio timing measurements like timing advance and chip offsets are not defined	
Global Navigation Satellite System (GNSS) measurements	Provides the basic types and constructs necessary to define assistance data and measurements for any GNSS system. Specifically, it defines structures for the Global Positioning System (GPS) and Galileo GNSS deployments	
Digital Subscriber Line (DSL) measurements	Layer 2 Tunnelling data Protocol (L2TP) measurements	Provides a schema to encapsulate the tunnel and session identifier for a data stream from the regional access provider network to the ISP
	RADIUS measurements	Provides a schema to encapsulate the access node, slot and port information often reported by a network access server (NAS) in accounting record streams or at access request time
	Ethernet VLAN tag measurements	Used to identify the source data stream in DSL networks where Ethernet is deployed to the DSLAM
	ATM Virtual Circuits	Provides a measurement schema that contains the VPI and VCI of the ATM data stream terminated at the network access server (NAS)

The measurement specification provides measurement definitions listed in Table 2.4. Measurements can be included in a HELD location request from the device to the LIS as shown in Figure 2.24.

Additional measurement schemas to cover other access technologies also exist, including one for WiMAX [82], which is included in the WiMAX Release 1.5 LBS specification.

2.4.6 HELD as a Dereference Protocol

Figure 2.5 showed the two location mechanisms, a literal location (location by value) and a location (by) reference. A location reference is useful in circumstances when the device might

```
<locationRequest xmlns="urn:ietf:params:xml:ns:geopriv:held">
   <locationType> geodetic </locationType>
   <measurements xmlns="urn:ietf:params:xml:ns:geopriv:lm"
                 time="2010-23-16T14:33:58">
       <wifi xmlns="urn:ietf:params:xml:ns:geopriv:lm:wifi">
         <nicType>Example \index{WiFi}WiFi Device</nicType>
         <ap serving="true">
           <bssid>00-12-F0-A0-80-EF</bssid>
      <ssid>wlan-home</ssid>
           <channel>7</channel>
           <type>a</type>
           <band>5</band>
           <antenna>2</antenna>
           <flightTime rmsError="4e-9" samples="1">2.56e-9</flightTime>
           <deviceSignal>
             <transmit>10</transmit>
             <gain>9</gain>
             <rcpi dBm="true" rmsError="9.5" samples="1">-98.5</rcpi>
             <rsni rmsError="6" samples="1">7.5</rsni>
           </deviceSignal>
         </ap>
         <ap>
       <bssid>00-12-F0-A0-80-F0</bssid>
           <ssid>crazy-home</ssid>
           <flightTime rmsError="4e-9" samples="1">5.78e-9</flightTime>
         </ap>
       </wifi>
   </measurements>
</locationRequest>
```


Figure 2.24 Measurements in Location Request. (From the HELD Measurements draft, http://tools.ietf.org/id/draft-ietf-geopriv-held-measurements-05.txt.)

be moving and the application requiring location cannot predict in advance exactly when it will need location information. In this situation, a location reference allows the application to request location information directly from the source, the LIS. Other uses for location references include the application of policy: for example, a PSAP may have privileges to get very precise location information, while a commercial application may not. By identifying the recipient, quality-of-service policy can be applied.

HELD supports location references in the form of location URIs, and these can be explicitly requested as a location type. HELD can also be used to dereference a location URI. A location URI with a scheme of http: or https: should be considered to point to a HELD service. Using HELD in this manner is described in the HELD dereference specification [83].

The location by reference requirements document RFC 5808 [84] describes two basic authentication methods for location URIs: those requiring explicit authentication, and those that grant authorization based on possession of the URI alone. These two methods are commonly referred to as the authentication and possession models. The HELD dereference specification allows for both methods to be used, though it does not define how user policy

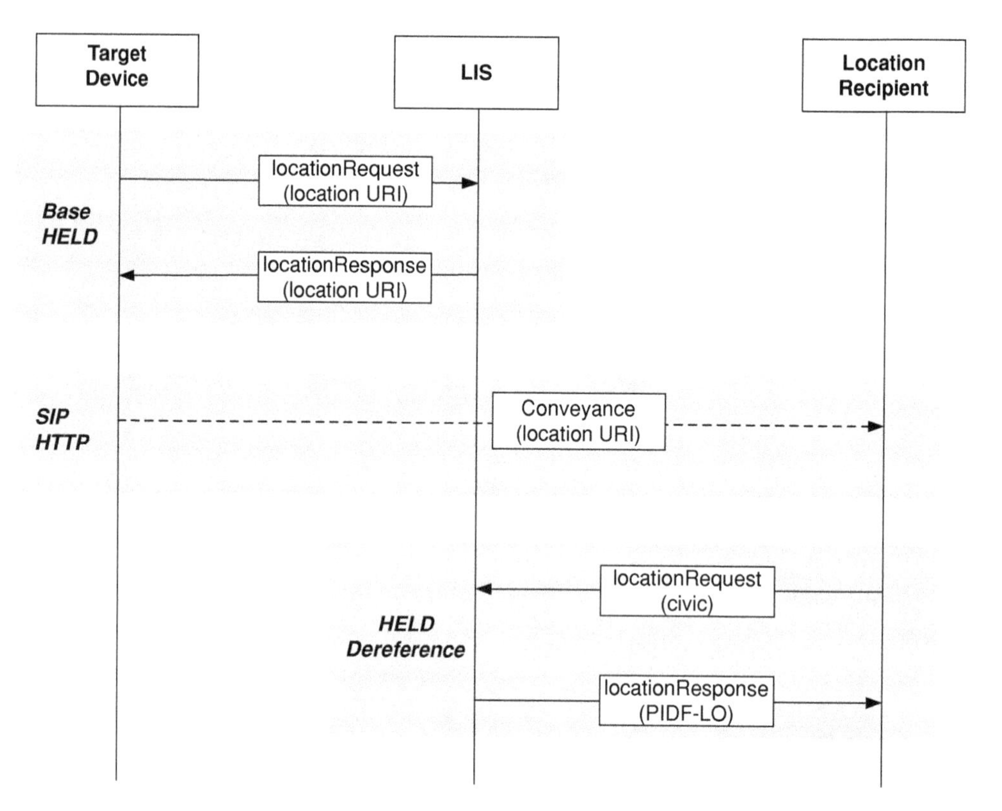

Figure 2.25 HELD dereference context diagram.

concerning who may dereference the URI is made available to the LIS. Since basic HELD does not define a mechanism for the user of a device to attach policy to a location URI, the user should assume that the URI adheres to the location by possession model described above, and take the necessary precautions to restrict access to this URI to the same degree as if it were a literal location value (Figure 2.25).

Identity extensions and measurements must not be included in a HELD URI dereference operation and result in an error being returned if they are. Furthermore, the LIS must not return a location URI in response to a HELD URI dereference operation. If a location type of locationUri is requested and the exact attribute is set to true, then the LIS returns a cannotProvideLiType error.

2.4.7 *HELD Policy URIs*

The previous section described how location URIs in the basic HELD specification follow the possession security model. In this model, location is always granted when a location URI is dereferenced because possessing the URI is all that is required in order to prove authorization to the information. Often this is not good enough, and a user wants to restrict who can use a location URI. In HELD this is done using the policy URI specification [85].

A LIS operator may provide their own policies around what information a location URI may reveal and to whom, in addition to the general possession model provided by default. The policy URI specification not only allows the user of a device to set, view, and change their own policy associated with a location URI, but also allows them to view operator policy as well.

A policy URI is associated with a location URI, and points to a place where policy is stored that grants access to location information for a particular device. If there is no stored policy associated with the policy URI, then a possession model is assumed. This ensures backwards compatibility with the basic HELD specification, where policy URIs are not considered.

A LIS that understands the policy URI specification may always return a policy URI when a location URI set is provided to the device. This gives the device the option of specifying their own specific policy if they understand this extension. Alternatively, a client that understands the policy URI extension may explicitly request a policy URI when it requests a location URI. The device needs to take care in this case, because, if a policy URI is not returned, it means that the LIS does not understand the extension and that a possession authentication model should be assumed for all location URIs. An example request and response to an LIS for a policy URI is shown in Figure 2.26.

The policy associated with a URI may be stored in the LIS itself, or in a separate policy server, the choice of which is a matter for implementation. The policy URI uses the http or https URI scheme; the policy is viewed and manipulated using the http GET, PUT, and DELETE operators. The policy server must accept a policy specified in accordance with the common policy specification RFC 4745 [86]. Alternative policy specifications may be used after successful content negotiation between the device and the policy server has occurred.

The general flow for acquiring location information when using a policy URI is shown in Figure 2.27.

To view the current policy associated with a URI, a GET is issued against the policy URI by the client device. The client device may then add, change, or delete rules in the policy, and

```
<locationRequest xmlns="urn:ietf:params:xml:ns:geopriv:held">
   <locationType exact="true">locationURI</locationType>
   <requestPolicyUri xmlns="urn:ietf:params:xml:ns:geopriv:held:policy"/>
</locationRequest>

<locationResponse xmlns="urn:ietf:params:xml:ns:geopriv:held">
   <locationUriSet expires="2010-12-29T14:33:00.0Z">
      <locationURI>
         https://lis.example.com:9692/38x983nbl5r0cjn2w
      </locationURI>
   </locationUriSet>
   <policyUri xmlns="urn:ietf:params:xml:ns:geopriv:held:policy">
     https://ps.example.com:9768/policy/79d8409s930s0930ds
   </policyUri>
</locationResponse>
```

Figure 2.26 HELD policy URI request and response.

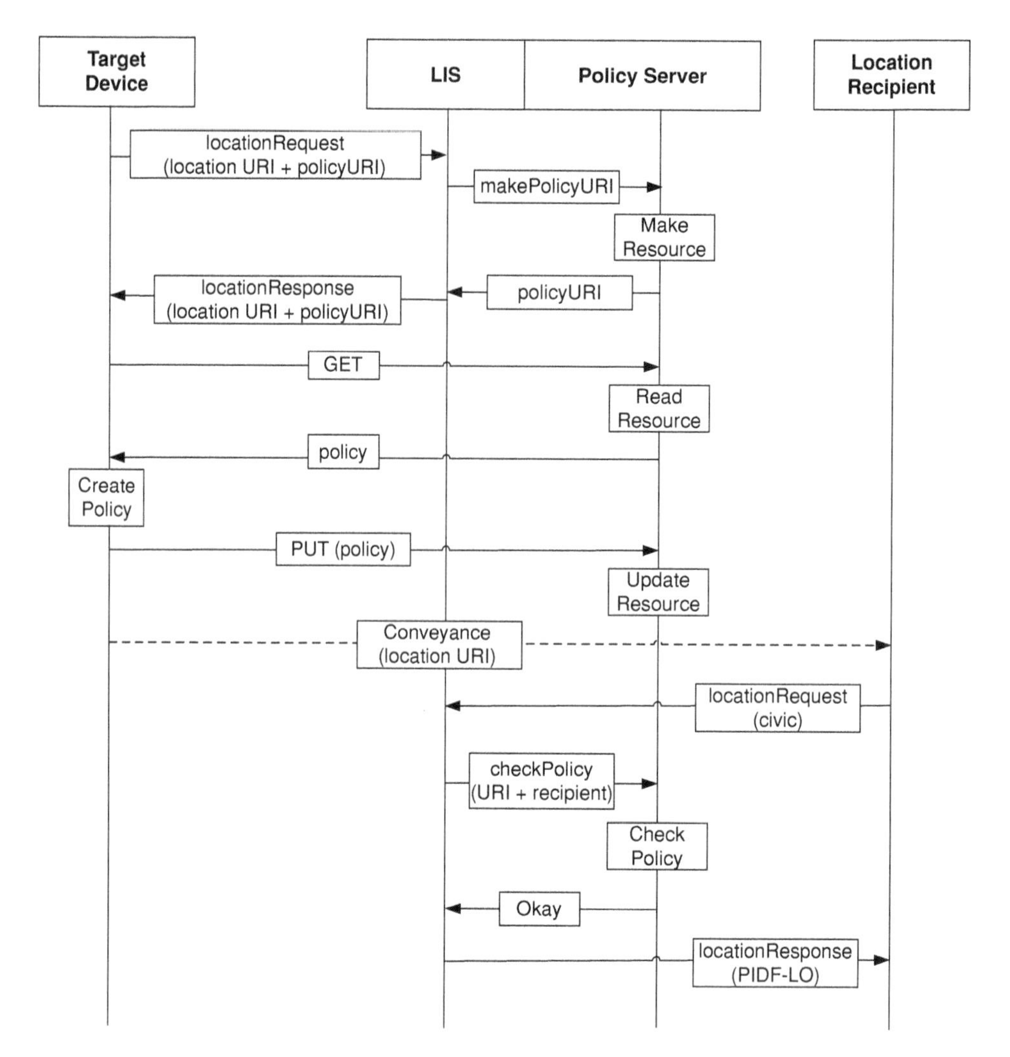

Figure 2.27 Policy URI Usage.

return the resulting policy to the policy server using a PUT command. When this occurs, the policy server checks the policy document to ensure that it is valid, and that any mandatory rules are still present. The policy server will return an HTTP error if it cannot process the policy document provided by the client.

Deleting the policy URI using a DELETE operation is not the same as putting an empty policy document in the policy server. Issuing a DELETE to a policy URI voids not only the policy URI, but also by proxy any associated location URI. This is because the DELETE destroys the resource associated with the policy URI, and any attempt to deference the URI will now result in a 404 NOT FOUND error. If an LIS has a link between a location URI and a policy URI, and the policy URI is not valid, the LIS must deny all access for location information via the location URI.

It is important to note that the policy URI and associated policy are related to a specific location URI set. If a device makes a request for another location URI set, it will be provided with a different policy URI, which is unique from any other policy URI.

2.4.8 HELD Device Capabilities

The measurements specification [81] allows a device to capture information about its local environment and include this information as part of a location request to an LIS. This information is helpful to the LIS, as it provides information that the LIS would not otherwise have, and improves the accuracy or time to determine the location of the device. But if the location request comes from an external location recipient via a location URI, then there is no way to invoke these measurement capabilities in the device and thereby gain the advantage of these device-based measurements. The HELD capabilities specification addresses this shortcoming.

The device capabilities specification [87], allows a device to tell the LIS what kinds of measurements and functions it can perform. It does this by including a list of the capabilities that it supports when requesting a location URI from the LIS (Figure 2.28). If the LIS understands this extension, then it responds with an agreed set of capabilities that it might use, along with an HTTP resource that the device must monitor for invocation of a specific capability by the LIS. This mechanism relies on HTTP long polling to provide a session over which capabilities can be invoked.

When the LIS requires measurements from the device, it returns a request for certain information from the device on the invocation URI, as well as a destination for the device to post the results to. In this way the LIS can invoke advertised device capabilities as required to ensure requests made via location URIs can be provided with accurate and fast location responses.

2.4.8.1 Capabilities Indication

Capabilities are advertised by the device to the LIS in a location request message containing a deviceCapabilities element. The deviceCapabilities element may contain a location element and/or one or more measurement elements. A location element indicates that the device can determine its own location. A measurement element indicates that the device can provide measurement data of a specific type that may be useful in determining the location of the device.

Each capability defines a capability id that is unique within the scope of the deviceCapabilities element. The value of this attribute is used as a handle by the LIS to invoke the capability in the device when required.

Capabilities will generally be invoked on the device in order for it to be of service to the LIS, either calculating its own location, or collecting measurements so that the LIS can calculate the location, and all of this takes time. When the device indicates a capability to the LIS, it should also indicate how long the capability takes to provide a result so that the LIS knows whether to invoke the capability or not for a particular location request. The responseTime attribute in the measurement and location elements provides this function, and it is an integer specified in milliseconds.

The capabilities extract in Figure 2.29 shows the device indicating two capabilities. The device is able to provide its own location, and the device is able to provide GNSS (in this case

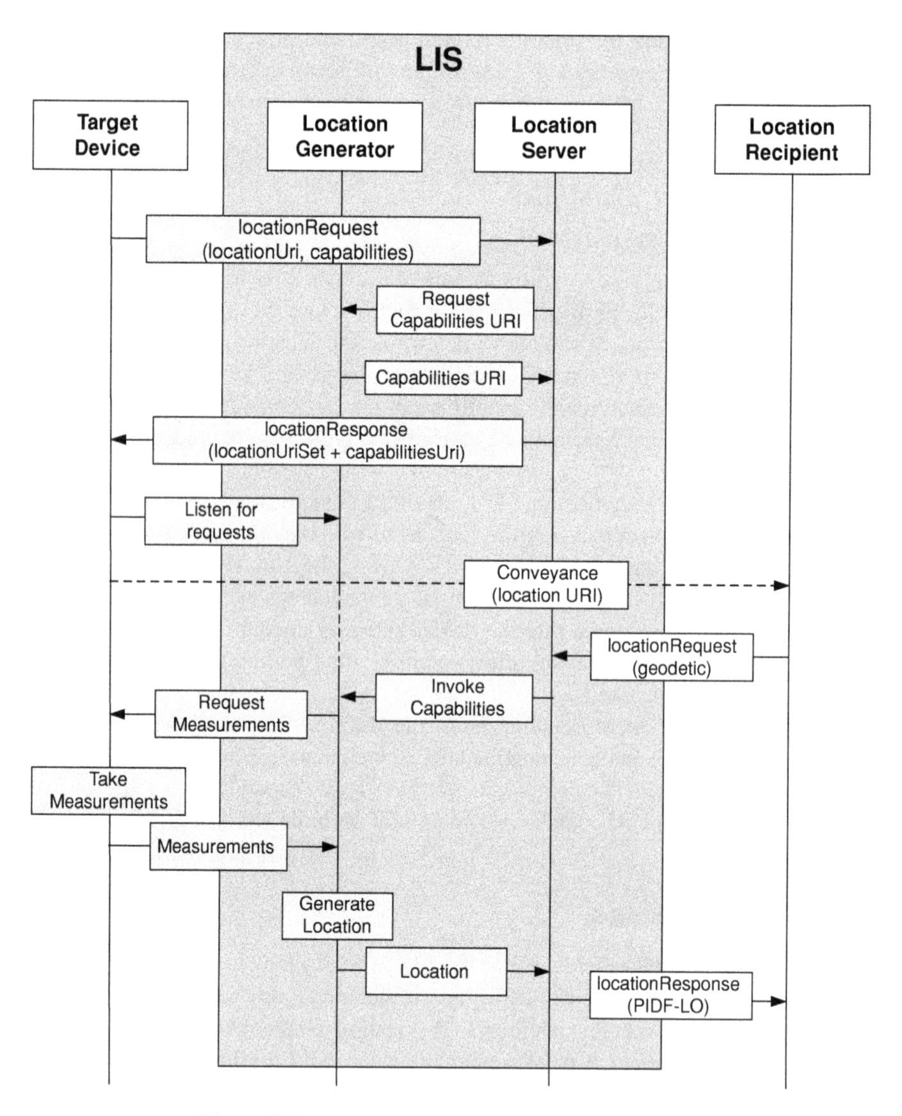

Figure 2.28 Device capabilities signaling diagram.

GPS) satellite measurement information. Additional capabilities are easy to specify, and can be done in any forum or organization that can register a namespace.

2.4.8.2 Capabilities Agreement

A device will generally indicate all capabilities that it supports to the LIS, but the LIS may not support or agree to use all of these, for a variety of reasons. The capabilities that the

```
<held:locationRequest
        xmlns:held="urn:ietf:params:xml:ns:geopriv:held">
      <held:locationType exact="true">locationURI</held:locationType>
      <requestPolicyUri
         xmlns="urn:ietf:params:xml:ns:geopriv:held:policy"/>

      <cap:deviceCapabilities
         xmlns:cap="urn:ietf:params:xml:ns:geopriv:held:cap">
        <cap:location id="loc" responseTime="45000">
          <held:locationType>geodetic</held:locationType>
        </cap:location>
        <cap:measurement
            xmlns:gnss="urn:ietf:params:xml:ns:geopriv:lm:gnss"
            type="gnss:gnss" id="gps" responseTime="12000">
          <gnss:gnss system="gps" signal="L1"/>
        </cap:measurement>
      </cap:deviceCapabilities>
</held:locationRequest>
```

Figure 2.29 Capability indication example. (From IETF HELD capabilities draft.)

LIS is prepared to use are included in the agreedCapabilities element provided in the location response message (Figure 2.30).

The agreedCapabilities element contains a child element, called monitor, that in turn contains a URI. Devices must connect to that URI in order to listen for when the LIS wishes to invoke the agreed capabilities. The agreedCapabilities also contains a list of the capabilities that the LIS may invoke when required.

```
<held:locationResponse xmlns:held="urn:ietf:params:xml:ns:geopriv:held">
   <held:locationUriSet expires="2011-01-06T5:55:00+10:00">
      <held:locationURI>
         https://lis.example.com/48c8nj3209sao\index{NENA!i2}i20sk3
       </held:locationURI>
   </held:locationUriSet>

   <cap:agreedCapabilities
       xmlns:cap="urn:ietf:params:xml:ns:geopriv:held:cap">
      <cap:monitor>
           https://lis.example.com/inv/c094jx983nd093d
      </cap:monitor>
      <cap:location id="loc"/>
      <cap:measurement id="gps"/>
   </cap:agreedCapabilities>
</held:locationResponse>
```

Figure 2.30 Capabilities agreement. (From IETF HELD capabilities draft.)

```
<invokeCapabilities xmlns="urn:ietf:params:xml:ns:geopriv:held:cap">
   <measurement id="gps" before="2010-12-25T12:00:00+10:00">
        <push>https://lis.example.com:9199/ggauwiz371873add</push>
   </measurement>
</invokeCapabilities>
```

Figure 2.31 Capability invoation. (Modified from the IETF HELD capabilities draft. Copyright © 2011 IETF Trust and the persons identified as the document authors. All rights reserved.)

2.4.8.3 Invoking Capabilities

The capabilities agreement that the device receives provides the device with a resource on which to connect and listen for capability invocation requests. This is an http or https URI, and HTTP long polling techniques are used to simplify procedures at the device end. When the LIS is ready to invoke a capability, it writes an invocation request to the resource provided to the device.

Figure 2.31 shows the invocation of the GPS measurement capability, requesting that the device provide GPS measurements to the LIS before noon on 25th December 2010. The LIS used specifies the before attribute to indicate that it requires the result before this specific time. The LIS also specifies a push URI as the place where the device needs to send the results. The device uses the HTTP PUT command to deliver the result to the push URI.

In addition to the before attribute, the LIS may request that the measurements be made again periodically and subsequently sent to the LIS. This is done using the periodic attribute, which specifies the number of milliseconds to wait between making measurements. The periodic interval starts at the time specified in the before attribute. When the device has taken the required measurements, or determined its location, it uses HTTP to PUT the result to the URI provided in the invocation request. Measurements are encoded in a measurements container defined in [81], while locations are sent as a PIDF-LO.

There are several advantages of this approach. First, it is very simple and easy to implement. Second, connections always occur from the device to the LIS, and this makes security much easier to manage. Third, using HTTP in this manner means that the protocol will easily transit most firewalls, meaning that it will work in most network environments.

2.5 OMA Enablers and Emergency Services

Khiem Tran
CommScope

The Open Mobile Alliance (OMA) [59] is a standardization body consisting of nearly 200 member companies, with a goal to produce open standards-based enablers for mobile services.

The OMA releases its specifications as part of "Enabler Releases," each of which may be either "Draft," "Candidate," or "Approved." An Enabler Release may contain several different specifications, and sometimes the same specification may appear in different Enabler Releases.

The OMA has a number of enablers related to either the calculation or the retrieval of location of a device attached to the mobile network. These enablers, SUPL, MLP, MLS (which contains the later versions of MLP), and LOCSIP, all have uses for emergency services, and a high-level overview of each is provided in the following sections.

2.5.1 SUPL

SUPL is the OMA's enabler for Secure UserPlane Location. At the time of writing, there were three separate Enabler Releases for SUPL: SUPL1.0 [88], which has been available as a Candidate Enabler since 2006 and upgraded to an Approved Enabler in 2007; SUPL2.0 [89], which was briefly approved as a Candidate Enabler in 2009, then subsequently reverted to Draft again; and SUPL3.0 [90]. This section gives an overview of SUPL and examines its applications for emergency services.

2.5.1.1 What is SUPL?

SUPL is an enabler that allows a location server to communicate with a mobile device over the User Plane (i.e., the normal data bearer provided by an access network) instead of the control plane (the signaling channels available for that access network). As such, it has a number of advantages over other location architectures that rely on control plane signaling.

The first is that SUPL can be relatively independent of access network. As long as there is User Plane connectivity between the device and the server, SUPL can be supported with little requirement for interaction between the SUPL server and the access network (there is still some necessary interaction in order to support the full SUPL authentication model). The drawback of this independence is that it is also harder for the SUPL Location Platform (SLP), a location server, to interact with network nodes to help calculate location – all measurements generally come from the SUPL Enabled Terminal (SET) and not from other network nodes.

The second advantage is that the SUPL architecture can be relatively simple and can be deployed on top of existing networks with minimal impact on other network nodes. In contrast, control plane location architectures, such as those defined by 3GPP in TS 23.271 [91], require the support of location functionality in a series of different network nodes and, in some cases, such as the UTRAN location architecture described in TS 25.305 [92], require multiple protocols. A typical SUPL location session, however, only involves direct communication between the SET and the SLP. This means that SUPL can be deployed rapidly without having to modify existing network nodes. It also makes it possible to deploy SUPL without any relationship at all with the access network, except for the security requirements.

The third advantage is that bandwidth on the User Plane is generally greater than that available on the control plane. This means that additional measurements that are not supported in control plane positioning protocols can be supported in SUPL.

2.5.1.2 Security in SUPL

As the "S" in the name suggests, security is an important aspect of SUPL. This is achieved in two ways. First, SUPL uses a secure data session for communication between the SET and the SLP, with a series of alternative authentication models to authenticate the SET and establish the secure session. Second, there is an implicit trust relationship between the SET and its Home SLP (H-SLP). This second aspect needs a bit more explanation, as it has implications for emergency services applications, as will be seen later on.

In SUPL, when a secure data session is required for communication between an SET and an SLP, it is always established by the SET. In the SUPL1.0 Enabler Release, it was a fundamental principle that the SET would always establish this session to its Home SLP (H-SLP). The address of the H-SLP could be pre-configured on the SET, or a default H-SLP address can be constructed based on the IMSI of the SET. In certain cases, as described in section 2.5.1.5, the SET may subsequently be instructed to establish a new secure session to a different node, but the basic principle remains that the SET always establishes a secure connection first with its H-SLP.

In SUPL2.0, with the introduction of emergency services support, this principle is modified to allow an SET to be ordered to connect to a new SLP (known as an Emergency SLP or E-SLP). To minimize the security risks of this change in approach, it is specified that the SET shall only do this when it is aware that it is in emergency mode (just how the SET determines that it is in emergency mode is undefined; see section 2.5.1.9 for more details).

2.5.1.3 Supported Access Networks

Although SUPL is nominally independent of the access network, the available security solutions limit which access networks can be used in practice.

SUPL1.0 supports GSM, WCDMA, CDMA, and CDMA-2000 deployments. SUPL2.0 supports all of those, plus LTE [93], UMB [94], I-WLAN (Interworking-WLAN where a 3GPP/3GPP2 subscriber uses WLAN to access 3GPP/3GPP2 services), WiMAX [95], and I-WiMAX (where a WiMAX subscriber uses WLAN to access WiMAX services).

Note that SUPL2.0 supports I-WLAN but not WLAN. Supporting WLAN access was an original requirement, but this was not met, as a suitable security solution was not provided.

2.5.1.4 The Components of the SLP

Architecturally, the SLP is divided into two separate elements, the SUPL Location Center (SLC) and the SUPL Positioning Center (SPC). The SLC is responsible for communication with external applications and coordinating location sessions. The SPC is responsible for the actual support of various positioning calculation methods.

In the SUPL1.0 Enabler Release, the protocol between the SLC and SPC is undefined, but in SUPL2.0, a new protocol known as ILP (Internal Location Protocol) has been created, allowing SLCs and SPCs supplied by different vendors to interwork using a standardized interface.

External to the SLP, the main impact of this split between the SLC and SPC is in the two supported modes of operation, Proxy Mode and Non-Proxy Mode, as described in the next section.

2.5.1.5 Proxy Mode Versus Non-Proxy Mode

SUPL supports two modes of operation for SLPs. In "Proxy Mode," all SET communication takes place over a secure session between the SET and its H-SLP. In "Non-Proxy Mode," once the secure session has been established between the SET and the H-SLP and mutual authentication has been performed, the H-SLP may instruct the SET to establish a separate SUPL Positioning Center (SPC) using credentials it has passed to the SET using the first secure session.

In SUPL1.0, only Proxy Mode is officially supported for 3GPP-based deployments (i.e., GSM or WCDMA), although, in practice, there is no reason why Non-Proxy Mode cannot also be used for these deployments. For 3GPP2-based deployments (i.e., CDMA or CDMA2000), both Proxy Mode and Non-Proxy Mode are supported. In SUPL2.0, both Proxy Mode and Non-Proxy Mode are supported for all deployments.

There are no hard requirements on which modes an SET must support in order to be SUPL compliant. For many access networks, therefore, there is a potential deployment issue in ensuring that an SLP and an SET both support the same mode. This was relatively easy to control in SUPL1.0, where the SET always deals first with the same H-SLP. But, with the introduction of the concept of an E-SLP for emergency services requests, SUPL2.0 deployments will require careful consideration to ensure that SETs and E-SLPs both support at least one common mode.

2.5.1.6 SET to SLP Communication

For communication between the SET and the SLP, SUPL defines a protocol known as the Userplane Location Protocol (ULP). This protocol enables the H-SLP to authenticate the SET and exchange information and instructions with it. It allows two separate mechanisms for the SLP and the SET to exchange measurements and assistance data to support various positioning methods. The first is via various encapsulated positioning protocols defined by other forums. The second is via the ULP layer itself. In some cases, a combination of the two methods is used, with assistance data not supported in some positioning protocols being carried in ULP.

The three supported positioning protocols for SUPL1.0 are the 3GPP defined the Radio Resource LCS Protocol (RRLP) [96], the Radio Resource Control (RRC) protocol [97], and the 3GPP2 defined TIA-801 [98]. For SUPL2.0, the LTE defined LTE Positioning Protocol (LPP) [99] has been added.

For each supported access network type, SUPL defines one mandatory positioning protocol and it is optional for the SET and SLP to support additional ones. This means that, for each supported access network, there should be at least one common positioning protocol that both the SLP and the SET support.

2.5.1.7 Overview of Positioning Procedures

Positioning procedures in SUPL can be most broadly categorized as either "SET Initiated," in which the SET initiates the location request via ULP, or "Network Initiated," in which the request is initiated by the SLP from an external agent (e.g., a location-based application, which could be an MLS). Note that the external agent in this case could still be co-located on the same device as the SET client. In SUPL2.0, a special type of request called an Emergency Services Request is introduced. This is a special type of Network Initiated Immediate request.

Within each categorization, there are a host of different variants resulting in a large number of slightly different procedures. Besides being SET Initiated or Network Initiated, a positioning procedure can use Proxy Mode or Non-Proxy Mode; it can be Immediate or Triggered; and it can be Roaming or Non-Roaming (and there are multiple ways in which roaming can be supported, with different procedures for each).

The following sections look at the most important procedures from an emergency services point of view. These are the SET Initiated Immediate location request and Network Initiated Immediate location request procedures in SUPL1.0 and SUPL2.0, and the Emergency Services Request introduced in SUPL2.0. Because of the similarities between the Emergency Services Request and the standard Network Initiated Immediate location request, these are discussed in the same section. Following this, the Triggered and Roaming variants will also be examined.

From an emergency services point of view, whether a procedure is Proxy Mode or Non-Proxy Mode does not make much difference once it has been ensured that the SET and the SLP both support at least one common mode, so the Proxy and Non-Proxy Mode versions of each procedure are covered in the same section.

SUPL2.0 also adds a number of different SET Initiated procedures to enable an SET to request delivery of its location to another SET, but the actual delivery of this location is out of scope for SUPL. These procedures are not covered.

2.5.1.8 SET Initiated Immediate Location Requests

For all SET Initiated procedures (Figure 2.32), the SET first needs a secure data connection to its H-SLP. Depending on the SET's implementation, it can either establish a separate data connection for each SUPL procedure or "session," or it can reuse the same secure data connection for multiple SUPL sessions.

As discussed in section 2.5.1.2, the initial secure data connection must always be to the SET's H-SLP and not some other SLP. This means that, if the SET is roaming on another operator's network, for example, it must still contact its H-SLP and rely on the H-SLP for service. Depending on the supported roaming scenarios, the H-SLP may then be able to identify other SLPs that can help in the positioning process, or it can attempt to provide the positioning itself.

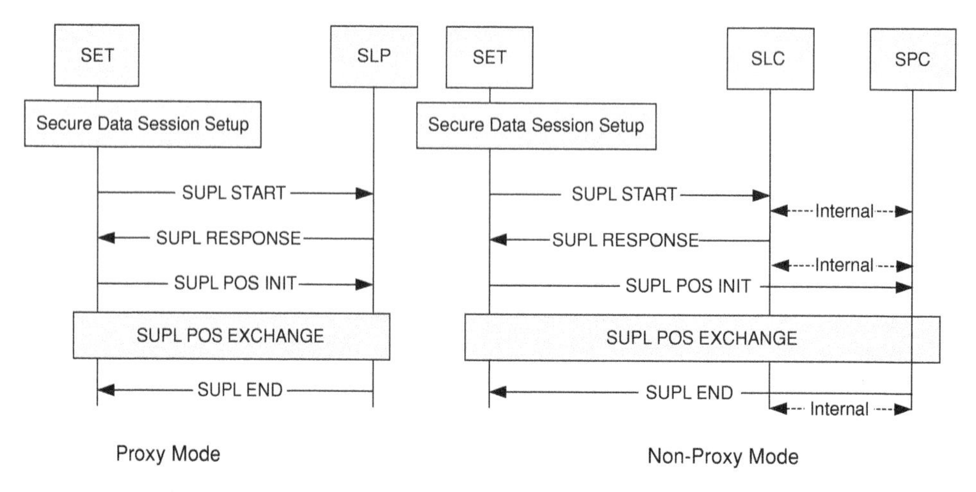

Figure 2.32 SET Initiated procedures for Proxy Mode and Non-Proxy Mode. (Reproduced with permission from © Open Mobile Alliance Ltd, 2013.)

After a secure data connection is available, the SET sends a ULP message known as an SUPL START to the H-SLP. This message includes various parameters that identify the capabilities of the SET and the requested Quality of Position (QoP).

The SUPL START also contains one or more Location IDs (only one in SUPL1.0, up to 64 in SUPL2.0). Location IDs provide location information that the SET has determined from its local environment. They are access-network-specific, that is, there is one type of Location ID for GSM networks, another for CDMA, another for UMTS, and so on (this is another reason why SUPL is only supported on a specific set of access networks – a new network cannot be supported until the ULP protocol has been updated to include an appropriate Location ID for it).

The Location ID generally contains information that identifies the point of access in the network (e.g., the identity of a cell or base station) and also various measurements specific to that access network.

In SUPL1.0, only one Location ID, for the current serving cell, is supported in the SUPL START, but in SUPL2.0, multiple Location IDs can be included. These can include both Location IDs for non-serving cells (and potentially those from other access networks) and "historical" Location IDs with associated timestamps, which indicate where the SET has been previously.

The inclusion of non-serving cells in SUPL2.0 also means that the SET can provide measurements from multiple access networks. So, for example, an SET connecting via I-WLAN could also report an observed UMTS base station even if it was not connecting via UMTS. This makes it easier for the H-SLP to try to locate an SET in an unknown network.

After it receives the SUPL START, the H-SLP determines whether it needs to invoke any of the various SUPL roaming procedures. These can include contacting other SLPs, which may be better able to locate the SET using the RLP protocol defined in the MLS enabler (see section 2.5.2). Just how the H-SLP determines which roaming procedures to use is out of scope of SUPL, and this remains an implementation problem. See section 2.5.1.12 for a more detailed discussion.

If the information included in the Location IDs is enough for the H-SLP to calculate a position that meets the request QoP, it can return a response and end the SUPL session then and there. If not, it needs to begin a process to prepare the SET to exchange positioning messages. For Proxy Mode, these messages are exchanged using the existing connection between the H-SLP and the SET. For Non-Proxy Mode, the SET is instructed to use a separate connection to a nominated SPC.

Based on the information in the Location IDs and the reported SET capabilities, the H-SLP is responsible for choosing the best positioning method to meet the requested QoP. This is another important aspect of the SUPL architecture. Based on its knowledge of the network, the H-SLP may be able to determine, for example, when A-GPS is not likely to work for a particular cell (say, a known pico-cell in a subway). Of course, a determined SET can still influence the choice of positioning method by simply lying about its own current supported capabilities (i.e., an SET which indicates that it only supports one method will always be told to use that method). The H-SLP is also responsible for choosing the positioning protocol to use, out of the available set indicated by the SET in the SUPL START.

One other function that takes place at the SLP on receiving the SUPL START is authentication of the SET. This generally involves some sort of interaction with the access network and is another reason why SUPL is not quite as network-independent as it might at first appear. In practice, it appears that some SUPL deployments have chosen to skip this step.

After choosing an appropriate positioning method, the H-SLP sends a ULP message called the SUPL RESPONSE. In Proxy Mode, the SET then responds with an SUPL POS INIT, which may include additional measurements and requested assistance data. If the selected positioning protocol allows the mobile device to send the first message (as TIA-801 and LPP do), the first message of the positioning protocol exchange may be included in the SUPL POS INIT as well. In Non-Proxy Mode, the SUPL POS INIT goes to a SPC identified in the SUPL RESPONSE. In non-roaming scenarios, this will be the H-SPC as it is still associated with the H-SLP.

Next, the SET and the H-SLP/H-SPC exchange messages using the chosen positioning protocol until a final position is determined. These positioning protocol messages are each encapsulated inside a ULP message known as the SUPL POS.

Supported positioning methods include various types of Global Navigation Satellite System (GNSS) and Assisted GNSS (A-GNSS) positioning (GPS is the only GNSS supported in SUPL1.0), Enhanced Observed Time Difference (E-OTD), Observed Time Difference of Arrival (O-TDOA), and Advanced Forward Link Trilateration (AFLT), as supported by the chosen positioning protocols.

Once a position has been determined, the H-SLP/H-SPC sends an SUPL END to the SET to end the session. The SUPL END may also include the final position if this was not already known to the SET after the last SUPL POS.

Emergency Services Implications

The SUPL SET Initiated Immediate procedure is theoretically compatible with any emergency services architecture that has the mobile device acquiring its own location and passing it on somewhere else. For example, in 3GPP TS 23.167 [100], describing IMS emergency sessions, the SUPL SET Initiated Immediate procedure can be used in Step 2 of Procedure 7.6.1 "Acquiring location information from the UE or the network," where the user equipment (UE) may acquire its own location before sending an SIP INVITE to the IMS Core. This is shown graphically in Figure 2.33.

There are some important considerations to be made, though. First, in SUPL, the SET needs to rely on its H-SLP either to find its location or to enlist other SLPs to help it. In roaming situations, this can mean a user making an emergency call might not be able to rely on their current access network being supported and might not be able to tell beforehand that this is the case.

Next, there is no mechanism to validate that the location used by the SET in the emergency scenario is actually the same location provided by the SLP (or indeed if the SLP was ever involved at all).

Finally, the time needed for the SUPL SET Initiated location procedure needs to be considered, and this may rule it out for some time-critical emergency applications.

2.5.1.9 Network Initiated Immediate and Emergency Services Location Requests

Both Network Initiated and Emergency Services location requests are generally initiated by an agent external to SUPL (Figures 2.34 and 2.35). One option for conveying these initial requests, and the subsequent responses, is via the MLP protocol defined as part of the OMA's MLS enabler. Another option is the OMA's LOCSIP enabler, which can serve the same purpose

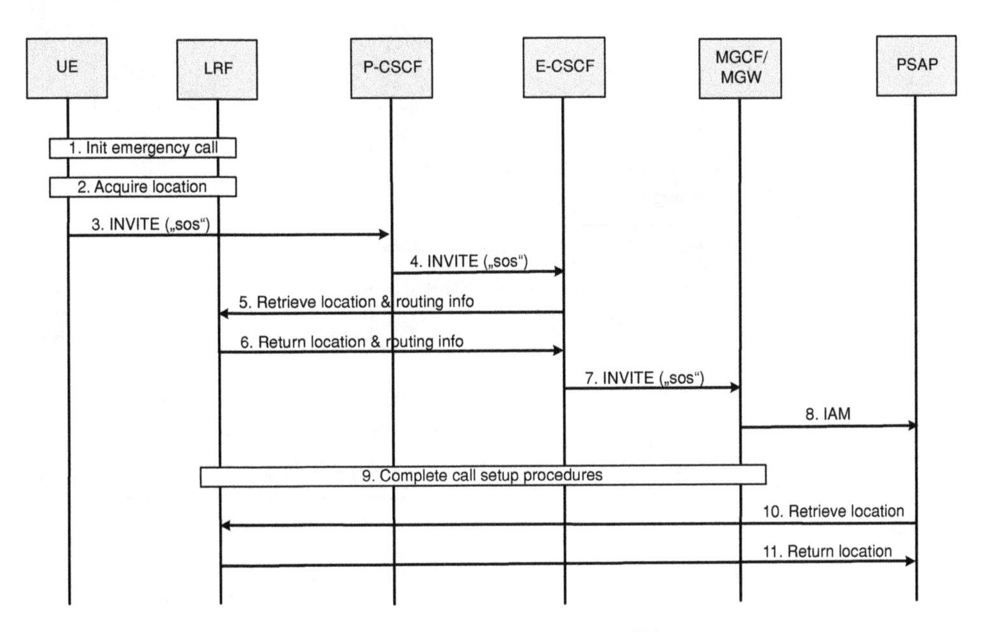

Figure 2.33 IMS emergency services call flow. (© 2009. 3GPP™ TSs and TRs are the property of ARIB, ATIS, CCSA, TTA and TTC who jointly own the copyright in them. They are subject to further modifications and are therefore provided to you "as is" for information purposes only. Further use is strictly prohibited.)

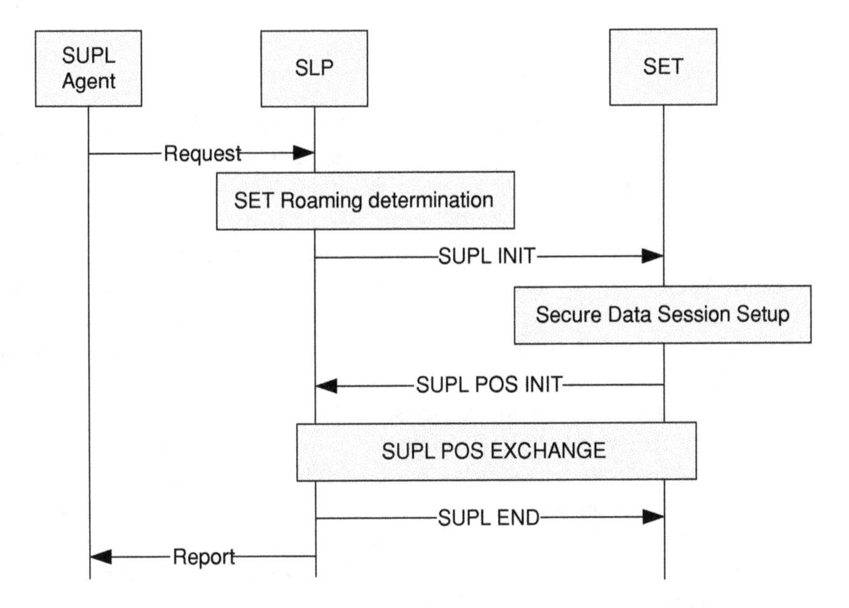

Figure 2.34 Network Initiated procedures for Proxy Mode. (Reproduced with permission from © Open Mobile Alliance Ltd, 2013.)

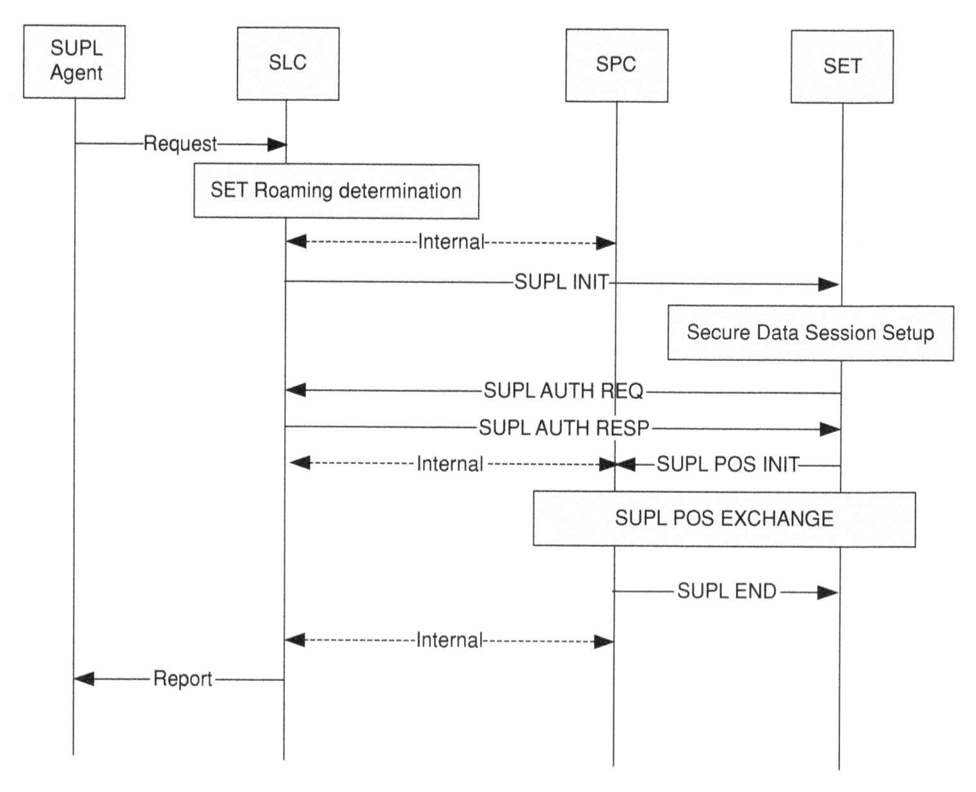

Figure 2.35 Network Initiated procedures for Non-Proxy Mode. (Reproduced with permission from © Open Mobile Alliance Ltd, 2013.)

(see section 2.5.4). Other protocols are also possible – the interface between the external agent and the SLP is functionally similar to the Le interface defined in 3GPP TS 23.271 [91]. Note that the external agent could also in fact be resident on the same device as the SET, so it is external only from the point of view of SUPL itself.

For Network Initiated location requests, including Emergency Services location requests, the SLP needs to send a ULP message called an SUPL INIT to the SET somehow prior to a secure data session being established. This process is somewhat problematic, as there are a number of different supported methods, each with their own drawbacks. Because this topic is rather complicated, it is discussed in more detail in section 2.5.1.10.

For normal Network Initiated location requests, the SET is only allowed to connect to its H-SLP, regardless of where the SUPL INIT originates from. For Emergency Services location requests, the SET is allowed to connect a special E-SLP indicated in the SUPL INIT itself. The conditions under which an SET is allowed to accept an Emergency Services location request are out of scope for SUPL and are expected to be determined by local regulatory requirements. This has implications, of course, for SETs roaming into regions with varying regulatory requirements.

To mitigate the risk of a deliberate attack on an SET by sending it spurious SUPL INITs for Emergency Services location requests, the concept of an Emergency SLP whitelist is introduced in SUPL2.0. This list allows an SET to prioritize requests from known E-SLPs over those from unknown ones (although requests from unknown E-SLPs still need to be responded to if there are no pending requests from known E-SLPs). Delivery of the E-SLP whitelist to an SUPL SET is also out of scope for SUPL.

The SUPL INIT includes a selected positioning method to be used for the session. This is somewhat anomalous, as the SLP is at this point still unaware of the SET's capabilities and environment and so the SLP has no idea if the chosen positioning method can actually be supported. For this reason, the SLP is allowed to change to a new positioning method when the actual positioning protocol exchange takes place. The main justification for the positioning method to be included in the SUPL INIT is to give the SET time to start up any resources needed for the positioning method before the first SUPL POS arrives.

The SUPL INIT also indicates whether Proxy Mode or Non-Proxy Mode should be used. As discussed in section 2.5.1.5, this is also potentially problematic if the target SET does not support the mode selected by the SLP.

For both standard Network Initiated location requests and Emergency Services location requests, once the SET receives the SUPL INIT, it needs to make sure a secure data session exists to the relevant SLP. As with the SET Initiated procedures, the SET may either open a new session or reuse an existing one.

In Non-Proxy Mode, the SET then sends a ULP message called an SUPL AUTH REQ to the H-SLP, which returns an SUPL AUTH RESP, indicating an SPC to connect to.

The SET then sends an SUPL POS INIT (to the H-SLP for Proxy Mode or to the indicated SPC for Non-Proxy Mode), which contains an indication of the SET's capabilities and one or more Location IDs (as with the SET Initiated case, only one Location ID is supported in SUPL1.0, while up to 64 Location IDs are supported in SUPL2.0).

For a standard Network Initiated location request, as with the SET initiated case, the H-SLP has a responsibility to authenticate the SET, typically using an interaction with the access network. For Emergency Services location requests, this requirement is waived, and, furthermore, location of "unregistered" SETs (e.g., 3GPP or 3GPP2 devices without UICCs or UIMs) is allowed.

Note that the arrival of the SUPL POS INIT is the first point at which the H-SLP or H-SPC is made aware of the SET's access network using SUPL-defined mechanisms. The SUPL specification indicates that the H-SLP should determine whether the SET is roaming prior to sending the SUPL INIT, but does not specify how this should be done. Various network-dependent methods are possible, but this remains another problematic area for SUPL. See section 2.5.1.12 for more details.

After the SUPL POS INIT, the SET and the H-SLP/H-SPC exchange messaging using a positioning protocol chosen by the H-SLP/H-SPC. As with the SET Initiated case, these are encapsulated within SUPL POS messages. Once a position has been calculated, an SUPL END message is used to terminate the session.

Emergency Services Implications

The SUPL 2.0 Emergency Services location request is intended for use in any emergency services architecture that requires the location of an SET to be provided on request to an external entity, such as a PSAP. As discussed above, there are a number of considerations that need to be borne in mind for a particular deployment.

First of all, there needs to be a mechanism outside of SUPL to ensure that the device to be located is actually an SUPL SET and one that is compatible with the E-SLP. Then, there needs to a mechanism to ensure that the SET meets local regulatory requirements for when an Emergency Services location request can be accepted. This could include, for example, a means to determine if the SET is in the process of an emergency call if that is part of the criteria. Also required is a mechanism to identify and deliver an E-SLP whitelist if that is to be deployed.

On the SUPL INIT delivery side, as discussed in section 2.5.1.10, there needs to be a mechanism outside of SUPL to ensure that the E-SLP knows which SUPL INIT delivery mechanisms the SET supports. This is especially problematic for the E-SLP scenario, as the E-SLP might not have any knowledge of the SET prior to the location request, or even be aware of which access network it is on (one of the few mechanisms that it has to predict is which SUPL INIT mechanism the SET might support). Whichever SUPL INIT delivery mechanisms are used, care must also be taken to be sure the SUPL INIT can be delivered reliably and in a timely fashion, even to a device that is in an emergency calling mode where it might be rejecting other traffic.

For emergency services applications, an additional issue to be considered is that of location validation and spoofing. Because the SLP relies heavily on information supplied by the SET for location calculation, there is a risk that a malicious SET could send false measurements, leading to a false location being passed to the PSAP. This is more of a risk in SUPL than it is with control plane location because there is less interaction with the access network where the location of the device could be confirmed with other measurements (e.g., by confirming that the SET really is connecting from the cell it claims to be). This risk can be mitigated by cross-checking locations from a number of different methods and using "SET Assisted" locations (i.e., locations calculated on the SLP, rather than on the SET) wherever possible. For any applications where it is critical that a malicious SET cannot lie about its location, SUPL is not suitable without some other method of verifying location, for example, by correlating the location given with a control plane query.

For roaming scenarios, a mechanism outside of SUPL is also needed first of all to determine which E-SLP the PSAP should contact, and then how the E-SLP determines if the SET is roaming and whether the assistance of another SLP is required. As with the SET Initiated case, there is also no easy way for the SET user to know in advance whether or not they are within a network supported by SUPL. However, for Emergency Services location requests, the problem is slightly different, because it is not the SET's H-SLP that needs either to provide coverage or to locate another SLP that can, but a potentially unknown E-SLP.

The use of the standard Network Initiated Immediate location request is not recommended because it is an SUPL requirement that an SET in an SUPL Emergency Session must immediately terminate non-emergency SUPL sessions. Note that this recommendation applies for SUPL1.0 too because an SUPL2.0 SET may refuse to handle a non-emergency SUPL INIT from an SUPL1.0 E-SLP during an emergency session.

2.5.1.10 SUPL INIT Delivery

SUPL INIT delivery is one of the key design challenges of the SUPL architecture. While all the other ULP messages can be passed via the secure connection between the SET and the

SLP, the SUPL INIT must travel via some other means, since it must go from the H-SLP to the SET while the secure connection is not present.

In SUPL1.0, two mechanisms for SUPL INIT delivery are supported. The SUPL INIT may be contained in an SMS or a WAP message delivered to the SET. Unfortunately, neither of these methods offers guaranteed delivery and both are prone to congestion. These problems can be alleviated somewhat by the use of a dedicated SMSC for SUPL INIT, but this is still less than ideal for emergency services applications.

The SUPL INIT delivery problem is compounded by the various roaming scenarios supported by SUPL. When sending the SUPL INIT, the mechanisms that the H-SLP has available to determine whether the SET is roaming or not are all beyond the scope of SUPL and the H-SLP may in practice be unaware of where the SET is and how long the message containing the SUPL INIT will take to deliver.

In SUPL2.0, two additional mechanisms are added for SUPL INIT. These are UDP/IP Push, in which the SUPL INIT is sent to the SET via a UDP datagram, and SIP Push, in which the SUPL INIT is sent via an SIP Push through the SIP/IP Core (and in the case of emergency services queries via the Emergency IMS Core). The WAP Push delivery mechanism has also been renamed OMA Push and expanded to include any delivery mechanism that uses the Push Access Protocol (PAP). This allows the SLP to send the SUPL INIT via PAP to a WAP PPG, which can then use an SIP Push to ultimately deliver the SUPL INIT to the SET.

In SUPL2.0, therefore, there is a choice of four different SUPL INIT delivery methods, each with their own special considerations, which need to be borne in mind during a deployment. As mentioned above, SMS and WAP delivery are both prone to congestion and do not offer guaranteed delivery. UDP/IP Push requires that the IP address of the SET is known to the SLP, but the mechanism for doing this is out of scope of SUPL. For emergency services queries, it also requires that a UDP datagram will not be blocked from delivery to an SET while an emergency call is in progress. SIP Push requires that the SET supports SIP, and also that the Emergency IMS Core supports delivery of the SIP Push to an SET in an emergency session.

Another deployment consideration is determining how the SLP knows which SUPL INIT delivery mechanisms are actually supported by the SET. In SUPL1.0, there is no requirement on the SET to support all methods of SUPL INIT delivery and it remains out of scope how the H-SLP determines which method to use. In SUPL2.0, new requirements have been added so that an SLP can deduce at least one method of SUPL INIT that an SET will support if it connects via certain access networks. For example, for GSM, WCDMA, and TD-SCDMA, it is a requirement that the SET must support at least OMA Push; and for CDMA or CDMA2000, the SET must support at least SMS delivery. At the time of writing, there is nothing defined for the other types of supported access networks for SUPL2.0. For emergency services, there is also a special requirement that an SET that is capable of supporting an emergency services call using an IP bearer shall support UDP/IP or SIP PUSH (which, unfortunately, still leaves the SLP with the problem of determine *which* of the two to use).

2.5.1.11 Immediate Versus Triggered Location Requests

In SUPL1.0, only "Immediate" positioning is supported. That is, each request results in a single location being calculated as soon as the request in received. SUPL2.0 introduces the concept of "Triggered" positioning, in which a single request may result in a series of positions being returned at regular intervals (a "periodic trigger") or when a particular event occurs, for example, the SET determining that it has entered or left a particular target area (an "event trigger").

Triggered location requests can be either SET Initiated or Network Initiated, although they are not currently supported for Emergency Services location requests. Because of this, their utility for emergency services applications is limited.

2.5.1.12 Roaming Considerations

As was mentioned previously, there are a number of different roaming mechanisms supported in SUPL, but they all generally rely on mechanisms outside the scope of SUPL to let the H-SLP or E-SLP identify Visited SLPs (V-SLPs) that can help to locate an SET.

In Proxy Mode, an H-SLP or E-SLP can choose between "V-SLP positioning" (in which RLP, the Roaming Location Protocol defined in the OMA's MLS enabler, is used to tunnel SUPL messages between the V-SLP and the H-SLP or E-SLP serving the SET), and "H-SLP positioning" (in which RLP is simply used to let the H-SLP pass measurements, such as Location ID to the V-SLP for conversion into a location).

In Non-Proxy Mode, an H-SLP or E-SLP can choose between the aforementioned H-SLP positioning and "V-SPC positioning" (in which the H-SLP or E-SLP instructs the SET to connect to the SPC of the V-SLP, and RLP is used to pass various security credentials to the V-SLP to make this possible).

All of these mechanisms require some sort of interoperator roaming arrangements and a way to identify which roaming modes are supported.

In practice, it appears that a much easier approach is simply to try to serve as many access networks as possible through the H-SLP. This is possible even for networks where there is no direct relationship by building or acquiring from a third party a database mapping cell information to coarse location. This is good enough for A-GNSS applications, but is not suitable for emergency services application as the database cannot be guaranteed to be accurate enough for fallback if A-GNSS fails.

For Network Initiated location requests, another aspect to roaming is how a location-based application, which might be in a visited network, identities the H-SLP of the SET in order to launch a query. RLP provides a mechanism for a Remote SLP (R-SLP) to inject a Network Initiated location request to an H-SLP, but, as you might have guessed, the mechanism for how it determines that it needs to do this is out of scope for SUPL.

2.5.2 MLS

MLS is the OMA's enabler for Mobile Location Services. At the time of writing, there are three Enabler Releases for MLS: MLS1.0 [101], MLS1.1 [102], and MLS1.2 [103].

2.5.2.1 What is MLS?

The MLS enabler contains a set of protocols for use on interfaces defined by 3GPP in 3GPP TS 23.271 [91]. As the name suggests, these protocols are all to do with location services.

The most significant of these, from an emergency service viewpoint, is the Mobile Location Protocol (MLP). The other protocols are the Roaming Location Protocol (RLP), used between two location servers (including SLPs during SUPL Roaming and two GMLCs in the 3GPP model), and the Privacy Checking Protocol (PCP). These last two protocols are not discussed in this chapter, but the MLP protocol is discussed in detail in section 2.5.3. Note that the first versions of MLP were separate enablers all by themselves.

2.5.2.2 MLS Enabler Releases

There are currently three different MLP enabler releases. MLS1.0 contained two protocols, MLP3.2 and RLP1.0. PCP1.0 was intended to be included, but it was not ready. MLS1.1 contained the same versions of MLP and RLP, but added PCP 1.0. MLS1.2, intended to be the final release of MLS, consisted of MLP3.3 and RLP1.1 and did not include PCP.

2.5.3 MLP

2.5.3.1 What is MLP?

The OMA's Mobile Location Protocol or MLP is an application-level protocol that allows a location-based application (LBA) to query a location server for a mobile device's location. It is intended to work for multiple location architectures, regardless of whether the location server being queried is a 3GPP-defined GMLC, a 3GPP2 defined MPC, or an OMA-defined SUPL SLP.

From an emergency services point of view, MLP is relevant because it can be used by a PSAP to request location from a location server.

MLP started out as a Location Interoperability Forum (LIF) specification, becoming an OMA enabler when the Location Interoperability Forum was consolidated in the OMA. The first OMA version is still known as LIF TS101 v3.0.0 [104], and the later OMA versions are known as MLP3.1 [105], MLP3.2, and ML3.3, continuing the sequence of LIF version numbering. In OMA terms, MLP 3.2 and MLP3.3 are not enablers in their own right, but are contained with different versions of the MLS enabler (see section 2.5.2).

MLP uses XML and is intended to support multiple transport layers, including the Hypertext Transfer Protocol (HTTP), the Wireless Session Protocol (WSP), and the Simple Object Access Protocol (SOAP), although, in practice, there are only currently standardized mappings for using it over HTTP. HTTP has the advantage that it is widely supported. The defined HTTP binding, however, does not allow multiple requests to be piggybacked in a single request, and the sender has to wait for a response before issuing a new request. This is partially alleviated by the use of MLP's "asynchronous" mode, as described in (section 2.5.3.2).

MLP supports a number of different "services" (Table 2.5). These are discussed in detail in the following sections. Of particular interest to emergency services are the Emergency Location Immediate Service and the Emergency Location Reporting Service. Combined with these are the Standard Location Immediate Service and the Standard Location Reporting Service. Finally, there is the Triggered Location Reporting Service. Each of these services are supported for all MLP versions from LIF TS101 v3.0.0 onwards, although there are minor variations for different MLP releases.

Table 2.5 MLP services

Service	Messages
Standard Location Immediate Service	Standard Location Immediate Request Standard Location Immediate Answer Standard Location Immediate Report
Emergency Location Immediate Service	Emergency Location Immediate Request Emergency Location Immediate Answer Emergency Location Immediate Report (MLP 3.2 and later only)
Standard Location Reporting Service	Standard Location Report
Emergency Location Reporting Service	Emergency Location Report
Triggered Location Reporting Service	Triggered Location Reporting Request Triggered Location Reporting Answer Triggered Location Report Triggered Location Stop Request Triggered Location Reporting Stop Answer

2.5.3.2 Immediate Location Services

There two Immediate Location Services, the Standard Location Immediate Service and the Emergency Location Immediate Service, both of which allow a location client to initiate a request and receive location information back from a location service.

The requested location information can be delivered in either a "synchronous" or an "asynchronous" response, as requested by the location client. Note that, for MLP versions prior to MLP3.2, "asynchronous" responses are not supported for the Emergency Location Immediate Service, and emergency requests are considered to be implicitly asynchronous. From MLP3.2 onwards, both synchronous and asynchronous response types are supported for the Emergency Location Immediate Service.

For synchronous location responses (Figure 2.36), the location server sends back all the requested location information in a single response message via the same connection from which the request arrived. The locations of multiple mobile stations can be requested in the same request and reported in the same response.

For asynchronous location responses (Figure 2.37), the location server sends back a single response to the client to indicate that the request has been accepted and to provide a request id, which will allow the client to match later responses to the original request. The location server then sends the requested location information in one or more location report messages via one or more separate connections. This allows a location server under high-load conditions to continue serving requests at a high rate without needing to process the last request before receiving the next.

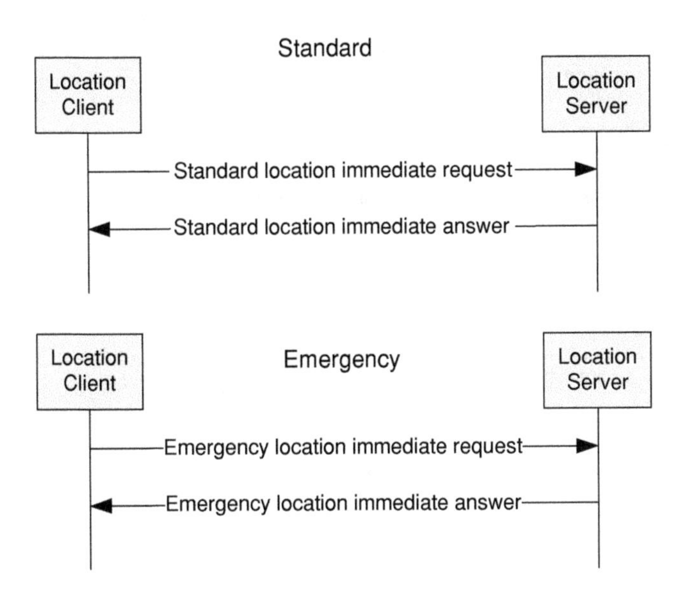

Figure 2.36 Synchronous responses for Standard and Emergency Location Immediate Services. (Reproduced with permission from © Open Mobile Alliance Ltd, 2013.)

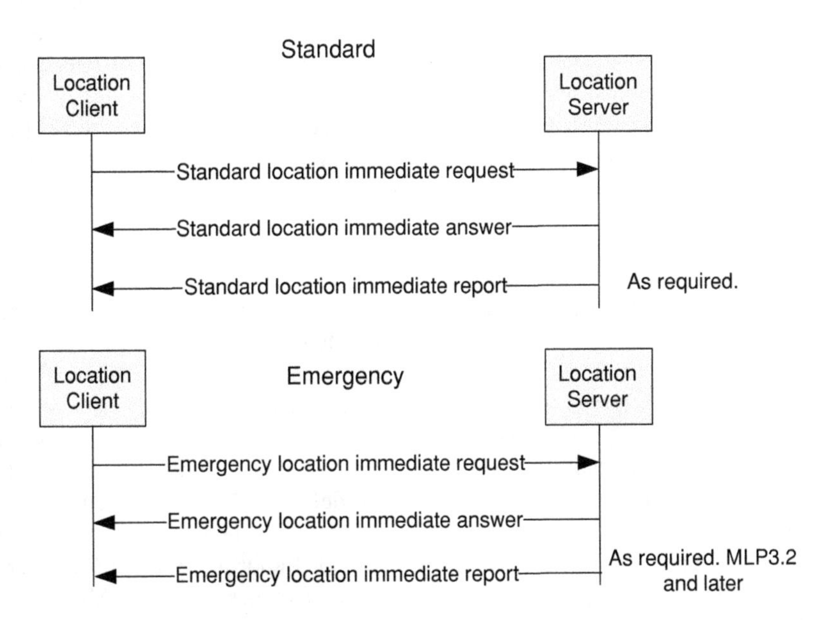

Figure 2.37 Asynchronous reporting for Standard and Emergency Location Immediate Services. (Reproduced with permission from © Open Mobile Alliance Ltd, 2013.)

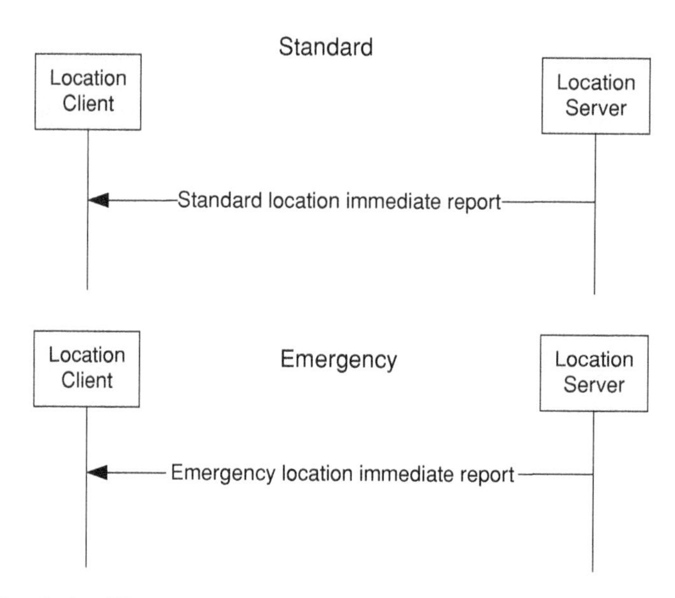

Figure 2.38 Standard and Emergency Location Reporting Service. (Reproduced with permission from © Open Mobile Alliance Ltd, 2013.)

Another point of different between the Emergency and Standard Location Immediate Services prior to MLP3.2 is in the parameters that the location client is able to use to request a location. Prior to MLP3.2, the Emergency Location Immediate Request only supported location request by horizontal accuracy, vertical accuracy, and altitude accuracy, while the Standard Location Immediate Request also allowed a maximum location age to be specified, as well as an indication of response time (either via a choice of "No delay," "Low delay" or "Delay tolerant," or via the specification of a response time in seconds).

2.5.3.3 Location Reporting Services

As well as the Immediate Location Services, there are also Standard and Emergency Location Reporting Services (Figure 2.38). These allow unsolicited location information to be pushed from the location server to the location client. For emergency services, a practical example where this can be used is when a caller has initiated or released an emergency call and triggered a network-induced location request. The Emergency Location Reporting Service allows the location server to push the resulting location directly to a PSAP, rather than waiting for the PSAP to request it. If the location server is a 3GPP defined GMLC, for example, it might receive a Subscriber Location Report containing the location and could then push it to a predefined address using the MLP Emergency Location Reporting Service (the alternative would be for the GMLC to simply cache the location until a request from a PSAP was received using the Emergency Location Immediate Request service).

2.5.3.4 Triggered Location Reporting Service

The Triggered Location Reporting Service allows a location client to request location information for a mobile station at particular intervals, or when particular trigger conditions are met. As of MLP3.3, there is no way to indicate that a Triggered Location Reporting Service is for an emergency request.

2.5.3.5 Mobile Station Identifiers

MLP supports a number of different ways of identifying a mobile station. These include standards-defined identifiers, such as MSISDN, IMSI, IMEI, MIN, and MDN, and also implementation-specific identifiers, such as "EME MSID" (Emergency Mobile Station Identifier) and "OPE ID" (Operator Specific Identity). Note that, in practice, sometimes the MSISDN identifier is also used in a non-standard way. For example, in the Canadian Radio-television and Telecommunication Commission's (CRTC) Phase 2 E9-1-1 specification [106], the use of the MSISDN field is mandated, but with only 10 digits, instead of the full international format mandated by MLP.

MLP also supports IP addresses and SIP URIs, and also various forms of anonymous subscriber identification.

For emergency services, MLP also allows Emergency Services Routing Digits (ESRD) or Emergency Service Routing Key (ESRK) fields to be included with a position estimate from the location server. The ESRD field allows the inclusion of a North American ESRD (NA-ESRD), which is used to identify a particular North American emergency services client and the base station and cell sector from which the emergency call was made. Likewise, inclusion of the ESRK field allows MLP to support the North American ESRK (NA-ESRK), which is a temporary number assigned to an emergency call in the North American emergency services architecture. As of MLP3.3, only North American encodings of ESRD and ESRK have been defined.

2.5.3.6 Position Estimates and Civic Address Information

Up until MLP3.3, the position estimates supplied by the location server always took the form of geographic data represented by different shapes (points, circles, arc bands, polygons, etc.). MLP3.3 introduced support for civic addresses, in which a location client can request either shape-based information, or a civic address, or both. The civic address format is based on that used in the IETF's RFC 4119 [17] and the Revised Civic Location Format for PIDF-LO [21], which was still a draft at the time MLP3.3 was being standardized. The resulting structure allows civic addresses to be composed of a sequence of elements, starting at the country level and moving down all the way to an individual desk within a building. Support for multiple languages is also provided with each element.

2.5.4 LOCSIP

LOCSIP is an OMA enabler for support of location in the SIP/IP Core.

2.5.4.1 What is LOCSIP?

The main role of LOCSIP is to provide an SIP-friendly interface between application in an SIP/IP Core and a location server. In this regard, it duplicates somewhat the role played by the

MLP protocol (see section 2.5.3), but it makes it easier for SIP-based applications to access location information. LOCSIP also makes use of other OMA enablers, such as SIP for Instant Messaging and Presence Leveraging Extensions (SIMPLE) [107], the Global Permissions Management (GPM) [108], the OMA XML Document Management (XDM) [109], and many IETF RFCs, including RFC 4119 [17].

In LOCSIP, a location client subscribes to a location server to request location information about a particular target or set of targets. Originally, it was intended that the location client would be independent of the target device, just as it does not matter for MLP whether a location client is on the target device or not. During the specification work, however, it was decided to add specific support for a location client that was on the target device, to allow for any resulting special requirements needed to support this.

2.5.4.2 Subscription to Location Notifications

When a LOCSIP location client wishes to subscribe to location notifications from a LOCSIP location server, it simply sends an SIP SUBSCRIBE toward the location server. The SIP SUBSCRIBE contains the SIP URI of the target, or a target list if there are multiple targets, and any required Quality of Service[16] (QoS) or filter criteria for when the location notifications are to be sent.

The SIP SUBSCRIBE makes its way through the SIP/IP Core, as shown in Figure 2.39, with authorization being performed along the way by the Home Subscription Agent (HSA).

On receiving the SUBSCRIBE, the location server responds with a 200 OK, and then an SIP NOTIFY. One thing to note here is that, in order to be compliant to RFC 3265 [33], the SIP NOTIFY needs to be sent immediately, even though the location might not be available yet or the location request might not yet be authorized by the location server. To get around this, it is possible to send an "empty" SIP NOTIFY and then another update when the actual location is available.

The QoS parameters that the location client can use include location type (geospatial, civic address or both), maximum uncertainty (both horizontal and vertical), maximum response time, and maximum location age. If a civic address was requested, the required civic address elements can also be specified (i.e., the location client can indicate that it wants to see both country and postal code elements in any civic address it receives).

The location client can also indicate whether the requested QoS parameters should be treated as "best effort," "assured" or "emergency" via a parameter known as QoS class. For "best effort," the location server should do its best to meet the requested parameters, but may return the best it has if it cannot. "Assured" means the location will only return a response if it can meet all the requested QoS parameters. A QoS class of "Emergency" indicates to the location server that this is an emergency services location request and that any other QoS parameters may have no effect on the location server.

For emergency location subscriptions, the location client can also include an "emergency" Priority header in the SIP SUBSCRIBE. This allows the location server to give the request higher priority and may also allow it to skip enforcement of any local location access control rules.

---

[16] In some organizations the term Quality of Service (QoS) is used while in others Quality of Positioning (QoP) is preferred. These two terms often refer to the same concepts.

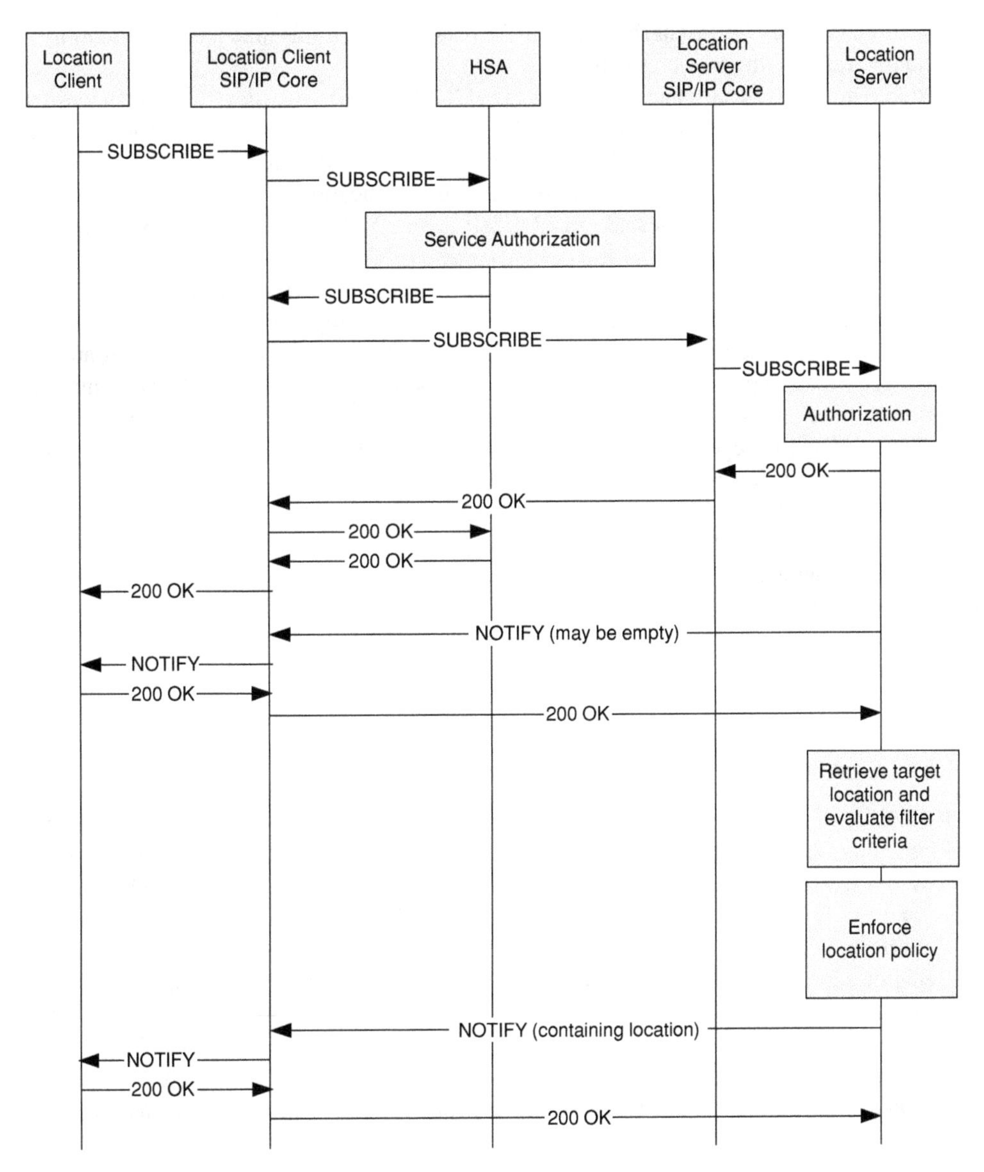

Figure 2.39 Subscription to Location Notifications. (Reproduced with permission from © Open Mobile Alliance Ltd, 2013.)

2.5.4.3 Subscription to Notifications of a Periodic or Area Events Trigger

In addition to subscribing to immediate location notifications, a LOCSIP location client can also subscribe to periodic updates of location or notifications of a particular area event (i.e., leaving or entering a given geographic area). This is done with a very similar procedure to that shown in section 2.5.4.2.

As before, the location client sends an SIP SUBSCRIBE, but this time it also includes filter criteria that indicate the trigger conditions to be used.

Periodic notifications are handled as per the IETF RFC for event throttling [110] and the location client is allowed to specify both a "throttle" and a "force" time period. The "throttle" time period is the minimum time allowed between location notifications, and the "force" time period is the maximum time period allowed between notifications. This means that the way periodic notifications are requested via LOCSIP is subtly different from the way they are handled in SUPL or MLP.

If both "throttle" and "force" periods are supplied and they are equal in value, then the result is a regular set of reports. But if the "force" value is higher than the "throttle" value, the location reports may arrive at some point between the two periods, especially if an area event trigger is also specified. The net result is that the location server may be configured to report each time a particular event happens *as long as* it has not reported as recently as the throttle period and to report *regardless of whether the event has occurred* if the force period has elapsed since its last report.

The available area event triggers, based on the IETF specification for location filters [34], are also quite different from those defined in MLP and SUPL2.0. As well as triggering on entering and leaving a given area, it is also possible to trigger on the target traveling a certain distance, exceeding a certain speed, changing a particular value in its address (e.g., changing the city or postal code in its civic address) or moving in to or out of a specified range from another target.

2.5.4.4 Emergency Services Implications

LOCSIP is an obvious candidate for providing location information to support IMS emergency sessions, although there are also other methods that can provide the same access. As per 3GPP 23.167 [100], there is already an Ml reference point between the E-CSCF in the IMS Emergency Core and the Location Routing Function, and the 3GPP was involved in the definition of this interface.

Being designed for SIP from the ground up, it is neater and cleaner than the existing MLP specification and certainly more flexible. On the other hand, it is also useful only for SIP, while other protocols, such as MLP, can be used for other types of user sessions as well as SIP. At this point in time, it is unfortunately too early to tell how widely and how soon LOCSIP will be adopted and whether it will supplant existing deployments of protocols, such as MLP.

2.6 3GPP Location Protocols

Jan Kåll
Nokia Siemens Networks

2.6.1 Introduction

The 3GPP has standardized service capabilities and network solutions to deliver information about the location of mobile phones to be used for value-added services, location-based applications, and emergency services. Location information may be available for any User Equipment (UE), but there are restrictions on choice of positioning method and sending notification of a location request to the UE depending on the capabilities of the network and the UE.

The support for location services has been developed stepwise, with enhancements added in newer 3GPP releases. For example, the support for location services (and emergency services) in 4G access networks (i.e., Long Term Evolution, LTE) was developed for 3GPP Release-9.

The development of new location capabilities and features is based on service requirements that have been identified on the market either as enhancements to existing features or as new features found to be useful or necessary, for example, to improve location privacy. It has been an essential requirement from the start to be able to report the location of 3GPP terminals used for emergency services to the Public Safety Answering Point (PSAP). The service requirements for 3GPP location services are specified in 3GPP TS 22.071 [111].

Based on the location service requirements, the network and terminal location functions and corresponding signaling between the network and the terminal were developed in the 3GPP. In general, it is possible to determine the current location of a specific user's terminal and to report the location information, for example, geographical coordinates, in a standard format to the users, network operators, and value-added service providers that, for example, offer navigation services or inform about local attractions.

The user and network operator may set limitations how the user's location information is shared externally in order to ensure location privacy. In most countries, the location privacy settings of a user do not apply to emergency calls because it is of crucial importance to know the location of an emergency caller. However, in some places it is possible for the emergency caller to stay anonymous toward the PSAP, and in some other places the location of a terminal can only be stored for the duration of a call.

The emergency call is routed to the appropriate PSAP, which is serving the area of the emergency call, based on the approximate location (radio cell), or optionally based on a more accurate location information of the emergency caller.

The main functional specification for location services in 3GPP is TS 23.271 [91]. The location information available from 3GPP networks can consist of the following:

- geographic location and optionally velocity information;
- timestamp of the given location;
- quality of positioning; and
- user-specific privacy requirements regarding the distribution of the user's location information.

The 3GPP network can also deliver emergency-related location information to the PSAP without any explicit request from the PSAP when recognizing that the call was to an emergency number.

2.6.2 Location Technology in 3GPP Networks

The 3GPP network entity that supports location services is termed "LCS Server." The "LCS client" requests the location of a specific UE from the LCS Server and can use the received location information for location-based services. Using these terms, the PSAP is one LCS client that can request and get emergency-related location information from 3GPP networks. The functionality of how to obtain the mobile's location information from a 3GPP network is described in more detail below.

2.6.2.1 3GPP System Architecture for Location Services

The "system architecture" in 3GPP terms describes the functional split between the network entities and the interconnecting interfaces, the reference points, between the network entities used for signaling. Figure 2.40 is based on 3GPP TS 23.271 [91] with some simplifications and shows the general arrangement of the Location Service (LCS) feature in 3GPP GSM, UMTS, and EPS systems, with the corresponding access technologies GERAN, UTRAN, and E UTRAN. The Core Network (CN) entities communicate with the Access Networks (AN) across the A, Gb, Iu, and S1 interfaces.

The LCS Client requests the UE's location information from a Gateway Mobile Location Center (GMLC). The GMLC authenticates the LCS client and determines how to handle the location request based on the user's privacy requirements and the UE's registration state, for

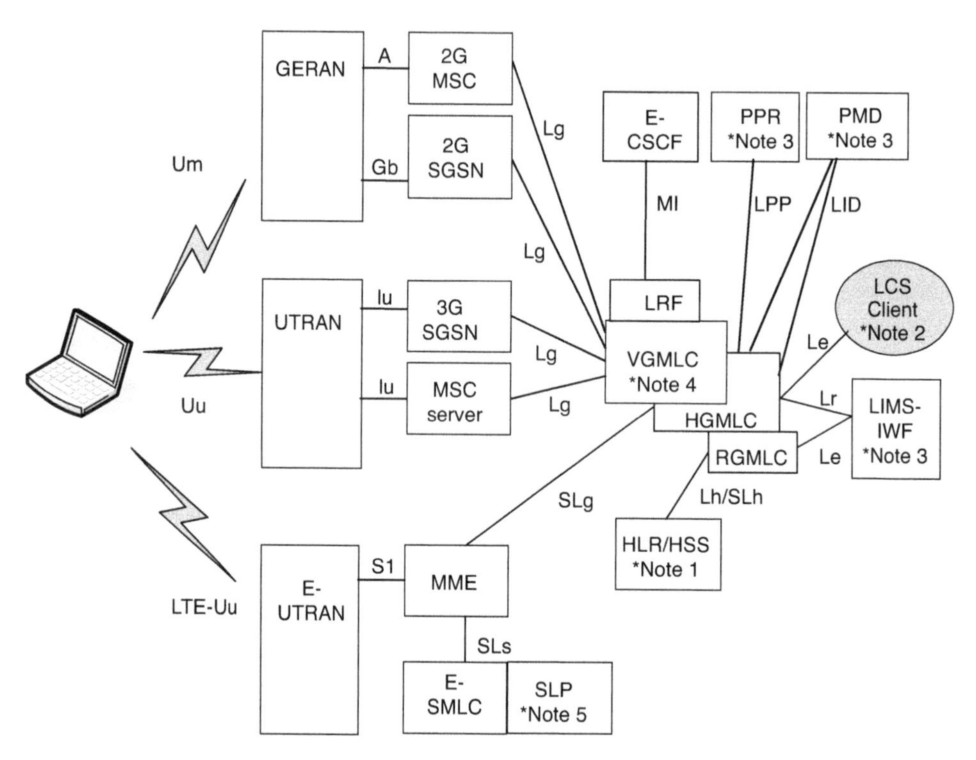

Figure 2.40 Location services architecture in 3GPP™ networks (see notes in Table 2.6). (© 2009. 3GPP TSs and TRs are the property of ARIB, ATIS, CCSA, TTA and TTC who jointly own the copyright in them. They are subject to further modifications and are therefore provided to you "as is" for information purposes only. Further use is strictly prohibited.)

Table 2.6 Notes to Figure 2.40

Note 1:	The HSS includes both 2G-HLR and 3G-HLR functionality
Note 2:	The LCS client may contain OSA Mobility SCS with support for the OSA user location interfaces
Note 3:	The PPR, PMD or LIMS-IWF functions may be integrated in GMLC
Note 4:	The LRF may interact with a separate GMLC or may contain an integrated GMLC
Note 5:	The SLP, which may be an H-SLP, V-SLP or E-SLP, may optionally be associated with an E-SMLC in order to share assistance data for support of both control plane LCS and OMA SUPL for an operator who deploys both solutions. The interaction between the E-SMLC and the SLP is outside the scope

example, if the UE is idle or making a call. The GMLC can have different roles depending on whether the UE is in its home network (HPLMN) or roaming in a Visited PLMN (VPLMN) and depending on where the LCS client is located.

In a complex scenario, the UE is roaming in the VPLMN and the LCS client is in some third country, that is, neither in the VPLMN nor in the HPLMN. The LCS client sends the location request to the so-called Requesting GMLC, R-GMLC, which authenticates the LCS client and checks with the H-GMLC in the HPLMN what privacy requirements apply for the mobile in question. The R-GMLC finds out with the help of the Home Location Register (HLR) in which country (VPLMN) the UE is roaming and forwards the location request to the corresponding V-GMLC in the VPLMN. The V-GMLC finally sends the location request to an appropriate core network entity, namely to the MSC for call-related services and to the SGSN/MME for packet data services. The core network entities handle the request together with the access network entities serving the mobile. After a location estimate has been determined for the UE, the core network entity returns the location estimate to the V-GMLC, which forwards the result via the R-GMLC to the LCS client.

The LCS client can define what kind of location information is requested for the target mobile, and the location information delivered to the LCS client contains the corresponding location information, together with the accuracy of the location estimate, the time of day the measurement was made, used positioning method, and so on.

The network entities shown in Figure 2.40 are listed in Tables 2.7, 2.8, 2.9, and 2.10, grouped according to their function.

Table 2.7 General-purpose 3GPP network entities

E-UTRAN	Evolved UTRAN
GERAN	GSM EDGE Radio Access Network
GPRS	General Packet Radio Service
HLR	Home Location Register
HSS	Home Subscriber Server
LTE	Long Term Evolution
MME	Mobility Management Entity
MSC (server)	Mobile services Switching Center, for circuit-switched telephone services
SGSN	Serving GPRS Support Node, for packet-switched (IP) services in the PS domain
UTRAN	Universal Terrestrial Radio Access Network

Table 2.8 Network entities dedicated to location services

E-SMLC	Evolved Serving Mobile Location Center
GMLC	Gateway Mobile Location Center, R = Requesting, H = Home, V = Visited GMLC
PMD	Pseudonym Mediation Device functionality
PPR	Privacy Profile Register
SLP	SUPL Location Platform
SUPL	Secure User Plane Location

Table 2.9 Network entities for location and emergency services in IMS

E-CSCF	Emergency Call Session Control Function in IMS
LIMS-IWF	Location IMS – Interworking Function
LRF	Location Retrieval Function in IMS

2.6.2.2 Positioning Architecture in 3GPP Access Networks

The general architecture of 3GPP location services was described in the previous section. Here we give some more details about the positioning functions and architecture in 3GPP UTRAN and E-UTRAN access networks.

UTRAN

The access network elements related to location services in UTRAN access networks are shown in Figure 2.41. The Serving RNC receives the location request originating from the LCS client and GMLC from the core network over the Iu interface. The RNC interacts with the Node Bs, LMUs, and possibly with the Standalone SMLC (SAS) in order to generate a location estimate for the UE, and then returns the resulting location information to the Core Network (CN).

Table 2.10 Interfaces, and reference points

Gb	2G-SGSN – BSS
Gs	MSC – SGSN
Le	External LCS Client – GMLC (external interface)
Lg	GMLC – VMSC, GMLC – MSC Server, GMLC – SGSN (gateway MLC interface)
Lh	GMLC – HLR (MAP-based)
Lid	GMLC – PMD
Lpp	GMLC (HGMLC) – PPR entity
Lr	Between GMLCs and e.g. between LIMS-IWF – HGMLC
Ml	E-CSCF – LRF
SLg	GMLC – MME
SLh	GMLC – HSS (Diameter-based)
SLs	MME – E-SMLC
Um	GERAN Air Interface
Uu	UTRAN Air Interface
LTE-Uu	E-UTRAN Air Interface

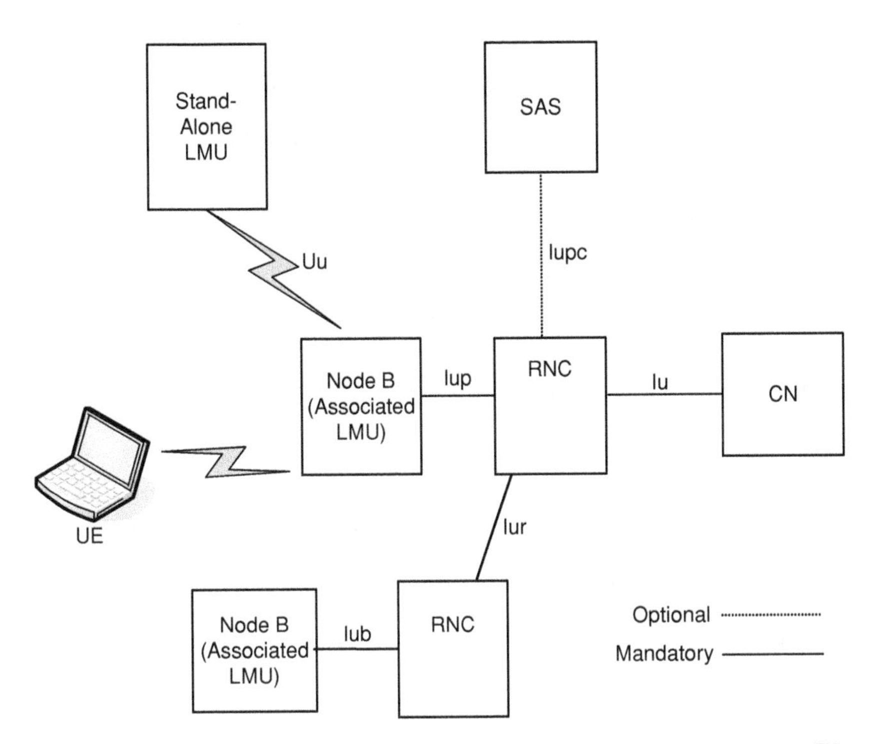

Figure 2.41 Positioning network elements for UTRAN access networks. (© 2010. 3GPP™ TSs and TRs are the property of ARIB, ATIS, CCSA, ETSI, TTA and TTC who jointly own the copyright in them. They are subject to further modifications and therefore provided to you "as is" for information purposes only. Further use is strictly prohibited.)

In order to generate the location estimate, the UE, Node B, and/or LMU are requested to perform measurements and return the measurement results for location calculations.

The positioning functionality in UTRAN can reside in the RNC or in a Standalone SMLC. With a Standalone SMLC in the access network, the RNC only has a kind of relay functionality of positioning-related signaling toward the core network and the UE. The SAS receives measurement data from the RNC over the Iupc interface and also sends positioning assistance data to the UE via the RNC.

The Location Measurement Unit (LMU) makes radio measurements and calculations and sends the measurement results to the RNC/SAS. The LMU is standalone or integrated with Node B.

E-UTRAN

New network elements and interfaces have been introduced for positioning of a UE in E-UTRAN access networks, as shown in Figure 2.42

The positioning functionality in the E-UTRAN differs from GERAN and UTRAN because the MME sends the location request to the E-SMLC after receiving the location request from the GMLC, the eNodeB or the UE. The MME itself initiates the location request when the UE

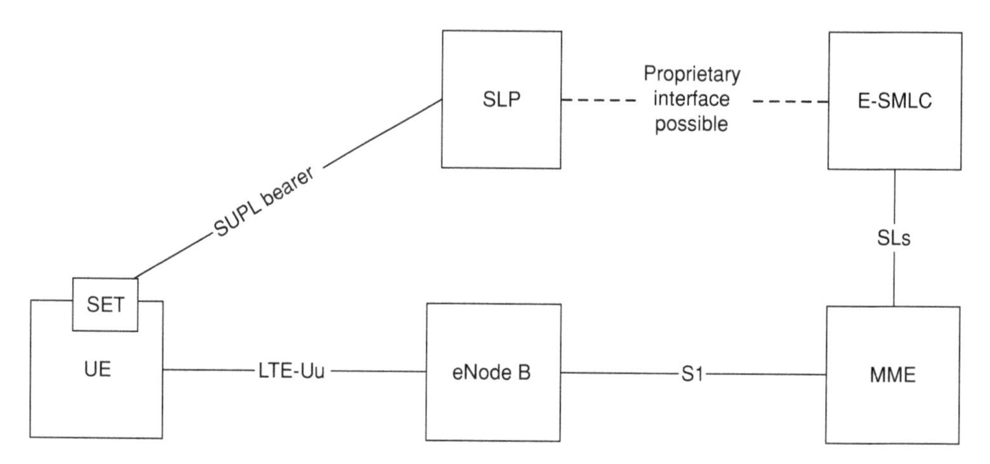

Figure 2.42 Positioning network elements for E-UTRAN access networks. (© 2010. 3GPP™ TSs and TRs are the property of ARIB, ATIS, CCSA, ETSI, TTA and TTC who jointly own the copyright in them. They are subject to further modifications and therefore provided to you "as is" for information purposes only. Further use is strictly prohibited.)

has initiated an IMS emergency call. The E-SMLC sends the resulting location information to the MME, which forwards the information to the requesting entity.

If location is requested for a mobile that is idle and not connected to the network, the MME starts establishing a signaling connection with the UE and assigns a specific eNodeB for the connection, because the UE needs to be in connected mode before the positioning procedure is started.

The E-SMLC decides which positioning method should be used and executes the positioning with the UE and/or serving eNodeB. The E-SMLC interacts with the eNodeB via the MME in order to get location-related information from the eNodeB, for example, timing information and measurement results, either determined by eNodeB or measurement results delivered by the UE to the eNodeB. The E-SMLC also sends assistance data to the target UE to assist with positioning, if needed.

The E-SMLC combines the received measurement results and other information to generate the location estimate for the target UE and sends the location information to the MME.

The SLP in Figure 2.42 is the SUPL server that handles positioning of the UE over the user plane, which is a peer-to-peer data connection between the SUPL server and the UE. Positioning in the E-UTRAN can be made more efficient by combining OMA-SUPL's user plane solution with E-UTRAN control plane positioning. OMA-SUPL is further described in section 2.5.

2.6.2.3 Positioning Methods in 3GPP Access Networks

The location of a mobile terminal can be detected and estimated based on the radio signals emitted by the base stations in the radio access network or by the mobile terminal itself or by using satellite signals received by the mobile. The satellite-based positioning methods can be improved by delivering so-called assistance data to the mobile terminal over cellular. The positioning methods are quite similar in the access networks standardized by 3GPP, that is, GERAN (2G), UTRAN (3G), and E-UTRAN (4G), but IMS emergency services are only available on 3G and 4G access networks, not over 2G, GERAN.

- *Cell coverage-based positioning.* The terminal's location is estimated according to the location of the serving radio cell, possibly enhanced by timing advance, round-trip time, neighbor cell signal strength, or angle of arrival measurements. The Enhanced Cell ID (E-CID) positioning refers to techniques that use the UE's and/or access network capabilities and measurements to improve the location estimate.
- *Enhanced Observed Time Difference (E-OTD) in GERAN access networks only.* The E-OTD method is based on measurements in the mobile terminal of the Enhanced Observed Time Difference of arrival of bursts of nearby pairs of Base Transceiver Stations (BTSs).
- *Idle Period Downlink Observed Time Difference of Arrival (IPDL-OTDOA).* This positioning method in UTRAN access networks is based on the measurements of frame timing in the terminal and possibly in the LMU. There is an idle period downlink because the serving base station can optionally stop radio transmission for short periods of time in order to improve the terminal's reception of neighboring base stations (hearability problem). In E-UTRAN, OTDOA suffers from the same neighbor cell hearability problem, which can be mitigated by using specific positioning subframes periodically and simultaneously sent by all base stations. This effectively reduces the inter-cell interference as well as increases the energy of the detectable signal.
- *Uplink Time Difference of Arrival (U-TDOA) in UTRAN.* The location of the terminal is estimated based on the differences in time at which the terminal's signal arrives at multiple, geographically distinct receiving Location Measurement Units (LMU), for example, co-located in the base stations.

The Global Navigation Satellite System (GNSS)-based methods can use one or a combination of selected satellite systems as listed below and is applicable to all 3GPP access networks, if supported by the mobile terminal:

- GPS L1 C/A
- Modernized GPS
- Satellite Based Augmentation Systems
- Galileo
- Quasi Zenith Satellite System (QZSS)
- GLONASS

The satellite-based methods are improved by delivering assistance data to the mobile, and there is a 3GPP standardized support for Network Assisted GPS and Network Assisted Galileo.

In LTE access networks, the large amount of assistance data can be delivered peer-to-peer to the UE using the LTE Positioning Protocol (LPP), which is transparent to the radio access subsystems. But in UTRAN the assistance information needs to be processed in the RNC and the eNodeB when taken out of one protocol and put to another. Still, the information is just relayed.

The OMA-SUPL user plane solution can also be used to deliver satellite assistance data from an SUPL server to the mobile. The 3GPP specifications TS 25.305 [92] and TS 36.305

[112] describe the details of the positioning architecture and positioning methods in UTRAN and E-UTRAN access networks. (3GPP TS 43.059 [113] describes positioning in GERAN (2G) networks.)

2.6.3 Emergency Location Information in 3GPP CS Domain, Control Plane

When someone makes an emergency call using a mobile phone, the 3GPP network provides the location of the caller to the PSAP. The PSAP may also request updated location information for the mobile after the call is established, even if the call was lost. The emergency call handling in the CS domain is described in section 3.6.

3GPP TS 23.271 [91] specifies the so-called network-induced location procedure to provide the location of emergency callers. When the network is recognizing that an emergency call is being made, the core network starts to request the location of the caller from the SMLC/E-SMLC. The emergency call is normally routed to the PSAP that serves the area of the base station that connected the emergency call. In some countries, however, there is a requirement to acquire a more accurate location estimate before selecting the PSAP to which the call should be routed. This will cause some delay for the emergency call, because it takes some time for the network to generate a possibly better location estimate for accurate routing.

When location information is available, such as the serving cell identity and call correlation information, then this information is sent to the GMLC associated with the PSAP serving the emergency call. The GMLC stores the location information for later retrieval by the PSAP and may optionally forward the emergency location information immediately to the PSAP. The GMLC can also record charging information for the emergency location service, if relevant.

When the emergency call is released, the core network may send a report to the GMLC that the emergency call was completed. The GMLC deletes any stored emergency location data, if required by national regulation.

2.6.4 Emergency Location Information in the IMS

When an IMS network can be used for emergency services over a Packet-Switched (PS) network, it is by regulation necessary to provide the location of the caller to the PSAP that serves the area where the caller is. Section 3.6 gives the general description of emergency calls in IMS. Several new IMS core network entities are introduced to support emergency calls. The Emergency CSCF (E-CSCF) routes the emergency call and handles emergency location information together with the Location Retrieval Function (LRF). The Proxy CSCF (P-CSCF) needs to be enhanced to support location information and the so-called "local break-out" functionality for emergency calls. Details are provided in section 3.6.

The location information of the mobile is used for the same purposes in IMS as in the CS domain. The IMS must be able to route the emergency call to the appropriate PSAP, and the PSAP needs to know the location of the emergency caller. In IMS the mobile has to include its own serving cell identification, DSL location information, and other available location information in the very first SIP INVITE when initiating an emergency call. The IMS network can use such location information for routing the call and also delivers the location information to the PSAP. If the mobile is equipped with a GPS receiver, it may know its own location quite accurately if the GPS receiver happens to be activated and is able to receive satellite signals at the time when the emergency call is made. The mobile can also use a so-called location identifier (i.e., P-Access-Network-Info header) to indicate which radio cell it is using for the emergency call. If such mobile-provided location identifier can be trusted, it is often sufficient for routing the call even though less accurate location-wise. The location identifier is access-network-specific and referenced in 3GPP TS 24.229 [114] and standardized in RFC 3455 [115].

In some cases the mobile can request more specific location information for itself or assistance data from the network, if the access network (and mobile) supports some advanced positioning method. The mobile can also use OMA-SUPL for this purpose (see section 2.5).

If the location information delivered by the mobile is insufficient for a given purpose, or if it needs to be validated by the network for correctness, the E-CSCF sends a location request to the LRF. For example, UE-provided location information may not be trusted and may need to be verified for liability reasons. In such a case, a request to the LRF for validation includes the location sent by the mobile, information identifying the packet-switched access network, information identifying the mobile terminal, such as the IP address, and information about the positioning methods supported by the mobile. The E-CSCF can also request the LRF to map the location identifier received from the UE into the corresponding geographical location information. The LRF uses the input data to determine the location of the mobile and to validate the location information provided by the mobile terminal as far as possible. Then, the E-CSCF uses the information returned by the LRF to select the PSAP to which the call should be routed.

The PSAP can also request the mobile's location information from the LRF during the initial emergency call setup or updated location during the call. The LRF knows the initial location of the mobile because the PS network requests the location of emergency callers immediately when an emergency service request is detected.

Hence, the LRF receives the mobile's location information in two separate ways: (i) directly from the mobile terminal via IMS, and (ii) delivered from the PS network after the emergency service request was detected there. The network-generated location information is obtained in somewhat different ways in the different 3GPP access networks. However, as can be seen from the description below, the same principles apply to how to deliver the emergency location information to the PSAP; only the network entities performing the tasks differ from case to case.

2.6.4.1 Emergency-Related Location in UTRAN Access Networks

When a UTRAN mobile initiates an emergency call, the SGSN in the PS core network requests the location of the emergency caller from the RNC/Standalone SMLC, which initiates the positioning procedures in the radio access network (RAN) and the UE as described in section 2.6.2.2. If cell-based position accuracy is not sufficient, the SMLC should initiate other positioning methods, if supported by the network and the UE.

The SGSN selects to which GMLC the location information should be sent using the Service Area Identity (SAI), the serving cell identity or the received location estimate. The location report sent by the SGSN contains the MSISDN identifying the mobile, the identity of the selected GMLC, the reason for the report, and finally the location estimate, its age, and an estimate about the achieved accuracy of the location information. Normally, the SAI or serving cell identity is included in the location report.

The GMLC/LRF accepts the location report only after checking that it serves the identified PSAP and the PSAP is available. The GMLC/LRF then sends the location information to the PSAP immediately or only after the PSAP has asked for it. When the GMLC/LRF does not serve the identified PSAP, the location report will be forwarded to the requesting GMLC/LRF.

2.6.4.2 Emergency-Related Location in E-UTRAN Access Networks

When a mobile initiates an emergency call in an E-UTRAN access network, the MME first selects an E-SMLC network entity and sends a Location Request message to that E-SMLC as described in section 2.6.2.2. The E-SMLC produces the location information together with the access network and mobile, and returns the location information to the MME, including an indication about the estimated accuracy of the information.

The MME uses the reported location information, or the information about which E-UTRAN cell is serving the mobile, when the MME selects the GMLC to which it should send the location information, and further when determining which PSAP is serving the area where the emergency call was initiated.

The GMLC checks whether the LRF is accessible and is able to serve the PSAP before accepting the location information from the MME. The LRF stores the location information and can also use the received information to determine which PSAP is appropriate. The LRF delivers the emergency location information immediately or only after the PSAP requests it.

2.6.4.3 Location for IMS Emergency Services Over Fixed Broadband Access

IMS emergency services can be supported also for terminals using fixed broadband access when the access network contains a Network Attachment Subsystem (NAS) with a Connectivity Session Location and Repository Function (CLF), as specified in ETSI ES 282 004 [116]. In addition to the location services architecture shown in Figure 2.40, the P-CSCF has a direct interface to an LRF in this case.

The terminal using fixed broadband access may know its own network location or geographical location, for example, provided by the user of the terminal when taking the terminal in use. When the terminal knows its location, it inserts the location information, or connection line information, in the SIP INVITE it sends to request an IMS emergency service. If the terminal does not know its own location, it can request its location from the access network, which may know the location of the terminal, for example, based on connection line information.

If the terminal did not deliver any location information, the P-CSCF may request location information from the LRF using the interface dedicated for this purpose. IMS can also request the LRF to validate location information provided by the terminal, if it is not sufficiently trusted.

Location information for a fixed broadband access terminal can also be obtained from a DHCP server using the DHCP option for coordinate-based geographic location, as specified in RFC 6225 [27], and the DHCP option for civic location, as specified in RFC 4776 [16]. More information about the civic location format in general can be found in the introduction of Chapter 2 and in section 2.1. A description of the geodetic location in DHCP can be found in section 2.2.

3

Architectures

This chapter on architectures is probably the most important part of this book, since it illustrates how various building blocks work together. The individual sections are driven by different requirements imposed on the end-to-end communication.

In NENA i2 the baseline assumption is that VoIP emergency callers want to place an emergency call and the existing emergency services infrastructure has not been updated. Quite naturally, such a design decision will lead to changes in other parts of the architecture, as Anand Akundi, the Chairman of the group that developed the i2 solution, describes in section 3.1. As will be seen in section 4, various deployments borrow concepts and ideas from the NENA i2 architecture.

For NENA the i2 architecture was only a first step towards the long-term vision, NENA i3. The assumptions made in NENA i2 are relaxed in NENA i3, and the design assumes an IP-enabled PSAP. Needless to say, such an upgrade is necessary to unlock the full potential of Internet-based multimedia communication. Brian Rosen, the lead author of NENA i3 and the Chair of the Long Term Definition (LTD) Working Group, describes the architecture in section 3.2. The design of NENA i3 was used as a blueprint for the work in EENA on the European version of the i3 standard, called the Next Generation 112 Long Term Definition (NG112 LTD) [117]. The work on the NG112 LTD document was done with input from the members of the NENA LTD Working Group. Regular conferences organized with members from both groups helped to advance both specifications, and to ensure proper alignment.

In section 3.3 we describe the IETF emergency services architecture. While it comes with a few variants that allow one to deal with legacy nodes, it focuses on an end-to-end IP-based communication model. A core component of the final deployment vision is the usage of the Location to Service Translation (LoST) protocol that allows PSAPs to publish their service boundaries and information for how to get emergency calls closer to them. The architecture truly enables multimedia and also offers a story for how to obtain location information and to convey it to a PSAP using the formats described in Chapter 2. The IETF architecture also

Internet Protocol-Based Emergency Services, First Edition. Hannes Tschofenig and Henning Schulzrinne.

provides a solution for emergency calling of over-the-top providers, not only for those network operators who also operate voice services to their customers.

The remaining three architectures take their access-network-specific technologies into account. In section 3.4 Gabor Bajko, an IEEE standardization participant, explains the minimalistic WiFi emergency services architecture. While the write-up is quite short, this is actually very desirable, since radio technologies should be (at least in theory) independent of the application layer protocols in the layered Internet architecture. This book does not, for example, cover a description of emergency services in the Broadband Forum [118] (formerly known as the DSL Forum) and also of CableLabs [119], an organization that standardizes lower-layer equipment used in cable networks, since there is nothing in emergency services that is unique to these environments; these two organizations took the view that this is better left to the higher-layer protocols developed in other organizations.

In section 3.5 Dirk Kroeselberg, a WiMAX standardization participant, illustrates the design of the WiMAX emergency services architecture. The IEEE 802.16 radio and link layer standard, on which the WiMAX architecture is based, is considerably more sophisticated than the IEEE 802.11-based WiFi standards. IEEE saw themselves in competition with 3GPP for a new radio technology, and therefore some of the architectural design decisions found in the 3GPP architecture have been incorporated into the WiMAX architecture.

Finally, in section 3.6 Hannu Hietalahti, who was chairman of the 3GPP Core Networks and Terminals (CT) Technical Specification Group for several years, explains the design decisions leading to the 3GPP emergency services architecture. Since 3GPP is also responsible for the SIP-based IP Multimedia Subsystem (IMS) communication architecture, a description of SIP signaling is also included in the same section. IMS has been the telecommunication operator-preferred application-layer communication system, so offering emergency services support is logical since the switch to IP-based communication may not be visible to most end users and they will assume the same level of functionality from the new communication infrastructure.

3.1 NENA i2

Anand Akundi
Telcordia

3.1.1 Background

Enhanced 9-1-1 (E9-1-1) is a North American telecommunications system that routes emergency calls to the appropriate Public Safety Answering Point (PSAP) and associates a physical address with the calling party's telephone number. Calls made to 911 from wireline phones get automatically routed to the appropriate PSAP for that location, and the physical location associated with the calling number gets passed to the PSAP call-taker. This enables the PSAP to dispatch help even in the event of the caller not being able to convey their location to the call-taker. In wireless networks, the Federal Communications Commission (FCC) requires wireless carriers to provide mechanisms to allow the delivery of both location and callback number to PSAPs. The support for wireless E9-1-1 in the United States was provided in two phase. In Phase 1 Wireless E9-1-1, calls to PSAPs were required to be delivered to the PSAP with callback number and coordinates of the cell tower that is handling the call. This would give the PSAP a general idea of where the call originated. In Phase 2 Wireless E9-1-1, calls to PSAPs are required to be delivered with the latitude and longitude of the caller along with the callback number. The FCC also provides requirements on the accuracy that must be achieved when providing the caller location.

The introduction of VoIP and its popularity with consumers brought forward a new challenge for providing an acceptable E9-1-1 solution for VoIP consumers. VoIP, in its infancy, was a mainly a computer-to-computer type service (e.g., Yahoo chat, etc.) and typically a service that was a closed user group. A VoIP caller could only call other VoIP users who were logged on to the VoIP service. With these early VoIP services, it was very clear that one could not use the service to call emergency services. But as VoIP became a more mainstream service and Cable Providers, Telcos, and other Service Providers starting rolling out VoIP services to consumers, the need to have a solution to support calls to emergency services became critical. Consumers who replaced traditional wireline service with VoIP services expected their VoIP service to provide the same services as they were getting with their previous wireline service, including support for emergency services.

The National Emergency Number Association (NENA) kicked off an effort in August 2003 to start defining solutions to support calls to emergency services from VoIP devices. NENA realized that there were two separable problems that needed to be solved. One was the issue of how to support calls made from VoIP devices to the traditional E9-1-1 network. This issue dealt with the near-time issue of supporting the thousands of consumers who were switching to VoIP and needed a way to contact emergency services. The second issue that NENA decided to tackle was the evolution of the E9-1-1 infrastructure from a TDM to an IP-based infrastructure. The near-time solution was termed the i2 solution and the evolutionary long-term solution was defined as the i3 solution.

NENA formed the VoIP Migratory Work Group (VMWG) to create the i2 solution. The Charter of the VMWG was to create the i2 solution in such a way that VoIP calls could be routed to the appropriate PSAP with location and callback number without requiring any changes to the existing E9-1-1 infrastructure. The WG understood that it was important to ensure that the i2 solution did not require any changes to the traditional E9-1-1 infrastructure, because any changes to the existing infrastructure would increase the cost of deploying the solution and potentially it may never be deployed. Any new functionality needed to support calls to emergency services from VoIP devices would have to become the responsibility of the VoIP network. The VMWG is composed of a diverse group of members, including Service Providers (both VoIP and TDM), vendors, State and local emergency services authorities, and PSAPs.

3.1.2 The i2 Architecture

The i2 solution was designed with the primary goal of enabling VoIP call originations to be routed to the appropriate PSAP with both location and callback number. The goal was also to enable this capability with no new functionality requirements on the PSAP and with little or no new functionality required on the existing E9-1-1 infrastructure. The basic functionality required of the i2 solution is the following:

- detection in the origination network that an emergency call has been placed;
- routing of the emergency call, based on the location where the call originated, to the appropriate E9-1-1 network;
- selective routing of the call by the E9-1-1 network to the appropriate PSAP; and
- allowing the PSAP to retrieve location information associated with the call.

The capabilities listed above are available in both wireline and wireless networks today, and needed to be provided in the i2 solution to ensure VoIP customers got the same set of services as their wireline and wireless counterparts. The first task of the VMWG was to determine what type of VoIP calls should be supported by the i2 solution. VoIP services were divided into three categories, fixed, nomadic, and mobile. Fixed VoIP devices are those devices that are not expected to be moved. The location associated with the device is expected to remain the same once the device is installed. VoIP services that do not allow the device to be moved fall into this category. Nomadic VoIP devices are those that may be moved. The location associated with the device is not expected always to be the same, and mechanisms are needed to allow updated location to be provided when the device is moved. The third category is mobile VoIP devices. These are devices that may move during the duration of a call. It is expected that the location associated with these devices may change during a call, and mechanisms are needed to provide updated location in real time. The VMWG decided to define the i2 solution to support both fixed and nomadic devices in its first iteration. It was felt that this would cover most other VoIP devices being offered to consumers at the time that the solution was being developed. These categories are important when defining how location is determined and provided to the PSAP when emergency calls are made.

Figure 3.1 depicts the i2 functional architecture diagram. The architecture depicted in Figure 3.1 shows several new functions that were introduced to support the following:

- validation of location information;
- routing of emergency calls to the appropriate interconnection point with the existing infrastructure; and
- providing the interconnection for the IP domain with the existing network elements and databases needed to support delivery of location information to the PSAP.

The left side of the diagram depicts the IP domain from which calls originate and the right side depicts the traditional E9-1-1 infrastructure with which the IP domain needs to interconnect. The architecture is developed such that any new functionality needed to support VoIP emergency calling is restricted to the IP domain, and efforts are made to ensure that no new functionality is needed in the traditional E9-1-1 network.

SIP was chosen as the protocol for call/session setup, and the interfaces supporting SIP are depicted as dashed lines in Figure 3.1. The logical interfaces for the exchange of location-related data between and among functional elements in the IP domain are represented by solid lines. The interfaces defined in this standard are labeled (vX) for convenience in referencing individual interface descriptions.

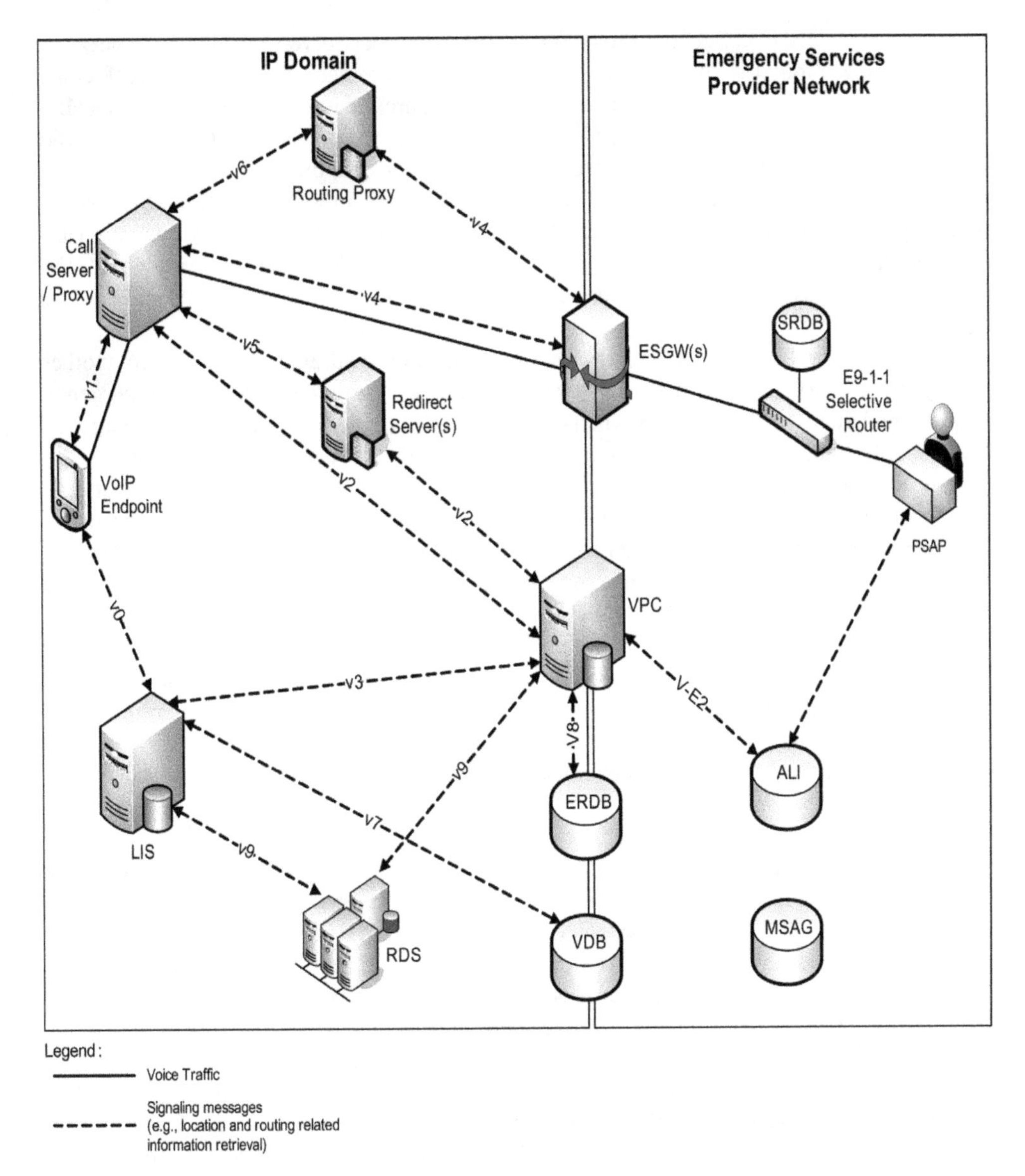

Figure 3.1 The i2 functional architecture diagram. (Reproduced and adapted with the permission of the Canadian Radio-television and Telecommunications Commission on behalf of Her Majesty in Right of Canada, 2013.)

3.1.2.1 Functional Elements

The i2 solution defines several new functional elements to support calls to E9-1-1 from VoIP callers. These new elements are needed both to support routing of VoIP emergency calls and for determination of location information.

VoIP User Agents, Call Servers, Proxy Servers, and Redirect Servers are all typical network elements that are found in the VoIP networks and are used to generate and route VoIP calls. These network elements are also part of the i2 network architecture. Some of the i2-specific functionality is described in the following paragraphs. There are no i2 architecture-specific requirements on the VoIP user agents other than the fact that they should allow the user to call E9-1-1. Voice Service Provider Call Server is always in the call path and needs the capability to recognize that an E9-1-1 call has been originated. Procedures for this may vary based on the end device capability. Once the call server recognizes an emergency call, the i2 solution provides support for various scenarios that a Voice Service Provider may choose to support.

A Voice Service Provider (VSP) may choose to hand over all emergency calls to another party. In this scenario, the VSP may contract with a third-party provider to handle all emergency calls on their behalf. In this scenario, when the call server recognizes the call to be an emergency call, it simply routes the call to a designated Routing Proxy for further call treatment.

In the second scenario, a VSP may choose to handle routing of emergency calls but may contract out the determination of the routing based on the location of the caller. In this scenario, the VSP Call Server would forward all emergency calls to the designated Redirect Server. The Redirect Server would access i2 functional elements that determine location and return a routing address back to the VSP Call Server. In this scenario, the VSP Call Server would not require any new or i2-specific interfaces to service emergency calls.

In the third scenario, a VSP may choose to handle the emergency call entirely in-house, and in this scenario the call server accesses the i2 functional elements that determine routing and location information.

Some of the i2-specific functional elements are described in the following paragraphs:

- *Emergency Services Gateway (ESGW).* The ESGW is the signaling and media inter-working point between the IP domain and conventional trunks to the E9-1-1 SR. It uses either Multi-Frequency (MF) or Signaling System #7 (SS7) signaling. The ESGW uses the routing information provided in the received call setup signaling (the ESGWRI) to select the appropriate trunk (group) and proceeds to signal call setup toward the SR, including the ESQK in outgoing signaling. The ESGW may signal only the 10-digit ESQK or both the ESQK and callback number to the SR.
- *Emergency Service Zone Routing Data Base (ERDB).* The ERDB contains the routing information associated with ESZs. It supports the boundary definitions for ESZs and the mapping of civic address or geospatial coordinate location information to a particular ESZ.
- *VoIP Positioning Center (VPC).* The VPC is the element that provides routing information to support the routing of VoIP emergency calls. It also cooperates in delivering location information to the PSAP using the existing ALI DB infrastructure. The VPC interfaces to the ERDB to obtain routing data. The VPC receives queries from the CS or RP/RS over the v2 interface.
- *Validation Database (VDB).* The VDB contains information that describes the current, valid civic address space defined by the Emergency Services Network Provider's MSAG. The VDB has the capability to receive a validation request containing a civic address consisting of data elements included in the civic Location Object (LO) and be able to determine if this civic address is a valid address. The VDB will return a response indicating that a given

location is a valid address or an error response. This process ensures that the address is a real address (i.e., the address exists) but does not ensure that it is the location of the caller.

- *Location Information Server (LIS).* A Location Information Server (LIS) is a functional entity that provides locations of endpoints. A LIS can provide Location-by-Reference, or Location-by-Value, and, if the latter, in geo- or civic forms. A LIS can be queried by an endpoint for its own location, or by another entity for the location of an endpoint. In either case, the LIS receives a unique identifier that represents the endpoint, for example, an IP address, circuit ID or Media Access Control (MAC) address, and returns the location (value or reference) associated with that identifier. The LIS is also the entity that provides the dereferencing service.
- *Root Discovery Server (RDS).* The RDS is the entity that stores the serving areas of ERDBs and VDBs. It accepts queries that contain geospatial or civic location information over the v9 interface. When queried with location information, it returns a list of URIs for ERDBs or VDBs serving that area.

3.1.2.2 Description of 9-1-1 Data Objects

This section identifies 9-1-1 data objects defined in the i2 solution architecture. These data objects are needed to support routing of emergency calls and delivery of location information to PSAPs. The use of the data objects and the protocols defined to carry them are described in detail in the specification of the various interfaces in the i2 solution architecture.

- *Emergency Services Gateway Route Identifier (ESGWRI).* The ESGWRI is used by the Call Server or Routing Proxy to route an emergency call to the correct ESGW, and by the ESGW to select the desired path to the appropriate SR for the call. The ESGWRI format is as follows:
  sip: +1 numberstring@esgwprovider.domain;user=phone
  where the "numberstring" is 10 numeric characters (e.g., *nnnnnnnnnn*).
- *Emergency Route Tuple (ERT).* The ERT consists of the elements from the Emergency Services Zone Routing Database that are used to select an ESGWRI. The ERT is communicated via the v2 and v8 interfaces and represents a route through a Selective Router for a collection of ESNs that are associated with a Selective Router. The components of the ERT are:
  - a Selective Routing Identifier consisting of a Common Language Location Identifier (CLLI) code of the form AAAABBCCDDD, where
    - AAAA represents the city/county
    - BB represents the State/Province
    - CC represents the building or location
    - DDD represents the network,
  - a 3- to 5-digit routing Emergency Services Number (ESN) that uniquely identifies the ESZ in the context of that SR,
  - an NPA that is associated with the outgoing route to the SR.
- *Emergency Services Query Key (ESQK).* The ESQK identifies a call instance at a VPC, and is associated with a particular SR/ESN combination. The ESQK is delivered to the E9-1-1 SR and may be delivered as the calling number/ANI for the call to the PSAP. The ESQK

is used by the SR as the key to the Selective Routing data associated with the call. The ESQK may be delivered by the SR to the PSAP as the calling number/ANI for the call, and subsequently used by the PSAP to request ALI information for the call. The ALI database includes the ESQK in location requests sent to the VPC. The ESQK is used by the VPC as a key to look up the location object and other call information associated with an emergency call instance. The ESQK is expected to be a 10-digit North American Numbering Plan (NANP) Number.

- *Last Routing Option (LRO).* The LRO is sent by the VPC to the Call Server or Routing Proxy and provides the Call Server or Routing Proxy with a "last chance" destination for the call. The LRO may be the Contingency Routing Number (CRN), which is a 24×7 PSAP emergency number, or it may contain a routing number associated with a national or default call center. The content of the LRO will depend on the condition that resulted in the providing of the LRO. Ultimately, the usage of LRO routing data for call delivery is based on logic internal to the Call Server or Routing Proxy.
- *Contingency Routing Number (CRN).* A 10-digit 24×7 number that could be used, when available, to route emergency calls to the PSAP in scenarios where a network (i.e., trunk or SR) failure results in the ESGW being unable to route the emergency call over the desired path to the SR. The CRN is expected to be a 10-digit NANP number.
- *Location Object (LO).* In the context of this document, the LO is used to refer to the current position of a VEP that originates an emergency call. The LO is expected to be formatted as a Presence Information Data Format – Location Object (PIDF-LO) as defined by the IETF in RFC 4119 [17] and RFC 5139 [21].
- *Location Key (LK).* This is an object that uniquely identifies an instance of a LO that is stored and/or managed by an LIS on behalf of a VoIP endpoint. The Location Key may contain a unique LIS identifier and a unique identifier for the endpoint. For example, in SIP this is represented as a URI and the term used is "Location by Reference." Details on the Location Key are for further study.
- *Location Information Element (LIE).* This is a protocol container for either or both of:
  - one LK,
  - one PIDF document.
- *Callback Number.* A Callback Number is an identifier for an emergency caller that can be used by the PSAP to reach an emergency caller subsequent to the release of an emergency call. In the i2 solution, the Callback Number is an E.164 number, and is represented in VoIP signaling by a tel-Uniform Resource Identifier (URI) [80]. The Callback Number, when available, must be sourced from the VSP's subscription data, which shall not be directly modifiable by the user.
- *Customer Name.* The Customer Name represents the name of the VSP's customer who subscribed to the VoIP service and is typically billed for it. The Customer Name, when available, must be sourced from the VSP's subscription data, which shall not be directly modifiable by the user.
- *Postal Address.* A Postal Address includes the following data elements in the VDB validation request and response:
  - HouseNum
  - PrefixDirectional (not included in all addresses – e.g., N 7TH)
  - StreetName

  - StreetSuffix (not included in all addresses – e.g., CONGRESS AVE)
  - PostDirectional (not included in all addresses – e.g., ACADEMY DR E)
  - PostalCommunity
  - CountyName
  - StateProvince
  - PostalCode
  - Country

  An address is considered Postal valid if it exists in the country's Postal Service authority data. For example, a valid Postal Address will conform to USPS abbreviations as specified in USPS Standard Publication No. 28 or the Canadian Postal Guide. The VDB may use a database or service based on USPS data to determine whether an address is a valid postal address.

- *MSAG Address.* An MSAG Address includes the following data elements in the VDB validation request and response:
  - HouseNum
  - HouseNumSuffix
  - PrefixDirectional (not included in all addresses – e.g., N 7TH)
  - StreetName
  - StreetSuffix (not included in all addresses – e.g., CONGRESS AVE)
  - PostDirectional (not included in all addresses – e.g., ACADEMY DR E)
  - MSAGCommunity
  - CountyID (not used by all MSAG addresses)
  - StateProvince

  An address is considered MSAG valid if it exists in the MSAG database. The MSAG database is created by the Addressing Authority for a region. It contains entries for valid Address Ranges for the Streets (within the Communities, Counties, and State/Province) for which the Addressing Authority is responsible. The Abbreviations, Street Names, and Community Names may not be the same as the valid Postal Address for the same address (in fact, they will most probably be different). An MSAG address may also require a County ID to be specified. The County ID may be an abbreviation of the County Name or it may be the Tax Area Rate (TAR) Code for the County.

3.1.2.3 Interface Definitions

This section provides brief definitions of the VoIP and data exchange interfaces included in the i2 Solution.

- *v0 – LIS to VoIP Endpoint.* The v0 interface is used to provide a means for a VoIP endpoint to receive information corresponding to a predetermined location. The information provided may be in the form of an LK, or it can be a PIDF-LO containing the actual location. It is expected that the PIDF-LO may include either or both Geodetic and Civic information, to enable routing of the emergency call, and a Civic Location, for example, addresses for non-mobile IP devices (mobile IP endpoints are not specifically addressed), for display to the PSAP.

- *v1 – VoIP Endpoint to Call Server/Proxy.* The v1 interface is between the VoIP Endpoint and the Call Server within the VSP's network. It is used to initiate an emergency call, which will ultimately be answered by the correct PSAP, and is also used to communicate the location information to the Call Server when an emergency call is initiated.
- *v2 – Call Server or Routing Proxy or Redirect Server to VPC.* The v2 interface is used to request emergency call routing information. The Call Server can invoke the v2 interface directly or utilize a Routing Proxy or Redirect Server, which requires forwarding the LIE, or sufficient information to construct the LIE to the Routing Proxy or Redirect Server. It is expected that the VSP or Routing Proxy Operator will have a business relationship in place that would allow the Call Server or Proxy to send requests for routing information to the appropriate VPC.
- *v3 – VPC to LIS.* The v3 interface provides a means for the VPC to obtain the emergency caller's location. This is used when the LIE, provided to the VPC via the v2 interface, contains an LK. The LK is used in obtaining the location from the LIS. The VPC uses the returned location information to obtain routing information from the ERDB.
- *v4 – Call Server/Routing Proxy to ESGW.* The v4 interface is used to forward the call to the appropriate ESGW. The Call Server or Routing Proxy uses the ESGWRI returned from the VPC, or derived based on the ERT returned by the VPC, to forward calls to the appropriate ESGW, and inserts the ESGWRI and ESQK into the signaling message.
- *v5 – Call Server/Proxy to Redirect Server.* The v5 interface is defined as an SIP interface to a Redirect Server so it supports a subset of the SIP specification. The Call Server sends an SIP INVITE containing callback information, the subscriber name, the LO and/or LK, and the VSP identifier to the Redirect Server. The Redirect Server interfaces to the VPC over v2 to obtain routing instructions for the emergency call based on the location of the caller.
- *v6 – Call Server to Routing Proxy.* The v6 interface is defined as an SIP interface to a Routing Proxy. The Routing Proxy supports the full SIP specification. The CS sends an SIP INVITE containing the callback information, the subscriber name, the LO and/or LK, and the VSP identifier to the RP. This interface is used when the CS (unconditionally) routes all emergency calls to a Routing Proxy.
- *v7 – LIS to VDB.* The v7 interface is used by the LIS provider to request validation of a given Civic Location as compared with the MSAG-based data stored in the VDB. The VSP may also use this interface from its LIS, when acting on behalf of its customers in the function of location provider and/or verifier.
- *v8 – VPC to ERDB.* The v8 interface supports queries from the VPC to the ERDB. The VPC sends location information for the emergency caller to the ERDB to obtain routing information (ERT), and other information to help in selection of an appropriate ESQK for the call and to support the delivery of call/location information in response to ALI database requests.
- *v9 – LIS/VPC to Root Discovery Operator.* The v9 interface allows an LIS/VPC to discover the appropriate VDB/ERDB. Several mechanisms have been defined to allow for the discovery of the VDB/ERDB.
- *v-E2 – VPC to ALI DB.* The v-E2 interface uses the E2+ protocol as defined in NENA Standards 05-001, with modifications required for support of i2. The ESQK is sent in a query by the ALI DB over the v-E2 interface. The VPC responds with the emergency caller's location, callback number, and VoIP Provider identifier and/or information it received in the VSP and/or source element.

3.1.2.4 Basic Call Routing of VoIP Emergency Calls to ESGW

Figure 3.2 illustrates a basic call routing scenario for a VoIP emergency call origination. In this scenario, a Call Server queries the VPC for routing information, and routes the call directly to the ESGW. The steps shown in the figure are as follows.

1. The endpoint is populated with a Location Object (LO) or a Location Key (LK). The figure illustrates the method whereby a PIDF-LO/LK is downloaded from the LIS to the endpoint using the v0 interface.

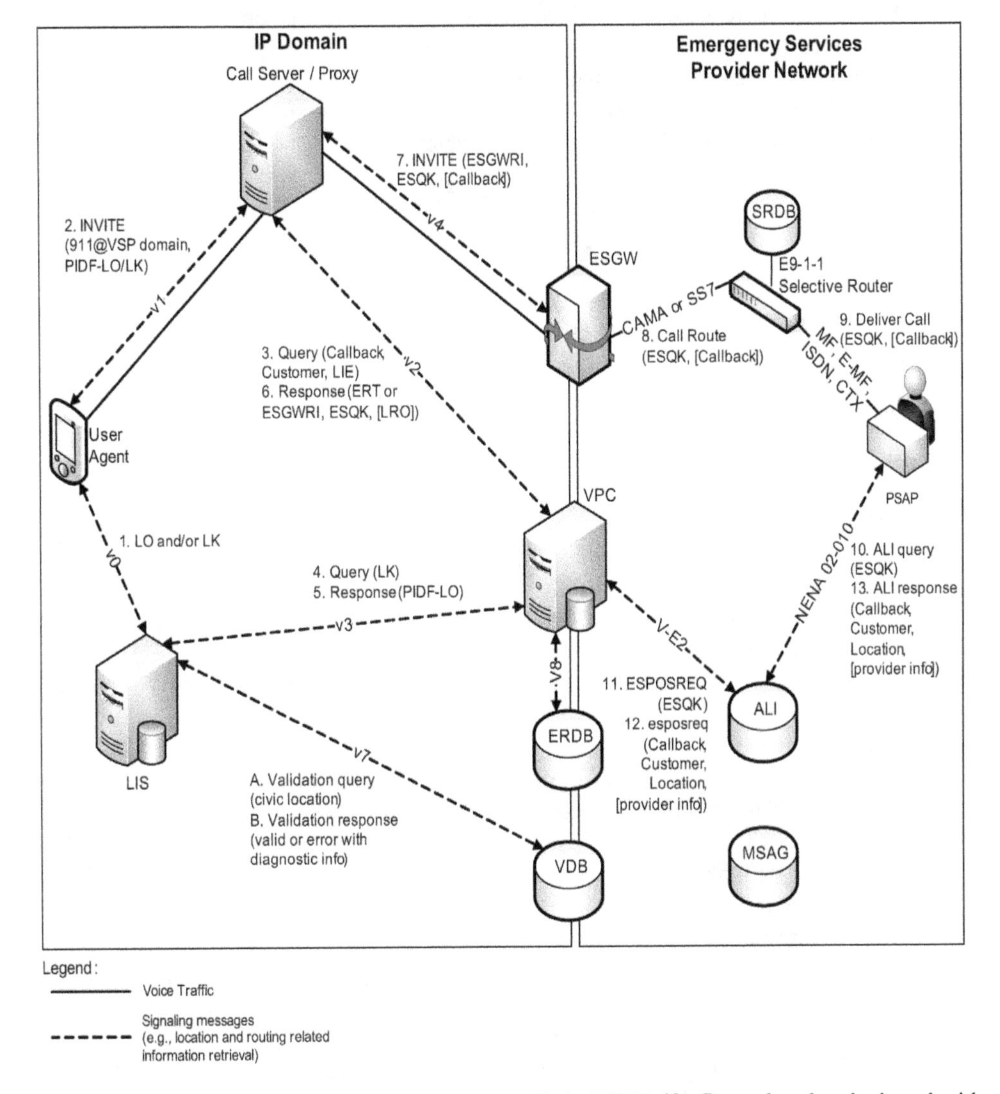

Figure 3.2 Basic call routing of VoIP emergency calls in NENA i2. (Reproduced and adapted with the permission of the Canadian Radio-television and Telecommunications Commission on behalf of Her Majesty in Right of Canada, 2013.)

2. The VoIP endpoint originates an emergency call by sending a call initiation request, designating 911@VSPdomain as the target destination, and including the LO and/or LK. Using SIP as the example, the User Agent would send an INVITE, including sip:911@VSPdomain in the Request URI header, the Address of Record (AoR) associated with the User Agent (possibly in the form of a tel uri in the From header, and including the PIDF-LO member body in the INVITE).
3. The Call Server receives the call initiation request and uses local procedures to determine the provisioned callback number and subscriber name associated with the Address of Record received in the incoming INVITE. The Call Server then sends a routing request to the VPC using the information received in, and determined based on, the call request. The VPC receives the routing request that includes the following key pieces of information:
   - a callback number associated with the emergency caller in the form of an E.164 telephony number (e.g., a tel URI),
   - a subscriber name associated with the emergency caller,
   - a LIE.
4. If an LK is received, the VPC queries the identified LIS over the v3 interface, including the LK.
5. The LIS returns to the VPC the requested location over the v3 interface.
6. The VPC uses the location (received from the Call Server or queried from the LIS) to obtain the ESZ-related routing information from the ERDB over v8 (not shown). The ERDB identifies the ERT and CRN that will facilitate routing via the appropriate ESGW to the SR that serves this ESZ. The VPC uses the received routing information to allocate an available ESQK from the pool of ESQKs appropriate for the SR/ESN associated with the caller's location/ESZ and sends a response to the routing request for this call. The routing response sent to the Call Server will contain the allocated ESQK and the appropriate LRO. In addition, routing information will be included in the response that consists of either the ERT (if the Call Server is responsible for performing the ERT-to-ESGWRI translation), or an ESGWRI (if the VPC is responsible for performing the ERT-to-ESGWRI translation). The VPC also determines the VSP either from the routing request contents or from the originator of the request. The VPC maps the caller's callback number, subscriber name, VSP, and the contents of the location into the appropriate fields necessary to populate the response to an Emergency Services Positioning Request (ESPOSREQ) and stores this information pending a subsequent query from the ALI DB.
7. If the Call Server receives an ERT in the routing response from the VPC, the Call Server will use the ERT to determine the target ESGW and the appropriate ESGWRI value for the call. If the Call Server receives an ESGWRI in the routing response from the VPC, the Call Server will take the ESGWRI received in the response and use it as the basis for selecting the appropriate ESGW toward which to route the emergency call. In either case, the Call Server routes the call to the ESGW, including the ESGWRI, the ESQK, and the callback number in outgoing signaling. The LRO will be retained at the Call Server and only used for emergency call routing if the ESGW detects a failure condition that makes routing based on the ESGWRI impossible.
8. The ESGW uses the received ESGWRI to select an outgoing route (i.e., trunk group) to the appropriate E9-1-1 SR. If the trunk group is available, the ESGW seizes a trunk and signals an emergency call origination to the E9-1-1 SR, using outgoing (SS7 or MF) 10 or 20-digit signaling. If the ESGW determines that it cannot route the call over a route

associated with the ESGWRI due to a failure condition, it shall inform the Call Server that a failure has occurred.

9. The SR receives the emergency call, uses the ESQK to query the SRDB for the associated Emergency Service Number (ESN), and uses the ESN to identify the appropriate PSAP for the call. The SR then delivers the call to the appropriate PSAP, signaling the ESQK, the callback number (if available), or both the ESQK and the callback number, based on local implementations.
10. The PSAP ANI/ALI controller receives the call setup signaling, and sends an ALI query to its serving ALI DB, using the ESQK or callback number as the query key, based on local implementations.
11. The ALI DB sends an ESPOSREQ to the VPC (identified in the shell record for the ESQK in the ALI DB), and includes the ESQK or ESQK plus callback number in the request.
12. The VPC receives the ESPOSREQ from the ALI DB, and uses the ESQK to retrieve the ALI record information it stored previously (in step 4). The VPC returns an Emergency Services Positioning Request response (esposreq) to the ALI DB, which includes the callback number, the subscriber name, the location information, and the VSP-provided information that can be supported by the v-E2 interface.
13. The ALI DB receives the esposreq from the VPC. It may also extract additional information from the shell record for the ESQK. The ALI DB returns an ALI response to the PSAP.
14. (not shown) When the VPC receives an indication that a particular instance of an emergency call is being cleared, the VPC de-allocates the associated ESQK and makes it available for subsequent emergency calls. Note that release of the ESQK may occur as a result of an indication of call release over the v2 interface from the Call Server or Routing Proxy, or the expiration of the ESQK guard timer, whichever occurs first.

3.1.2.5 Proxy Redirect Server

Figure 3.3 illustrates an example of an emergency call setup using SIP signaling to a Redirect Server. In this scenario, the Call Server uses a Redirect Server to obtain routing information, and then routes the call to the ESGW. The SIP Redirect Server performs a routing query to the VPC over the v2 interface. Before the emergency call is originated, the location is validated in steps A and B. The steps shown in the figure are as follows.

- 1, 2. The same as in the basic scenario described in section 3.1.2.4.
- 3. The Call Server/SIP Proxy Server sends an INVITE over the v5 interface to an SIP Redirect Server. The INVITE includes the callback number, subscriber name, and PIDF-LO.
- 4. The SIP Redirect Server queries the VPC over the v2 interface, including the following key information:
  - a callback number associated with the emergency caller in the form of an E.164 telephony number (e.g., a tel URI),
  - a subscriber name associated with the emergency caller,
  - a LIE.
- 5–7. The same as steps 4–6 in the basic scenario described in section 3.1.2.4.

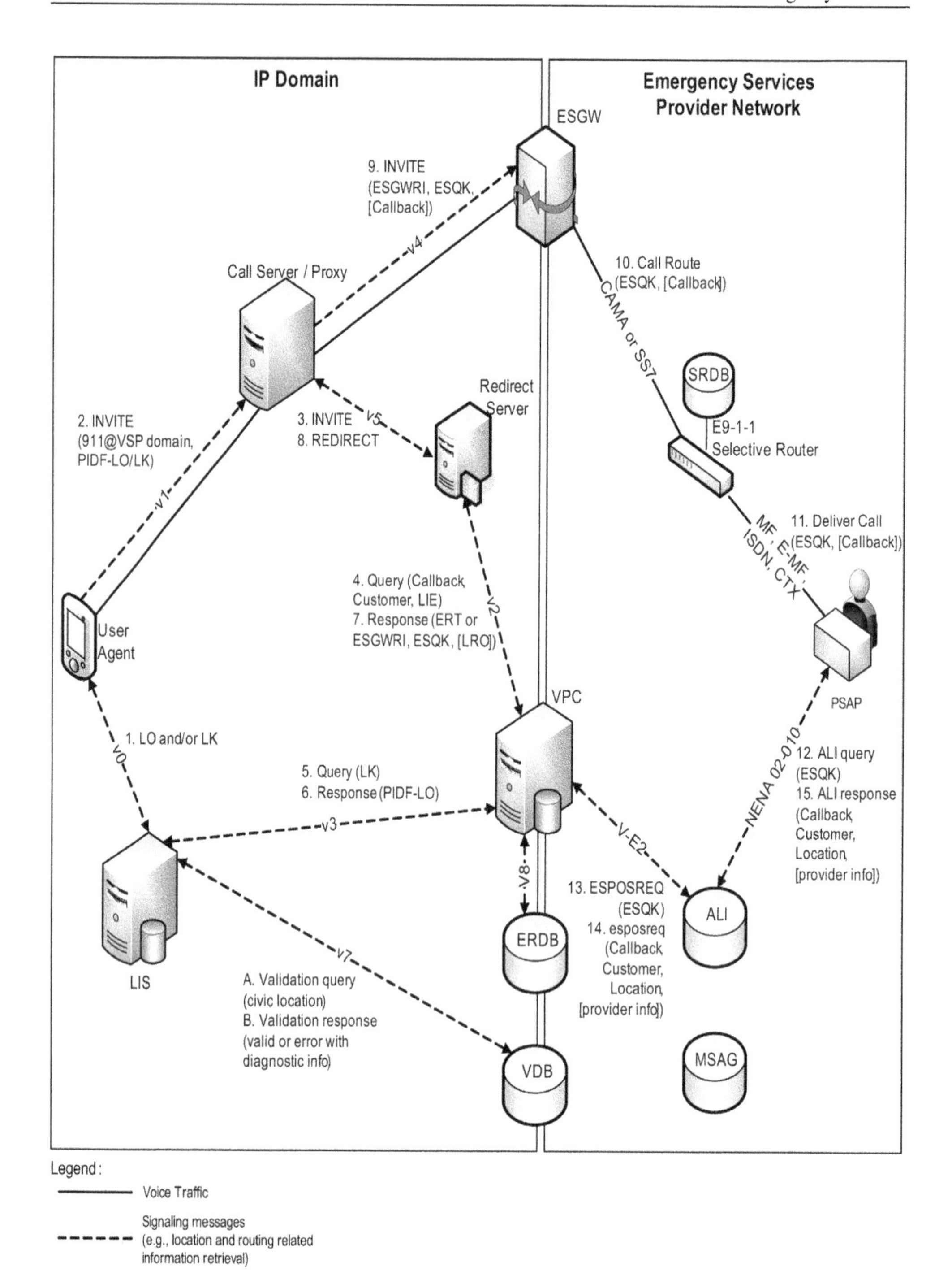

Figure 3.3 Call routing using SIP Redirect Server in NENA i2. (Reproduced and adapted with the permission of the Canadian Radio-television and Telecommunications Commission on behalf of Her Majesty in Right of Canada, 2013.)

8. The Redirect Server returns an SIP REDIRECT message to the SIP Proxy, including the ESQK and either the ERT or the ESGWRI provided by the VPC. It may also include a LRO, if provided by the VPC.
9. The Call Server/SIP Proxy routes the call to the ESGW over the v4 interface, including the ESGWRI provided by the VPC or determined by the Call Server, the ESQK, and the callback number.

10–15. The same as steps 8–14 in the basic scenario described in section 3.1.2.4.

3.1.2.6 Routing Proxy Routing Scenario

Figure 3.4 illustrates an example of emergency call routing via a routing proxy. In this scenario, the emergency call is routed via a routing proxy, which initiates the routing query to the VPC, and then routes the call directly on to the appropriate ESGW. The steps are as follows.

1, 2. The same as in the basic scenario described in section 3.1.2.4.

3. The Call Server/SIP proxy sends an INVITE over the v6 interface to an SIP Routing Proxy. The INVITE includes the callback number, subscriber name, and PIDF-LO.
4. The SIP Routing Proxy queries the VPC over the v2 interface, including the following information:
   - a callback number associated with the emergency caller in the form of an E.164 telephony number (e.g., a tel URI),
   - a subscriber name associated with the emergency caller,
   - a LIE.

5, 6. The same as steps 4 and 5 in the basic scenario described in section 3.1.2.4.

7. The VPC uses the received PIDF-LO to request routing information from the ERDB over v8 (not shown). The ERDB uses the received location to determine the ERT and CRN for the call. The VPC uses the ERT to determine the ESQK associated with the ESZ of the caller, and allocates an available ESQK for the call. The VPC sends a response to the routing request for this call over the v2 interface to the Routing Proxy, including the allocated ESQK and either an ERT (if the Routing Proxy is responsible for performing the ERT-to-ESGWRI translation) or an ESGWRI (if the VPC is responsible for performing the ERT-to-ESGWRI translation). An LRO is also expected to be included in the routing response. The VPC also determines the VSP either from the routing request contents or from the originator of the request. The VPC maps the caller's callback number, subscriber name, VSP, and the contents of the location into the appropriate fields necessary to populate the response to a subsequent ESPOSREQ from the ALI DB and stores this information pending a subsequent query.
8. The SIP Routing Proxy routes the call to the ESGW over the v4 interface, including the ESQK provided by the VPC, the ESGWRI determined by the Routing Proxy or provided by the VPC, and the callback number.

9–15. These steps in the Emergency Services Provider Network are the same as steps 8–14 in the basic routing scenario described in section 3.1.2.4.

3.1.3 Regulatory Situation and Deployment Status

The i2 solution received a lot of attention as a viable solution to support emergency calls from VoIP endpoints. In June 2005, the Federal Communications Commission (FCC) in the

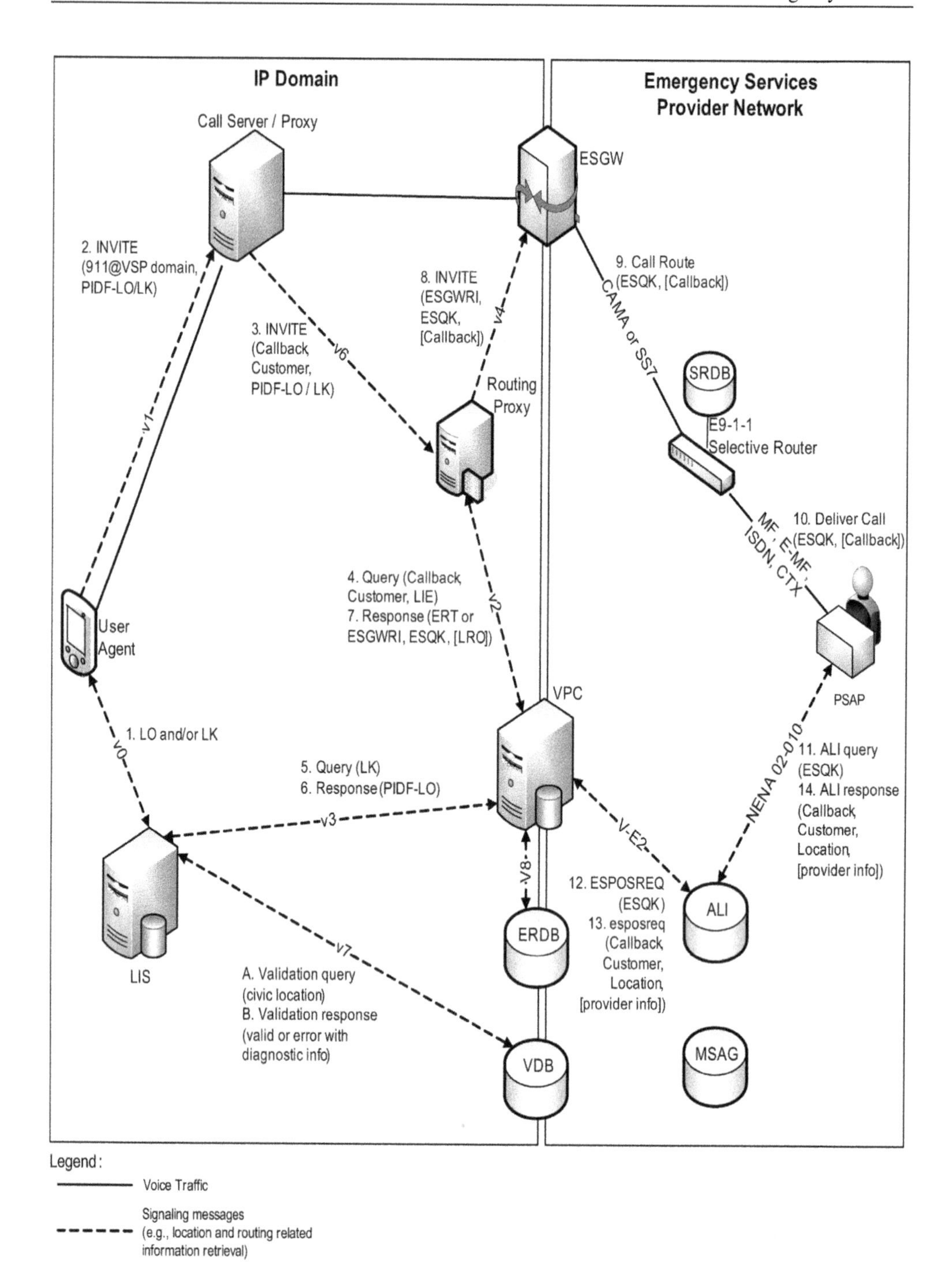

Figure 3.4 Call routing using an SIP Routing Proxy in NENA i2. (Reproduced and adapted with the permission of the Canadian Radio-television and Telecommunications Commission on behalf of Her Majesty in Right of Canada, 2013.)

United States passed rules that required "interconnected" VoIP Services Providers to support 9-1-1. Interconnected VoIP services are those VoIP services that allow users to make and receive calls from the Public Switched Telephone Network (PSTN). This made it critical for "Interconnected" VoIP Service Providers to deploy a solution that could support 9-1-1 services. In the midst of this regulatory turmoil, NENA published the i2 solution as a possible solution to support 9-1-1 calls from VoIP endpoints. While the FCC did not mandate the use of the i2 solution to support 9-1-1 from VoIP endpoints, a majority of the interconnected VoIP providers adopted some variation of the i2 solution to support their VoIP customers. Features like the automatic location determination, as defined in the i2 solution, are not currently deployed in the United States, but most of the other features have found their way into actual deployments.

The i2 solution has also received attention outside the United States. Regulators in both Canada and some countries in Europe have looked at i2 as a potential solution to support emergency calling from VoIP endpoints.

3.2 NENA i3

Brian Rosen
NeuStar

3.2.1 *History*

The North American Emergency Number Association (NENA) creates standards for the 9-1-1 emergency calling system in the United States, Canada, and some parts of Mexico. In 2003, PSAPs in North America started receiving calls from VoIP systems. Neither the PSAPs nor the VoIP carriers were prepared for these calls. It was assumed, and even sometimes explicitly stated, that VoIP services were not appropriate for 9-1-1 calling. This did not stop the users of these services, for whom it looked like a telephone, it acted like a telephone, and therefore it ought to support 9-1-1 calling.

To NENA, this was a repeat of the same story that led to mobile phone support in the 9-1-1 system: the carriers did not intend the service to be used to call 9-1-1, the users thought otherwise, and eventually 9-1-1 support was mandated by the regulator. The history of getting support for mobile phones was difficult and contentious, with NENA (and another organization, APCO) having a somewhat adversarial relationship with the mobile phone industry, and taking over a decade before most of the system was upgraded to handle automatic reporting of location of mobile phones calling 9-1-1. They further realized that the Internet brought not just VoIP, but a panoply of devices and services for which the same problems would occur repeatedly.

The 9-1-1 system, like most extant emergency calling systems, is built on a wireline TDM network, designed around the needs of regulated wireline carriers. Extensions were made to support automatic location reporting for wireline networks, and further extensions were made to support location reporting (in the form of latitude and longitude positions) for mobile wireless networks, but the latter was force-fit into the existing network. NENA knew that its network was not really capable of supporting many of these new services.

Meanwhile, a group in the IETF working on VoIP standards realized that, until VoIP supported emergency calling, it would not become a mainstream technology, and had begun discussions of what would be needed to support emergency calls on all kinds of IP services.

Participants from the IETF and the NENA groups were introduced to each other, and a meeting was held in Atlanta in 2003. At that meeting, a three-step process for introducing IP to the 9-1-1 system was defined. These three steps are as follows:

- *i1* was a small project to document what some VoIP systems had begun doing to support 9-1-1. An attempt was to be made to document best practices. This effort was completed quickly, but the work product was not of much value. While the NENA i1 specification was never published, a write-up can be found, for example, in [120].
- *i2* was a project in NENA to define a mechanism to support VoIP networks in the existing 9-1-1 system. Its guidelines were that, as much as possible, VoIP callers were to have identical service as wireline callers, and *no* upgrades to PSAPs were to be needed. Fortunately, the mechanisms put in place to support mobile networks were able to be adapted to VoIP networks. Unfortunately, the US regulator acted to require support of VoIP before the NENA work was completed, and what is actually deployed in the United States for VoIP support is close to, but not exactly, what NENA defined in its i2 work. It is, however, close enough, and is sometimes referred to as "pre-i2." A detailed description of NENA i2 can be found in section 3.1.
- *i3* defined a new 9-1-1 system, now called Next Generation 9-1-1, based on an IP network. Technically, i3 is the first version of the technical standards for NG9-1-1. The NENA work group writing the standards has close cooperation with the IETF ECRIT [120] and Geopriv [120] work groups, and i3 is based on the IETF work. The core publications relevant to NENA i3 are the stage 1 requirements [121], the stage 2 functional architecture [38], and the long-term vision solution specification (NENAi3 stage 3) [39]. NG9-1-1 is being deployed in the United States as of this writing.

3.2.2 Emergency Services IP Networks

NG9-1-1 is an all-IP network, with gateways to connect legacy TDM networks to the IP network. Public safety agencies are connected to a private, managed IP network called an Emergency Services IP Network (ESInet). Attempting to have a single IP network (or interconnected transports forming a single logical network) of *all* public safety agencies is difficult, because of the political difficulties of getting police, fire, EMS and PSAPs to agree on technical, operational, management, and governance issues, but the benefits of having all agencies on one network is seen as key to integrated response in significant incidents. These networks are usually organized at the county level, and are interconnected to form a state-wide network. The state networks are in turn interconnected to form a nationwide network, and international interconnections are anticipated between Canada, the United States, and Mexico. ESInets are conventional routed IP networks, using typical routing protocols used on the Internet, such as the Border Gateway Protocol (BGP) between different networks. The ESInets are built from a combination of wireline, wireless, commercial, and government-owned facilities. ESInetworks are managed, and access to them is controlled and restricted, but they are not assumed to be "walled gardens."

The ESInetworks are connected to carrier networks and the Internet. These connections are often accomplished with state-level router interconnects. All traffic from outside the ESInet is passed through a Border Control Function (BCF), which is a firewall and session border controller. Although PSAPs are reluctant to allow direct Internet connections to the ESInet, we have been able to convince them that it is impossible to offer service to any device on any network without Internet connectivity. An example we use often is of a person sitting in a coffee shop in Chicago who subscribes to a VoIP service in Sierra Leone. The visitor has a laptop, with a WiFi connection to the coffee shop's network, and has a soft client running on the laptop. An accident occurs right outside the window, and the visitor dials 9-1-1 using his soft client. Without an Internet connection, Sierra Leone VoIP Services Pty. would be unable to send the call to the Chicago PSAP.

3.2.3 Signaling and Routing IP-Originated Calls

NG9-1-1 follows the IETF standards for emergency calling. The "phonebcp" document [122] from the IETF defines the signaling interface for calls and messages to NG9-1-1. A phonebcp-compliant device will be able to initiate a session to 9-1-1. This means that *all* calls are signaled with SIP, and arrive at the ESInet routed by LoST, containing a Geolocation header with the location of the caller, and a valid AoR and Contact URI to facilitate a callback from the PSAP to the caller. Calls are marked with the "sos" service URN.

NG9-1-1 standards recommend, but do not specify, the use of IETF standardized Location Configuration Protocols to deliver location to IP connected endpoints. NENA standards start at the interconnect of a carrier network or the Internet to the ESInet, and, from that starting point, it does not matter *how* the location found in the Geolocation header got there, as long as it represents the location of the caller. NG9-1-1 accepts civic or geo forms of location, by reference or by value on all calls. We expect Internet Service Providers to deploy location infrastructure (and specifically an LIS) for their subscribers. Regulation may be needed to assure that all devices and all service can report accurate location to 9-1-1.

Calls must be routed to the correct ESInetwork using a LoST server provided by the local (or state) 9-1-1 authority. Within NENA standards, the LoST server is termed an "Emergency Call Routing Function" (ECRF). The ECRF queried by an endpoint or proxy typically routes to a state-level "Emergency Services Routing Proxy" (ESRP), which is located inside the ESInet, just past the Border Control Function. The ECRF can be accessed over the Internet. Alternatively, a carrier can obtain a feed from the local or state 9-1-1 Authority to maintain a local copy of the ECRF within the carrier's network. The 9-1-1 Authority also provides a "Location Verification Function" (LVF) that is also a LoST server but is specialized to perform location validation. The ECRF and the LVF may be co-located, but do not have to be. Civic location must be validated against the LVF before it is loaded into an LIS.

NENA standards extend the IETF standards for incoming calls in a very few ways. Carriers, some devices, and any other service providers in the path of an emergency call must add a "Call-Info" header to the initial SIP message that contains a URI that points to a NENA-defined data structure containing "Additional Information about a Call." This structure has elements for:

- identification and contact information about the carrier, device, or service provider;
- subscriber contact information;
- information on the service provided (analogous to what in the current system is called "Class of Service"); and
- a "hook" for device- or service-specific information (for example, vehicle telematics data, medical monitoring device data, or other sensor data would be supplied using this URI).

In addition, there can be an optional Call-Info header containing a URI for caller-specific data, where we intend "caller" to mean the human making the call. This structure contains elements for:

- caller contact information;
- emergency contact information, including a request to add the emergency contact to a call (for example, an elderly parent who is often confused may ask to have a son or daughter added to a 9-1-1 call if they make one); and
- a hook for an emergency medical health record.

NENA does not anticipate that most consumers only use their telecommunication providers to store and maintain, for example, medical or emergency contact information. A customer may select a different entity to store this additional data. The design of the "Additional Data about a Caller" mechanism is that the user arranges with some trusted entity, perhaps an insurance company, to maintain the data. The trusted entity supplies an opaque URI, which is attached to an emergency call. A call from a home, work or mobile phone for a given person would yield the same data. On a call from a residence on a wireline phone, for example, multiple URIs may be included, one for each family member.

NG9-1-1 accepts instant messages, signaled by SIP, routed by LoST, and containing a Geolocation header as well as AoR (in From or P-A-I). Pager or Session mode, or Message Relay Session Protocol, are accepted. NG9-1-1 mandates support for video and real-time text media streams as well as audio.

3.2.4 Legacy Wireline and Wireless Origination

NENA standards define how legacy wireline and wireless origination networks deliver calls to NG9-1-1 systems. They specify that gateways outside the ESInet interwork TDM (ISUP, MAP, IS-41) protocols to SIP, and how location is interworked. Such calls must be routed with LoST, and arrive meeting IETF phonebcp standards. In fact, NG9-1-1 routes *all* calls the same way, with no exceptions.

For legacy wireline origination networks, current E9-1-1 standards have location stored in a 9-1-1 Authority-contracted database called "ALI," keyed against the telephone number. In NG9-1-1, a similar function is provided by a special-purpose LIS, which uses telephone number as the key. The same processes currently used to maintain the ALI are adapted to maintain the LIS, which could be local to the wireline carrier. When the call arrives at the "Legacy Network Gateway" (LNG), it queries the LIS with the telephone number (which could be accomplished using HELD with identity extensions, or could be a synthesized location URI using the telephone number as the userpart), and retrieves location in the form of a PIDF-LO. The location is signaled with the call in a Geolocation header (and the body containing the PIDF-LO), typically by value. The LNG uses the location to dip the ECRF and routes the call in the same way that an IP-originated call would be routed. For legacy wireless origination networks, the LNG queries an LIS, using a similar process as described above, but the LIS returns an LbyR, which is included in the Geolocation header. When the ESRP or PSAP queries the location URI, it reaches the LIS, which interworks the SIP or HELD signaling to that needed by the mobile network to get a location update. As with the wireline origination, the LNG dips the ECRF (with what is usually a cell site/sector-based location) to obtain a route to the appropriate ESInet/ESRP.

Legacy (i.e., i2) VoIP systems can also use similar techniques to introduce calls to the ESInet before they are upgraded to i3/IETF standards.

3.2.5 *Emergency Events*

NG9-1-1 handles asynchronous events from devices and systems. We differentiate calls from events by the lack of two-way media and the presence or absence of a person. If there is two-way media, and a human, then it is a call. If there is not two-way interactive media, or no human, it is an event. Emergency events are signaled with a CAP message transported in an SIP MESSAGE transaction. Using SIP means that the event is routed in exactly the same way as a call, using all of the same mechanisms.

The policy routing mechanisms, specifically, can be used to route events differently from calls if desired. PSAPs can decide which events, or which senders, to allow to forward events to it using the policy routing mechanism. Location is handled in the same manner, although, since a CAP message includes the area affected, the location is effectively repeated, once in the CAP message and once in the SIP MESSAGE wrapper.

3.2.6 *Routing Calls Within the ESInet*

When the call reaches the initial ESRP inside the ESInet, it is often a state-level ESRP. This is primarily due to the need to have enormous amounts of bandwidth and BCF capacity to withstand serious distributed denial-of-service attacks. Several gigabits of sustained incoming traffic is needed to cope with realistic attacks at this time. The costs of bandwidth and BCF capacity is more realistically shared among all PSAPs in a state, although some large cities may have their own top-level BCF and ESRP, and some groups of smaller states may cooperatively purchase such services and devices. Since the ESRP is not the PSAP, it must make a routing decision. To do so, it queries the ECRF with the location obtained from the call signaling message. Using either the identity of the ECRF, and/or a distinguished service URN, the ECRF returns the next hop route URI.

This URI represents the nominal next hop in the route. The ESRP uses this URI to fetch a route policy created by the entity whose URI was selected by the ECRF. The policy is evaluated by the ESRP using state information it has available to determine what the actual next hop will be. This policy mechanism allows routes to be determined based on actual conditions. Inputs to the route policy include:

- downstream (e.g., PSAP) state (normal, out of service, stressed, overloaded, under attack, ...);
- any field in the call signaling (such as language preference, media, or call source);
- call suspicion level assigned by the BCF; and
- other state known to the ESRP, such as time of day.

Outputs include the final URI to use as the next hop and a mechanism to notify (using an SIP Asynchronous Event Notify) another agency of some condition.

The policy routing mechanism affords PSAPs a powerful way to control how their calls are handled in a variety of circumstances. The URI for the next hop may be the original intended destination, an alternate PSAP, or an Interactive Media Response (IMR) unit. Returning "Busy" is also a possibility.

This routing paradigm may be repeated. For example, a state ESRP may route to a county ESRP, which may route to a PSAP; and an ESRP in a PSAP may use the same mechanism to choose a queue of call-takers. The latter would allow, for example, PSAP management to draw a polygon around the location of a major traffic incident, and enable a routing rule which preferentially routed calls within that area to two call-takers, with an IMR program as a backup if the call-takers were busy, and allow all calls not within the polygon to route to the remaining call-takers.

It should be noted that NG9-1-1 requires *all* calls to be routed via the ECRF, and uses the ECRF to route calls within the ESInet. The latter uses the policy routing mechanism. Routing rules can, if necessary, be changed dynamically, and calls will route using the new ruleset immediately. Updates to the routing within the ECRF can change the initial call route as well, although caching may delay the effect of changes in the ECRF. Information returned from the ESRP may be cached and it contains a time-to-live field. This caching concept is useful to improve performance and resilience, as known from DNS, but it also impacts the rate at which updates can be made.

Because all ESInets are interconnected, a call can be diverted to any willing PSAP. This means that, in times of disaster or deliberate attack, all (2000 or more) call-takers on duty can handle calls from the affected area. On the other hand, the routing rules and other provisioning allow PSAPs to control whom it will accept diverted calls from.

The exact same mechanism is used by the PSAP to route calls to the appropriate responders. A distinguished set of service URNs are used to choose the type of responder (police, fire, EMS, coastguard, mountain rescue, ...) and the ECRF has a set of polygons defining the service boundaries for the responders. The PSAP queries the ECRF with the appropriate service URN, and the ECRF returns the next hop to that responder. The policy routing mechanism can also be engaged by the responder route mechanism. If a call arrives incorrectly routed because of an incorrect location, the PSAP can use the ECRF to get the correct route. Using aggregation of ECRFs and forest guides, a query to any ECRF will be redirected or recursively resolved regardless of where the actual location is.

3.2.7 Provisioning the ECRF

In order to adequately handle calls with geo location format, PSAPs must have accurate maps. These maps are housed in a Geographic Information System. The maps contain layers that define all political boundaries, subdivisions, and postal codes as polygons, streets as line segments ("centerlines") polygons, and addressable areas on streets with points or polygons. The data are sufficient to convert civic address to a geo location ("geocode") or convert a geo location to a civic address ("reverse geocode") accurately, and this capability is used to determine a dispatch civic address for responders to a call reporting geo location.

The ECRF is defined as housing a specific set of layers from the 9-1-1 Authority GIS system, and is maintained by a "push"-based layer synchronization protocol. Effectively, the ECRF is a subset of the 9-1-1 Authority's GIS system. Routes in this system are always polygons, and routing from a proffered civic address is accomplished by geocoding the address and performing a point in polygon operation for routing (although an implementation could optimize civic routing in most cases). The reason this is specified is so that the routing polygons can be changed dynamically, and as soon as any cache TTLs expire, calls will route on the new polygon. Experience suggests that such a capability is needed when unanticipated problems arise.

The same layer replication mechanism is used to create "replicas" of the ECRF to allow aggregation of multiple local ECRFs to county or state ECRFs and to provide replicas to carriers or other entities who desire a copy of the data. As soon as a change in the underlying map, features, addresses, boundaries or routing polygons are changed in the master GIS system (which is where the authoritative data are located), the change is immediately propagated to all ECRFs and ECRF replicas.

3.2.8 PSAPs

NG9-1-1 PSAPs receive SIP signaled calls, with location and callback URIs routed via the ECRF and the appropriate routing policies. As mentioned above, the PSAP may have an ESRP with similar capability to route within the PSAP based on location and policy. The PSAP uses SIP presence subscription or repeated HELD dereferencing to obtain location updates for mobile callers. As described above, the PSAP uses the ECRF to route calls to the appropriate responders. Standard SIP mechanisms provide bridging and media control for call-takers. NENA defined a new data structure to gather all the information the PSAP knows about a call, which is passed via the Call Info header when a call is transferred (usually via bridge) to another PSAP or responder. The same structure is used to pass data to a Computer Aided Dispatch (CAD) system. By standardizing the CAD interface, a call answered by a distant PSAP can see the same map view the original PSAP can see, can route the call to the right responder, and can insert a record in the right CAD system. This provides an effective mechanism to handle diverted calls no matter where, or why, the diversion occurs. Even in severe disaster or deliberate attack, calls can be answered by any available call-taker, appropriate information can be obtained by the call-taker who answers, appropriate information can be supplied to the caller, and the relevant information can be inserted into the right dispatch system.

In current 9-1-1 systems, TDM trunk group sizes are used to control congestion. The trunk group going into a PSAP is usually only a few calls larger than the number of call-takers, and the size of the trunk group from the originating switch toward the PSAP is smaller than that, to avoid having one switch monopolize all the call-takers in a localized incident. When a call exceeding the trunk group size is attempted, the caller receives a busy signal. The original rationale for this design, other than the fact that all telephone systems are designed similarly, was that, once all the call-takers were busy, there was not much point in accepting new calls. Even if a call could be answered elsewhere, there was no way to get relevant information from the caller to the responder, and in major disasters, the responders get busy anyway.

NG9-1-1 is designed with different principles. It is designed so that, if the local policy permits, and other PSAPs are willing, *all* calls can be answered by a call-taker who has the right information available, can take the right information from the caller, and can provide it to the right responder. For the responder, when its resources are overwhelmed, it attempts to triage: allocate responders to where the need is the greatest and the probability of being able to help is the largest. To triage, the responder needs all the information about all the callers, so that it can decide which ones to allocate resources to. If calls get a busy signal, no information can be obtained, and thus appropriate triage is impossible. Of course, if there is no access network, or no path (at all) between the access network and the ESInet or Internet, then there is no way to originate the call. If there is a path, NG9-1-1 can take the call, and effectively handle it.

3.2.9 Other i3 Features

It is possible to establish a three-way call with an i3 PSAP. This occurs with service providers in the path of a call. For example, some telematics services have a call center and the call center operator is usually in the call chain. The call center can originate a 9-1-1 call, routed on the location of the caller, and the call will be established as a three-way call involving the caller, the call center operator, and the PSAP call-taker. This mechanism is also used for relay interpreters for deaf and hard-of-hearing callers. Since i3 PSAPs accept video and text as well as audio, a caller who is deaf or hard of hearing will be able to see, and be seen by, the PSAP, as well as the PSAP being able to see, hear, and be seen and heard by the relay interpreter. Again, the policy routing function can control when such three-way calls are permitted.

In i3, there are standardized security mechanisms. A single sign-on authentication system with a uniform credential is specified, and a standardized rights management system is defined for all data in the system. All transactions within the ESInet are protected (authentication, integrity protected, and privacy), usually with TLS. Everything (significant events, state changes, media) is logged in a standardized logger.

3.3 IETF Emergency Services for Internet Multimedia*

Hannes Tschofenig[1] and Henning Schulzrinne[2]
[1]*Nokia Siemens Networks*
[2]*Columbia University*

3.3.1 Introduction

The ability to summon the police, the fire department or an ambulance in emergencies is one of the most important functions available when using the telephone. As telephone functionality moves from circuit-switched to Internet telephony, its users rightfully expect that this core feature will continue to be available and work as well as it has in the past. Users also expect to be able to reach emergency assistance using new communication devices and applications, such as instant messaging or SMS, and new media, such as video. In all cases, the basic objective is the same: the person seeking help needs to be connected with the most appropriate public safety answering point (PSAP), where call-takers dispatch assistance to the caller's

*Reprinted (with minor amendments) from *The Internet Protocol Journal* (*IPJ*), Volume 13, No. 4, December 2010. *IPJ* is a quarterly technical journal published by Cisco Systems. See http://www.cisco.com/ipj.

location. PSAPs are responsible for a particular geographic region, which can be as small as a single university campus or as large as a country.

The transition to Internet-based emergency services introduces two major structural challenges. First, while traditional emergency calling imposed no requirements on end systems and was regulated at the national level, Internet-based emergency calling needs global standards, particularly for end systems. In the old public switched telephone network (PSTN), each caller used a single entity, the landline or mobile carrier, to obtain services. For Internet multimedia services, network level transport and application can be separated, with the Internet Service Provider (ISP) providing IP connectivity service, and a Voice Service Provider (VSP) adding call routing and PSTN termination services. We ignore the potential separation between the Internet access provider, i.e. a carrier that provides physical and data link layer network connectivity to its customers, and the ISP that provides network layer services. We use the term VSP for simplicity, instead of the more generic term Application Server Provider (ASP).

The documents that are being developed within the IETF Emergency Context Resolution with Internet Technology (ECRIT) Working Group [120] support multimedia-based emergency services, and not just voice. As will be explained in more detail below, emergency calls need to be identified for special call routing and handling services, and they need to carry the location of the caller for routing and dispatch. Only the calling device can reliably recognize emergency calls, while the ISP is a reliable source of location information of the calling device based on its point of attachment to the network. Handling of emergency calls is, however, complicated by the wide variety of access technologies in use, such as virtual private networks (VPNs), other forms of tunneling, firewalls, and Network Address Translators (NATs).

The emergency services architecture developed by the IETF ECRIT Working Group is described in [40] and can be summarized as emergency calls are generally handled like regular multimedia calls, except for call routing. The ECRIT architecture assumes that PSAPs are connected to an IP network and support the Session Initiation Protocol (SIP) [123] for call setup and messaging. However, the calling user agent may use any call signaling or instant messaging protocol, which the VSP then translates into SIP.

Non-emergency calls are routed by a VSP, either to another subscriber of the VSP, typically via some SIP session border controller or proxy, or a PSTN gateway. For emergency calls, the VSP keeps its call routing role, routing calls to the emergency service system, to reach a PSAP, instead. However, we also want to allow callers that do not subscribe to a VSP to reach a PSAP, using nothing but a standard SIP [123] user agent (see [124] for a discussion about this topic); the same mechanisms described here apply. Since the Internet is global, it is possible that a caller's VSP resides in a jurisdiction other than where the caller and the PSAP are located. In such circumstances it may be desirable to exclude the VSP and provide a direct signaling path between the caller and the emergency network. This has the advantage of ensuring that all parties included in the call delivery process reside in the same regulatory jurisdiction.

As noted above, the architecture does not force or assume any type of trust or business relationship between the ISP and the VSP carrying the emergency call. In particular, this design assumption affects how location is derived and transported.

Providing emergency services requires three crucial steps, which we summarize in turn below: recognizing an emergency call, determining the caller's location, and routing the call and location information to the appropriate emergency service system operating a PSAP.

3.3.2 *Recognizing Emergency Calls*

Chapter 1 explained the current situation of a large number of emergency numbers being used worldwide. Because of this diversity, the ECRIT architecture decided to separate the concept of an emergency dial string, which remains the familiar and regionally defined emergency number, and a protocol identifier that is used for identifying emergency calls within the signaling system. The calling end system has to recognize the emergency (service) dial string and to translate it into an emergency service identifier, which is an extensible set of Uniform Resource Names (URNs) defined in RFC 5031 [11]. A common example for such a URN, defined to reach the generic emergency service, is urn:service.sos. The emergency service URN is included in the signaling request as the destination and is used to identify the call as an emergency call. If the end system fails to recognize the emergency dial string, the VSP may also perform this service.

Since mobile devices may be sold and used worldwide, we want to avoid manually configuring emergency dial strings. In general, a device should recognize the emergency dial string familiar to the user and the dial strings customarily used in the currently visited country. The Location to Service Translation Protocol (LoST) [68], described in more detail later, also delivers this information.

3.3.3 *Obtaining and Conveying Location Information*

In the IETF emergency services architecture, location information can be conveyed in SIP either by value ("LbyV") or by reference ("LbyR"). With an LbyV, the XML location object is added to the body of a SIP message. In case of a LbyR a reference to the location object is passed in SIP, which requires a separate lookup step.

When passed as a value, then location is encapsulated within the Presence Information Data Format – Location Object (PIDF-LO), an XML-based document to encapsulate civic and geodetic location information. The format of PIDF-LO is described in [17] with the civic location format updated in [21] and the geodetic location format profiled in [28]. The latter document uses the Geography Markup Language (GML) developed by the Open Geospatial Consortium (OGC) for describing commonly utilized location shapes. More information about the standardization efforts on location can be found in Chapter 2.

Location by value is particularly appropriate if the end system has access to the location information, for example if it contains a Global Positioning System (GPS) receiver or uses one of the location configuration mechanisms described below. In environments where the end host location changes frequently, the LbyR mechanism might be more appropriate. In this case, the LbyR is an HTTP/HTTPS or SIP/SIPS URI, which the recipient needs to resolve to obtain the current location. Terminology and requirements for the LbyR mechanism are available with [84].

An LbyV and an LbyR can be obtained via location configuration protocols, such as the HELD protocol [31] or DHCP [16, 27, 125]. Once obtained, location information is required for LoST queries, and is added to SIP messages [32].

The ISP is the source of the most accurate and dependable location information. It is, however, not the only source. A calling device may have built-in location capabilities, such as GPS, producing highly accurate location information or may obtain location information from third party location databases. For landline Internet connections such as DSL, cable or fiber-to-the-home, the ISP knows the provisioned location for the network termination, for example. The IETF Geographic Location/Privacy (Geopriv) Working Group has developed protocol mechanisms, called location configuration protocols, so that the end host can request and receive location information from the ISP. The best current practice document for emergency calling [122] enumerates three options that should be universally supported by clients: DHCP civic [16] and geo [27], and HELD [31]. HELD uses XML query and response objects carried in HTTP exchanges and is described in detail in Chapter 2.4. DHCP does not use the PIDF-LO format, but rather more compact binary representations of locations that require the endpoint to construct the PIDF-LO.

Particularly for cases where end systems are not location-capable, a VSP may need to obtain location information on behalf of the end host using the HELD identity extension [35].

Obtaining at least rough location information at the time of the call is essential, as the LoST query can only be initiated once the calling device or VSP has obtained some location information. Also, to speed up response, it is desirable to transmit this location information with the initial call signaling message. In some cases, however, location information at call setup time is imprecise. For example, a mobile device typically needs 15 to 20 seconds to get an accurate GPS location "fix" and the initial location report is based on the cell tower and sector. For such calls, the PSAP should be able to request more accurate location information either from the mobile device directly or the Location Information Server (LIS) operated by the ISP. The SIP event notification extension, defined in RFC 3265 [33], is one such mechanism that allows a PSAP to obtain the location from an LIS. To ensure that the PSAP is only informed of pertinent location changes, and that the number of notifications is kept to a minimum, event filters [34] can be used.

The two-stage location refinement mechanism described above works best when location is provided by reference (LbyR) in the SIP INVITE call setup request. The PSAP subscribes to the LbyR provided in the SIP exchange and the LbyR refers to the LIS in the ISPs network. In addition to an SIP URI, the LbyR message can also contain an HTTP/HTTPS URI. When such a URI is provided, the HELD location dereference protocol is used to retrieve the current location [83].

3.3.4 *Routing Emergency Calls*

In scenarios where the VSP and the ISP roles are offered by the same provider, routing of IP-based emergency calls can be based on statically pre-configured rules. However, the separation of these two roles to separate companies introduces challenges for emergency call routing. In addition, emergency services authorities are interested to use IP and SIP-based functionality to route calls more dynamically (for example to change call routing dynamically in case of overload situations). To decouple the routing enforcement point from the decision point, an automated procedure has been standardized. Two mechanisms are available: the service boundary exchange mechanism called LoST Sync [126] and the Location to Service Translation (LoST) protocol [68]. [126] has limited applicability for exchange of service boundaries between trusted entities, and LoST may make use of LoST Sync for distributing mappings. LoST is a generic mechanism that maps a location and a service URN to a specific PSAP URI and a service region. LoST, illustrated in Figure 3.5, is an HTTP-based query/response protocol where a client sends a request containing the location information and service URN to a server and receives a response, containing the service URL, typically an

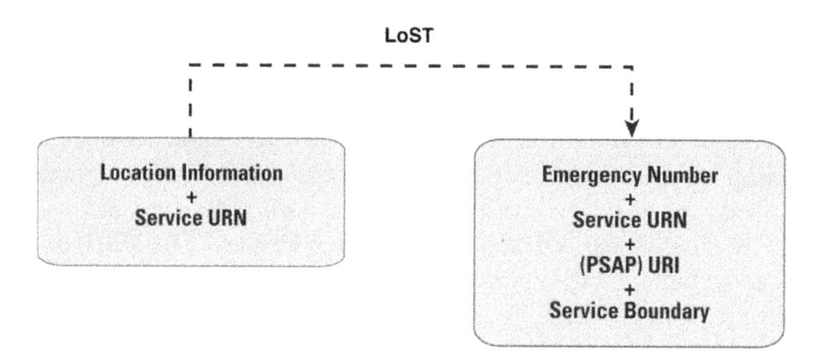

Figure 3.5 High-level functionality of Location to Service Translation (LoST). (Reprinted from The Internet Protocol Journal (IPJ), Volume 13, No. 4, December 2010. IPJ in a quarterly technical journal published by Cisco Systems. See www.cisco.com/ipj.)

SIP URL, the service region where the same information would be returned, and an indication of how long the information is valid. Both request and response are formatted as XML. For efficiency, responses are cached, as otherwise every small movement would trigger a new LoST request. As long as the client remains in the same service region, it does not need to consult the server again until the response returned reaches its expiration date. The response may also indicate that only a more generic emergency service is offered for this region. For example, a request for "urn:service:sos.marine" in Austria may be replaced by "urn:service:sos." Finally, the response also indicates the emergency number/dial string for the respective service.

3.3.5 Obligations

In this section we discuss the requirements the different entities need to satisfy, based on Figure 3.6. A more detailed description can be found in [122]. Note that this narration focuses on the final stage of deployment and does not discuss the transition architecture, in which some implementation responsibilities can be rearranged, with an impact on the overall functionality offered by the emergency services architecture. A few variations were introduced to handle the transition from the current system to a fully developed ECRIT architecture.

With the work on the IETF emergency architecture, we have tried to balance the responsibilities among the participants, as described below.

3.3.5.1 End Hosts

An end host, through its VoIP application, has three main responsibilities: it has to attempt to obtain its own location, determine the URI of the appropriate PSAP for that location, and recognize when the user places an emergency call by examining the dial string. The end host operating system may assist in determining the device location.

The protocol interaction for location configuration is indicated as interface (a) in Figure 3.6, and a number of location configuration protocols have been developed to provide this capability.

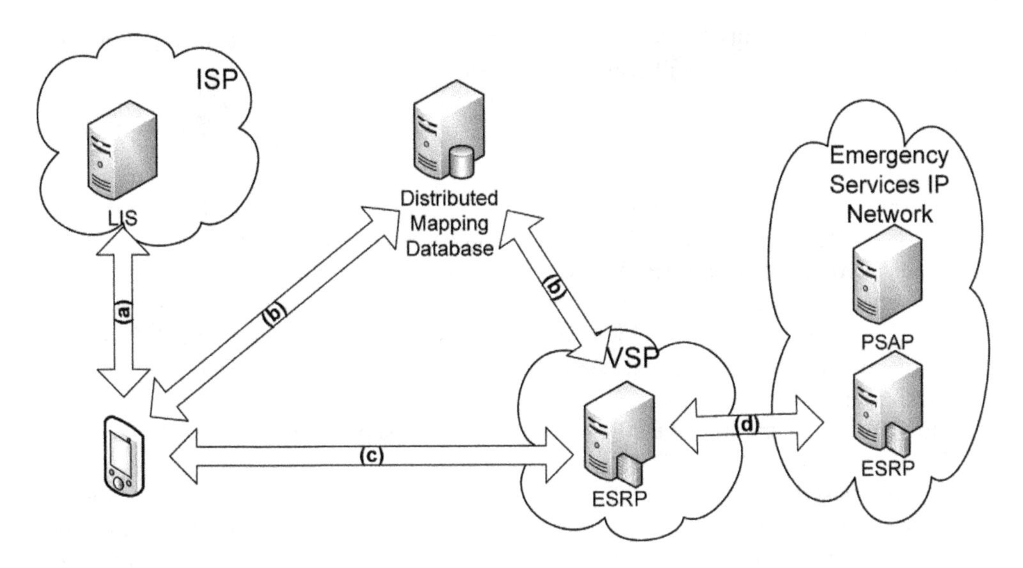

Figure 3.6 Main components involved in an emergency call.

A VoIP application needs to support the Location to Service Translation (LoST) protocol [68] in order to determine the emergency service dial strings and the PSAP URI. Additionally, the service URNs, defined in [11], need to be understood by the device.

As currently defined, it is assumed that PSAPs can be reached by SIP, but may support other signaling protocols, either directly or through a protocol translation gateway. The LoST retrieval results indicate whether other signaling protocols are supported. To provide support for multimedia, different types of codecs may need to be supported; details can be found in [122].

3.3.5.2 ISP

The ISP has to make location information available to the endpoint via one or more of the location configuration protocols.

In order to route an emergency call correctly to a PSAP, an ISP may initially disclose the approximate location for routing to the endpoint and more precise location information later, when emergency personnel are dispatched by the PSAP operator. The functionality required by the IETF emergency services architecture is restricted to the disclosure of a relatively small amount of location information, as discussed in [127, 128].

The ISP may also operate a caching LoST server to improve the robustness and the reliability of the architecture. This lowers the round-trip time for contacting a LoST server, and the caches are most likely to hold the mappings of the area where the emergency caller is currently located.

In the case where ISPs allow Internet traffic to traverse their network, the signaling and media protocols used for emergency calls function without problems. Today, there are no legal requirements to offer prioritization of emergency calls over IP-based networks. While the standardization community has developed a range of Quality of Service signaling protocols, their widespread deployment still remains to happen.

3.3.5.3 VSP

SIP does not mandate that call setup requests traverse SIP proxies, that is SIP messages can be sent directly to the user agent. Thus, even for emergency services, it is possible to use SIP without the involvement of a VSP. However, in terms of deployment, it is highly likely that a VSP will be used. If a caller uses a VSP, this VSP often forces all calls, emergency or not, to traverse an outbound proxy or session border controller (SBC) operated by the VSP. If some end devices are unable to perform a LoST lookup, VSP can provide the necessary functions as a backup solution.

If the VSP uses a signaling or media protocol that is not supported by the PSAP, it needs to translate the signaling or media flows.

VSPs can assist the PSAP by providing identity assurance for emergency calls, for example using SIP Identity (RFC 3325 [129]), thus helping to prosecute prank callers. However, the link between the subscriber information and the real-world person making the call is often weak. In many cases, VSPs have, at best, only the credit card data for their customers, and some of these customers may use gift cards or other anonymous means of payment.

3.3.5.4 PSAP

The emergency services best current practice document [122] only discusses the standardization of the interfaces from the VSP and ISP toward PSAPs and some parts of the PSAP-to-PSAP call transfer mechanisms that are necessary for emergency calls to be processed by the PSAP. Many aspects related to the internal communication within a PSAP, between PSAPs, as well as between a PSAP and first responders are beyond the scope of the IETF specification.

When emergency calling has been fully converted to Internet protocols, PSAPs must accept calls from any VSP, as shown in interface (d) of Figure 3.5. Since calls may come from all sources, PSAPs must develop mechanisms to reduce the number of malicious calls, particularly calls containing intentionally false location information. Assuring the reliability of location information remains challenging, particularly as more and more devices are equipped with Global Navigation Satellite Systems (GNSS) receivers, such as GPS, allowing them to determine their own location [130]. However, it may be possible in some cases to verify the location information provided by an endpoint by comparing it against network-provided location information.

3.3.6 LoST Mapping Architecture

So far, we have described the LoST protocol as it is specified in RFC 5222 [68], namely as a client–server protocol. A single LoST server, however, does not store the mapping elements for all PSAPs worldwide, for both technical and administrative reasons. Thus, there is a need to let LoST servers interact with other LoST servers, each covering a specific geographical region. The LoST protocol already provides the baseline mechanisms for supporting such a communication architecture, as described in RFC 5582 [131], an informational RFC providing terminology (in the form of different roles for LoST servers that distinguish their behavior) and explaining the basic concept of the LoST mapping architecture. RFC 5582 motivates the basic design decision for LoST to utilize it in a wide variety of architectures, but leaves the detailed instantiation to deployments in different jurisdictions.

The awareness of peering LoST servers determines the structure of the architecture rather than certain physical properties of a network, such as topology of a fiber installation, or the structure of a national emergency services organization. Two types of structures are used in combination, namely a mesh and a hierarchical structure. The mesh topology is envisioned for the top-level LoST entities, whereas the hierarchical structure reflects a parent–child relationship in a tree. Figure 3.7 shows this structure graphically, with the LoST servers acting in their roles of forest guides (FGs) and trees. A tree consists of a self-contained hierarchy of authoritative mapping servers (AMS) for a particular service. An AMS is a LoST server that can provide the authoritative answer to a particular set of queries. The topmost server in a tree is a tree root and this server peers with one or multiple FGs, that is the tree root announces its coverage region to FGs. In Figure 3.7, for example, the root of tree 1 interacts with FG A and makes the coverage area available. FG A also receives the coverage area from the root of tree 2. All tree roots receive information about the coverage area of their children in the tree. On the top level, all FGs (namely FG A, FG B, and FG C) form a mesh and synchronize their coverage areas.

Seekers and resolvers are two additional LoST entities in the LoST mapping architecture that are not shown in Figure 3.7. Neither seekers nor resolvers provide authoritative answers themselves but they may cache results. Particularly, the use of resolvers to cache mapping elements is expected to be very common.

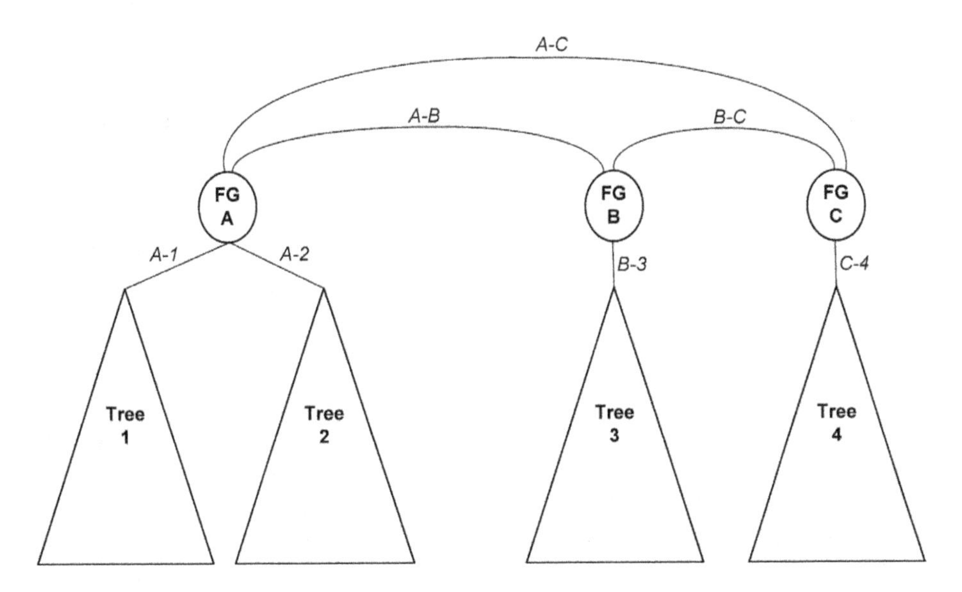

Figure 3.7 Trees and forest guides in the LoST mapping architecture.

To best understand the LoST mapping architecture, it is important to highlight the main design goals:

- *Robustness.* To ensure the stability of the system even if different people in different places of the LoST architecture make different decisions, the system will still function. It cannot be assumed that everyone has to agree with everyone else. The minimum level of agreement that has to be ensured is that AMSs are able to authoritatively answer mapping queries, i.e. only those LoST servers respond authoritatively if they indeed have the authority of a specific coverage area.
- *Consistent Responses.* Any device (called seeker) can issue a LoST query and it will get a consistent answer regardless of where the query enters the system. In some (rare) cases of territorial disputes, two AMSs may claim to be authoritative for the same region. In such a case, the answer received by a seeker will vary depending on the entry point into the mapping system.
- *Scalability.* Scalability of the LoST architecture is ensured by the use of caching and the distributed nature of the LoST servers in the architecture. Any LoST entity may support caching of received mapping elements. The mapping elements may be obtained as part of the ordinary operation of LoST (via query and responses) but also via separate replication of the mapping elements. LoST Sync [126] is one such protocol to exchange mappings between LoST servers (and other entities).
- *Minimal Seeker Configuration.* A seeker is a LoST client requesting a mapping. The only information a seeker needs to know is the address of a resolver; it does not need to know the structure of all forest guides nor does it need to maintain a global picture of LoST servers. To avoid having end user involvement in the configuration of LoST servers, Section 4 of the LoST specification provides a discovery technique based on DNS, and RFC 5223 [132] additionally offers a DHCP-based discovery procedure. Although LoST servers can be located anywhere, a placement topologically closer to the end host, for example in the access network, may be desirable in disaster situations with intermittent network connectivity. RFC 5223 offers this capability.

Even though it is technically possible to let seekers and resolvers enter their queries at any point in the LoST mapping architecture, a deployment choice is to configure resolvers with the addresses of the FGs. A query and response for an emergency caller located in Germany with a service provider in Finland could then be shown as depicted in Figure 3.8. In our example we assume that the VSP deploys a LoST resolver that is contacted by the their own customers, the seekers. We furthermore assume in this example that no caching takes place to illustrate the message flow (as shown with dotted lines). In message (1) the seeker contacts its pre-configured resolver with a recursive query providing its current location (somewhere in Germany). The resolver at this point in time does not have any information about the PSAP that has to be contacted for the given location in Germany (for the solicited service). Since the resolver knows the address of the forest guide (only one forest guide is shown in our example), it issues an iterative query to it, as marked with message (2). The FG responds with the entry point for the German LoST tree. The resolver then issues another query toward the provided tree root in message (3). For this example we assume that the root of tree 1 knows the address of the PSAP the seeker has to contact. This final response is then forwarded to the seeker via the resolver. The resolver would want to cache the intermediate and final results in order to

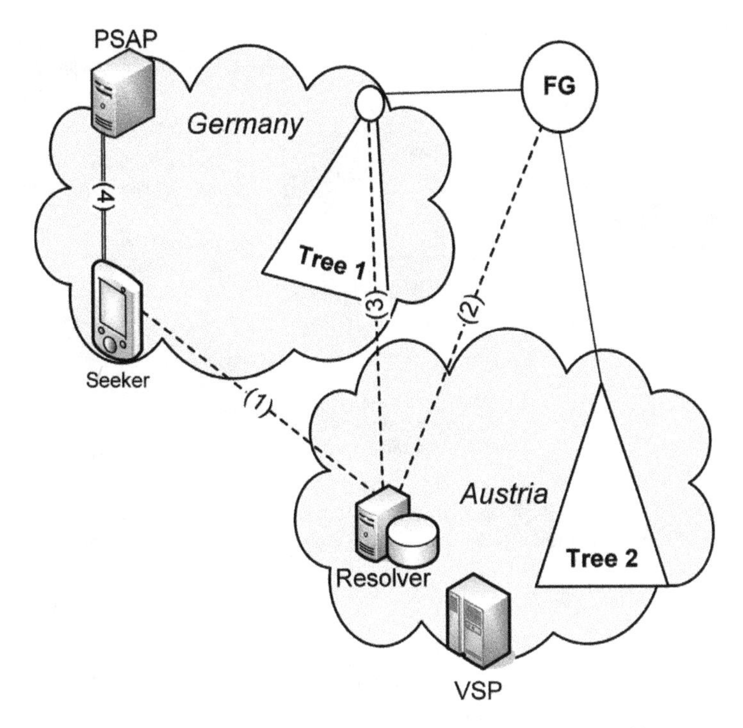

Figure 3.8 Example query/response in the LoST mapping architecture.

speed-up later lookups for the same geographical area and the same service. Once the seeker knows the final answer, it can proceed with the emergency call setup procedure to contact the PSAP, as shown in message (4) with the double line.

As illustrated, LoST servers form a distributed mapping database, with each server carrying mapping elements. These mapping elements are the main data structure that is communicated in the LoST protocol, synchronized between FGs and LoST servers in the tree, and cached by resolvers and seekers. Figure 3.9 shows the data elements of this important data structure graphically.

3.3.7 Steps Toward an IETF Emergency Services Architecture

The architecture described so far requires changes both in already-deployed VoIP end systems and in the existing PSAPs. The speed of transition and the path taken varies between different countries depending on funding and business incentives. As such, it is difficult to argue whether upgrading endpoints or replacing the emergency service infrastructure will be easier. In any case, the transition approaches being investigated consider both directions. We can distinguish roughly four stages of transition – the description below omits many of the details due to space constraints:

- Initially, VoIP end systems cannot place emergency calls at all. This is currently the case for many software clients, including popular over-the-top VoIP clients.

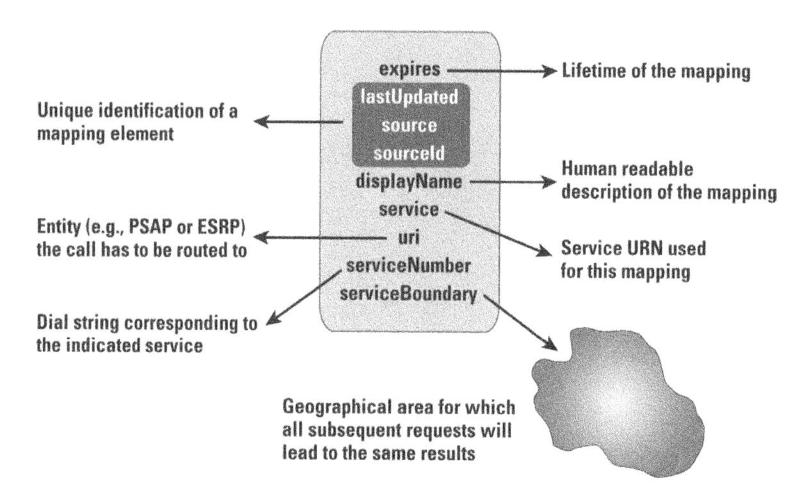

Figure 3.9 Mapping element. (Reprinted from The Internet Protocol Journal (IPJ), Volume 13, No. 4, December 2010. IPJ in a quarterly technical journal published by Cisco Systems. See www.cisco.com/ipj.)

- In a second stage, VoIP callers manually configure their location, and emergency calls are routed to the appropriate PSAP as circuit-switched calls via VoIP-to-PSTN gateways. This level of service is now offered in some countries for PSTN-replacement VoIP services, that is VoIP services that are offered as replacement for the fixed line Plain old telephone service (POTS). In the United States, this is known as the "NENA i2" service.
- In a third stage, PSAPs maintain two separate infrastructures, one for calls arriving via an IP network, and one for the legacy infrastructure.
- In the final stage, all calls, including those from traditional cell phones and analog landline phones, reach the PSAP via IP networks, with the legacy calls converted to VoIP emergency calls (fulfilling the requirements outlined in ref. [123] or by a translation gateway operated by the emergency service infrastructure.

3.3.7.1 Legacy Endpoints

Figure 3.10 shows an emergency services architecture with legacy endpoints. When the emergency caller dials the European-wide emergency number "112" (step 0), the device treats it as any other call without recognizing it as an emergency call, that is the dial string provided by the endpoint that may conform to RFC 4967 [133] or RFC 3966 [80] is signaled to the VSP (step 1). Recognition of the dial string is then left to the VSP for processing and sorting; the same is true for location retrieval (step 2) and routing to the nearest (or appropriate) PSAP (step 3). Dial string recognition, location determination and call routing are simpler on a fixed device and voice/application service provided only by the ISP than when they are provided via separate VSP and ISP.

If devices are used in environments without location services, the VSP's SIP proxy may need to insert location information based on estimates or subscriber data. We briefly describe these cases below.

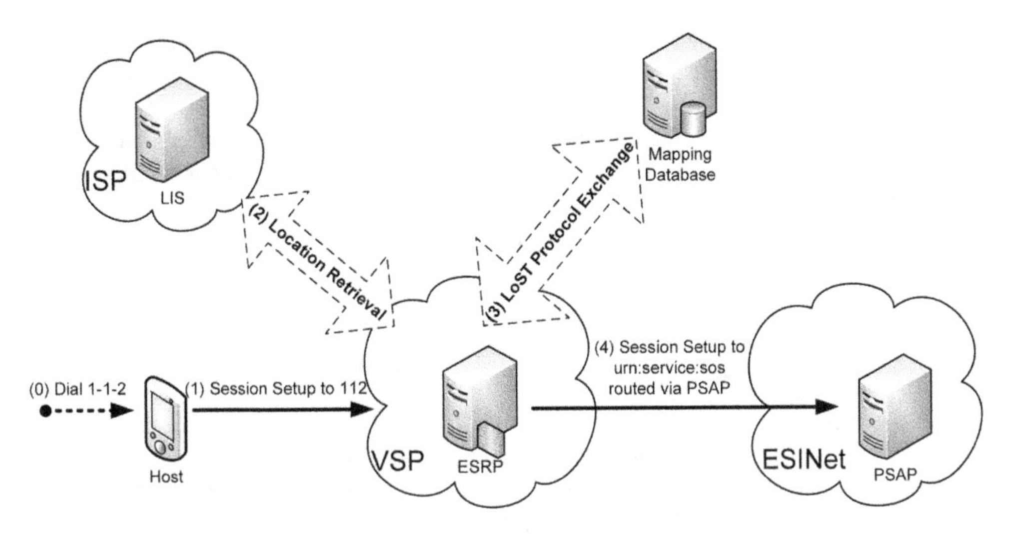

Figure 3.10 Emergency services architecture with legacy endpoints.

There are two main challenges to overcome when dealing with legacy devices. First, the VSP has to discover the LIS that knows the location of the IP-based end host. The VSP is only likely to know the IP address of that device, which is visible in the call signaling that arrives at the VSP. When a LIS is discovered and contacted, and some amount of location information is available, then the second challenge arises, namely how to route the emergency call to the appropriate PSAP. To accomplish the latter task, it is necessary to have some information about the PSAP boundaries.

[122] does not describe a complete solution for interworking with legacy PSAPs but instead offers building blocks to use. The following constraints exist when dealing with legacy endpoints:

- Only the emergency numbers configured at the VSP are understood. This may lead to cases where a dialed emergency number is not recognized or a non-emergency call is routed to a PSAP.
- Using the host's IP address to discover the ISP network to which the host is attached is challenging and may, in case of mobility protocols and VPNs, lead to wrong results.
- Security and privacy concerns may arise when a large number of VSPs/ASPs can retrieve location information from an ISP. It is likely that only authorized VSP/ASPs will be granted access. Hence, it is unlikely that such a solution would work smoothly across national boundaries.
- When the emergency call is not recognized by the User Agent, then functions like call waiting, call transfer, three-way call, flash hold, outbound call blocking, etc. cannot be disabled.
- The User Agent software may block callbacks from the PSAP calltaker since it does not have ways to recognize those as calls related to an earlier emergency.

- Privacy settings may not be respected and identity information may get disclosed to unauthorized parties. These identity privacy features exist in some jurisdictions even in emergency situations.
- Certain VoIP call features may not be supported, such as a REFER (for conference call and transfer to secondary PSAP) or the Globally Routable User Agent URIs (GRUU) for identifying individual devices issuing the emergency call.
- User Agents will not be able to convey location information to the VSP (even if it is available).

3.3.7.2 Partially Upgraded End Hosts

A giant step forward in simplifying the handling of IP-based emergency calls is to provide the end host with some information about the ISP so that LIS discovery is more reliable. The end host may, for example, learn the ISP's domain name, by using LIS discovery [71], or might even obtain a Location by Reference (LbyR) via the DHCP-URI option [125] or via HELD [31]. The VSP is then either able to resolve the LbyR in order to route the call or to use the domain to discover an LIS using DNS.

Additional software upgrades at the end device may allow emergency calls to be recognized based on some pre-configured emergency numbers (e.g., 112 and 911) and allow for the implementation of other emergency service-related features, such as disabling silence suppression during emergency calls.

3.3.8 Summary

The emergency services architecture developed in the IETF aims to offer a generic solution for emergency calling over the Internet by taking the split between the application service provider and the access network provider into account. By approaching this design the architecture is flexible with regard to developments of new software clients, including proprietary protocols offered by application service providers. The need for this flexibility has already proven to be useful since the development of new multi-media communication architectures is progressing at a fast pace. The recent developments related to the WebRTC architecture where Websites and browsers are turned into VoIP clients as well as the XMPP deployment can utilize the IETF emergency services architecture without significant changes. Furthermore, location information is available for many devices and services and can be re-used in emergency situations.

The full potential of the Internet Protocol stack can, however, only be leveraged when emergency services authorities rely on equipment providers that adhere to standards to avoid vendor lock-in. Since today's telecommunication infrastructure equipment is largely a software product it is essential to have the ability to upgrade software at a regular interval, not only to patch security vulnerabilities but also to benefit from new technological advances in the communication industry. These new developments will happen in the area of increased data sharing capability and multi-media communication, which both will increase the situational awareness for call takers as well as for first responders.

There are, however, challenges as well. The advances in Internet communication technology needs to be in sync with regulation and educational efforts so that emergency services can also benefit. Technology has made huge improvements over the last ten years: the standards are available, implemented and the Internet has formed the basis of today's communication society. Regulators will, however, have to catch up to tackle the global nature of the Internet instead of focusing on national structures. This process is slow and sometimes painful but it also offers the ability to re-think the legacy environment.

3.4 Emergency Services Support in WiFi Networks

Gabor Bajko
Nokia

3.4.1 Introduction

IEEE 802.11 [51, 134] is a set of specifications defining physical and MAC layer procedures for wireless LAN operations. The base specification and the draft amendments currently under development provide some limited support for location configuration and placement of an emergency call.

In WiFi networks, a STA (Station, an entity with 802.11 interface) is in state-1 when it is not yet associated with any Access Point (AP). In state-1, the STA is scanning the radio channels for beacons. The beacon is a periodic signal sent by the access point in a given channel, which allows a STA to identify an access point. Once the STA chooses an access point, it initiates an association procedure and after successful association it enters into state-3. Authentication may happen either before association (when a passcode is required for association) or after association (the so-called web-auth case, when the STA associates with an access point, but its traffic is redirected to a portal where it has to enter its credentials). Only in state-3 is a STA allowed to exchange data frames with the network.

There may be a large number of WiFi networks operating in one location: some of them, but not all, may provide assistance for location configuration; some may provide access only to local networks while some others may provide access to the Internet; some may provide support for emergency calls, but some others not; and so on. The recent spread of WiFi network deployments has shown how difficult and time-consuming the selection of the right WiFi network can be. A right access point is defined as one that best satisfies the user profile and the requirements of the running applications at the connection time.

Currently, a STA needs to perform a scan, select an AP based on the SSID or signal strength, associate with the AP, and only after association will know whether the services it expected are available or not (e.g., whether Web-based authentication is required or not, Internet access is available or not, emergency services are supported or not, etc.). If the initially selected AP was not the right one, then the STA performs a disassociation and starts the selection procedure from the beginning, resulting in a very time-consuming network selection. A selected AP may not be the right one if, for example, the user wants to browse the Internet but the AP requires a subscription, or the user wants to place an emergency call but the AP cannot assist the STA in the emergency call placement.

The amendments under development in the IEEE 802.11 task groups address this problem by defining procedures that allow STAs to advertise the capabilities they support and to exchange detailed service information with the APs while in state-1, i.e. before associating with the APs. This pre-associated state discovery procedure helps the STA to select the appropriate AP before it associates with it. In the context of Emergency Services, a STA will be able to find out during the network selection phase (i.e., in state-1, before association to an AP) which AP can help the STA in location configuration, or which AP has support for emergency services. This will enable a STA to perform a faster network selection and select a network that can assist in the placement of an emergency call, when the need for it arises.

3.4.2 Location Configuration

The IEEE 802.11 specification supports exchange of geodetic, civic, and URI location information between peer STAs.

When the location of an AP is configured using a civic location format, the AP will advertise that it has its own civic location configured in the beacon signal, in the form of a capability indication. The format of the civic location complies with the format described in RFC 4776 [16].

When the location of an AP is configured using a geodetic location, the AP will advertise that it has its own geodetic location configured in the beacon signal, in the form of a capability indication. The format of the civic location is almost identical to the format described in RFC 3825 [26], without the DHCP option and code fields.

When the STA is configured with a URI where its location information can be accessed, the AP will advertise that it has a URI pointing to its location in the beacon signal, in the form of a capability indication.

An AP may have its location configured using both civic and geodetic formats, and may advertise both in the beacon signal simultaneously. Any scanning STA is able to read whether location in any format is configured in the AP, but to get the location, the STA has to make a specific query. The type of the query depends whether the STA is in state-1 (not associated) or state-3 (associated).

A scanning STA in the pre-associated state-1 would query an AP, which advertises that its location is configured, to find out if the AP supports the feature of providing its location for STAs in pre-associated state. When the AP supports this feature, the STAs can request the AP's location without associating with it. The aim of this feature is to speed up location configuration by not requiring STAs to authenticate and/or associate with the AP to fetch the location. In many cases, an accurate location determination in the STA may require the STA to fetch the location of multiple APs. By providing their location to unassociated STAs, the APs may greatly reduce the time needed for determining location of a STA using information from multiple sources (e.g., STA-based trilateration or triangulation).

In pre-associated state, a STA can only request the location of an AP from the AP itself, that is, the STA is not able to request the AP to determine its own location using radio measurements.

Since information downloaded from an entity while in unassociated state is unreliable, STA implementations should carefully consider how to use the obtained information. Reliability of information downloaded from an AP belonging to a trusted network or in possession of a digital certificate is under consideration in the WiFi Alliance.

STAs in state-3, that is, already associated with an AP, may invoke the Location Measurement Request procedures to query either another STA or the AP to find out the location of that remote STA (or AP), that is, a response to the question "where are you?" The procedures also allow for the STA to request from the other STA its own location, that is, a response to the question "where am I?" This latter request may invoke some location measurements an AP would perform in order to locate the requesting STA, if the location of the requestor is not yet known by the AP. The Location Measurement Procedure may be used to obtain a civic location or a geospatial location, or a Location URI to the requested location.

STAs may also invoke the Timing Measurement Procedure to calculate their clock's offset relative to their peer's clock. STAs are also able to use this procedure to calculate the time it takes for the signal to travel between the peers (called the signal's flight time). Performing the procedure with multiple peers may help to determine a more accurate location than only relying on the location of the peer (in many cases that being the AP).

3.4.3 Support for Emergency Services

A WiFi AP may be configured to advertise multiple SSIDs. One such SSID may be configured for emergency call placements only. When an AP is configured with such an SSID, the AP will advertise in the beacon signal that this network is restricted to emergency services. Associating with this SSID would not require any authentication from the STAs, and the services the SSID would offer to the associated STAs would be restricted to emergency call placements only. IEEE does not specify how the restriction is to be implemented. One typical implementation would assign a specific VLAN to the traffic in the emergency services restricted SSID, and the traffic in that VLAN can then be filtered in the backend. A STA interested in placing an emergency call, but not having valid credentials to connect to any available network, would perform an active scan to identify any available emergency services restricted network. When one is found, the STA would be able to associate with the AP and place the emergency call.

If the operator of the AP does not wishes to configure a separate SSID for supporting emergency services, but it still wants its AP to support unauthenticated emergency services, then an AP that normally requires a passkey to associate with could be configured to allow association without a passkey, but for emergency services only. If an AP is configured to allow association without a passkey for emergency services purposes, it will advertise this in a form of a capability in the beacon signal. A STA interested in placing an emergency call, but not having valid credentials to connect to any available network, would perform an active scan to identify the networks that allow association for emergency service purposes without providing credentials. When one is found, the STA will request association by indicating in the association request its intention to associate without credentials, for emergency service purposes only. The AP will allow association of the STA and will apply the necessary restrictions. Again, how the AP restricts the traffic from unauthenticated STAs is considered deployment-specific and not addressed by the IEEE specifications.

3.4.4 Support for Emergency Alert Systems

When a WiFi network has Alerts typically issued by authorities, it will advertise the presence of such alerts in the beacon signal. When scanning STAs notice the presence of an Alert at one specific AP, the STAs can download the Alert from the APs using the so-called GAS procedure. The download of the Alert message does not require a STA to associate with the AP: the messages can be downloaded while in state-1. A STA interested in downloading an Alert from an AP should first check the Alert Identifier (a 64-bit hash of the Alert), to see if that specific Alert was downloaded previously or not.

3.5 WiMAX

Dirk Kröselberg
Siemens

Today's deployments of Mobile WiMAX technology mostly target the use case of providing wireless broadband access to the Internet. The service is often not bound to, or customized for, any particular application or application class. Mobile WiMAX network deployments are based on the IEEE-defined 802.16 radio interface [135] that supports a number of enhanced features like Quality of Service (QoS). The network architecture [136] designed by the WiMAX Forum [137] supports a set of network enablers that help to optimize the support of certain classes of applications. In addition, the 3GPP-defined IP Multimedia Subsystem (IMS) offers one common way to enable VoIP or other multimedia services for a WiMAX operator. WiMAX specifications are available to allow for a smooth integration of IMS services [138] into a WiMAX operator's infrastructure.

Independent of the above, WiMAX typically operates in regulated spectrum. This clarifies why different considerations apply to the support of emergency calls than for WiFi, which has a number of technical similarities otherwise. As soon as an operator offers commercial services across such a wireless medium, local regulation will apply and the resulting requirements must be met. As a result, the standards defining WiMAX networks provide appropriate building blocks to provide support for emergency services. Operators can deploy these building blocks based on their specific needs.

In the Release 1.5 of the WiMAX Forum network specifications [136, 139], this support is limited to citizen-to-authority emergency use cases and focuses on enabling emergency calls. The overall approach to support emergency services in WiMAX is one that

- maintains a clear split between the access network as well as the ISP infrastructure, on the one hand, and the VoIP application or other services that provide emergency support, on the other, and
- provides individual building blocks rather than a tight architecture for the support of emergency services.

The above are the key design criteria for WiMAX emergency services support. The former takes into account that, when taking the example of emergency calls, there is no single standardized VoIP emergency application available on the broad range of already deployed, or new, WiMAX devices. So the building blocks provided by the network and access technologies can be decoupled from the VoIP-specific parts that typically vary between the different applications.

The latter aspect of providing individual building blocks instead of a closed architecture for emergency services in WiMAX addresses the fact that emergency requirements are subject to local laws and regulations. These can be significantly different between countries or even between the different regions of the same country. The affected operator typically knows best what those actual requirements are that apply to the operator's network. Standards, and subsequently network equipment and device vendor implementations, must ensure that the appropriate building blocks to match the local requirements are available to the operator's deployment.

With the above general motivation, the following subsections provide a brief overview about the key aspects of the WiMAX network architecture and the main network functions and elements therein. Based on this, the WiMAX-specific aspects and building blocks for the support of emergency services are highlighted.

3.5.1 The WiMAX Network Architecture

This section gives an overview of the basic design principles of the WiMAX network architecture. It explains the main components of the network reference model defined by the WiMAX Forum Network Working Group (NWG).

3.5.1.1 Architecture Overview

For understanding the overall design of WiMAX networks, it is important to consider the influencing factors like use cases and involved business entities.

WiMAX networks provide broadband access to the Internet and to any kind of IP-based services. Operators can enable additional support in the network for certain classes of services over the WiMAX-provided IP connectivity to improve user experience. Such support includes QoS and device location. The toolkit that the WiMAX Forum developed and that actual network deployments and product developments use integrates architectural approaches from two worlds:

- fixed wireline access;
- mobile cellular networks.

The resulting Mobile WiMAX architecture is flexible and suits a range of different deployment scenarios, including fixed access like a simple wireless "DSL replacement," nomadic mobility, or even fully mobile access comparable to the broadband IP connectivity provided by today's 3G cellular systems. This takes into account both technical aspects and the typical involved business entities that are different in fixed and mobile offerings. The term "Mobile WiMAX" reflects the fact that, as long as it follows the radio standard, the same chipset can be used in any type of network independent of whether or not mobility is supported. That is, even devices used with nomadic mobility only are, in principle, capable of being fully mobile.

Looking at WiMAX from the mobility perspective, it allows the devices called mobile stations (MSs) to move between the base stations (BSs) of the same operator, by defining a number of network- or IP-based mobility mechanisms. The operator handing out the subscriptions for a WiMAX user is called the Network Service Provider (NSP), and roughly speaking

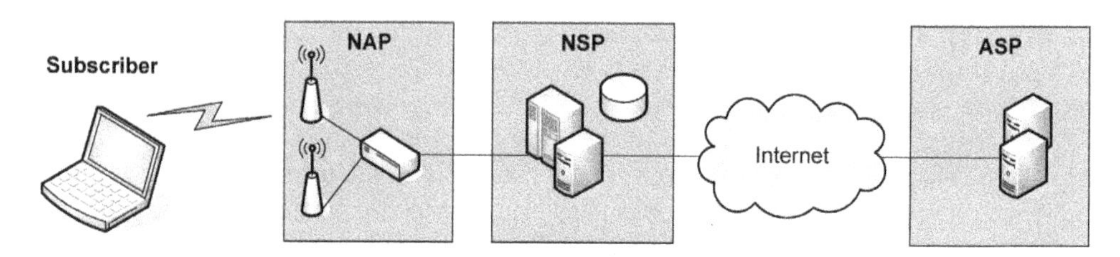

Figure 3.11 Business entities in WiMAX.

takes the role of an Internet Service Provider (ISP). The access part of the WiMAX network includes all the base stations. Logically, it is part of a different business entity (Figure 3.11) called the Network Access Provider (NAP). Of course, a single operator can own both NAP and NSP, which would, for example, be the case for a 3GPP mobile cellular operator, where the access and core functions of the network belong to the same operator.

There is mobility support between different NSPs, so, based on existing roaming agreements between NSPs, their subscribers can use the infrastructure of the other operator for gaining access. For this, the WiMAX network architecture introduces the logical split common to roaming. The home NSP has the contractual relationship with the actual user, that is, it owns the subscription. The visited NSP, which may reside in a different country, is the operator where the user's device is roaming.

In fixed-line access, roaming has no importance today. Typical business relations are not between operators entering roaming agreements. They are, for example, between access providers owning the copper in the ground and ISPs providing the IP connectivity. Hence, WiMAX introduced the fundamental split between NAP and NSP. The same NAP operating the WiMAX access infrastructure and owning the spectrum resources, that are, for example, spanning a whole country, can share its resources with several NSP operators.

A major value of the WiMAX network standards is to group the required network functions as logical clusters that map to the different business entities. The access serving network (ASN) is owned by an NAP and the connectivity serving network (CSN) is owned by an NSP. Between these clusters, reference points are defined that describe all required interaction in terms of signaling and user traffic in a standardized way.

3.5.1.2 Network Reference Model

The above subsection focused on the business entities behind a WiMAX deployment and the possibly different relations depending on the actual type of service. In this subsection, we look at the logical structuring of WiMAX networks from a technical perspective. This will also provide a first insight about where the support of emergency services impacts WiMAX. The focus is on the ASN and CSN, and on the way they interact with each other and with the WiMAX device.

Figure 3.12 shows the basic network reference model (NRM) for Mobile WiMAX networks, which consists of the following:

- The logical network clusters of ASN and CSN as well as the mobile station (MS) itself. The MS is also called subscriber station (SS). It covers both mobile and fixed WiMAX devices.

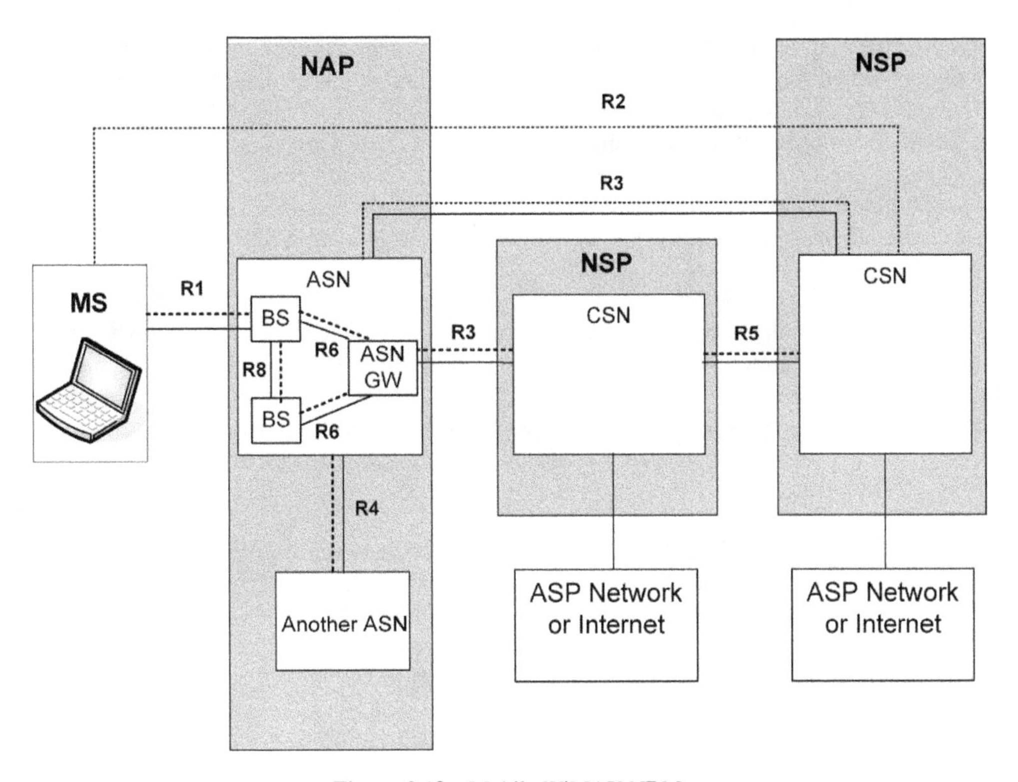

Figure 3.12 Mobile WiMAX NRM.

- Network elements that are allocated to either ASN or CSN. These network elements are logical groupings of individual network functions. They map to physical network equipment for many vendors. For the ASN, these elements are the base stations (BS) and the ASN gateways (ASN-GW) that control a set of base stations.
- Reference points that connect network elements within a network cluster, such as the ASN, or between two different network clusters, such as ASN and CSN, two different ASNs or CSNs, or MS and ASN or CSN, respectively. In Figure 3.12 a dotted line identifies a control path and a solid line identifies a data path.
- In addition, the corresponding business entities NAP and NSP, as well as external domains such as Application Service Providers (ASPs) that are largely out of scope from the perspective of the WiMAX-related specifications. Any WiMAX network logically consists of at least one ASN and one CSN cluster. To cover both roaming and non-roaming scenarios, the NRM shows two chained CSNs that serve as visited CSN and home CSN, respectively, in roaming (otherwise the visited CSN does not exist). Such CSN networks are connected by the R5 (roaming) reference point.
- A CSN comprises the functionality to perform authentication, authorization, and accounting in a central authentication, authorization, and accounting (AAA) server. The CSN enables IP connectivity by assigning IP addresses and providing connectivity with the Internet or any IP-based services through DHCP, a core router, or a mobile IP home agent (if IP mobility

is used between ASN and CSN). In addition, the CSN is in charge of managing the QoS for any connected MS. It covers roaming support and can optionally deploy dedicated server functions to provide location information, or over-the-air provisioning, of devices.

- An ASN comprises all functionality to operate the 802.16 wireless access infrastructure, including the base stations and the ASN-GWs. Common functions include the network support for network detection and selection, access control, handover and mobility support within the ASN and between ASNs, as well as forwarding to the CSN.

To ensure standardized signaling between the different ASNs and CSNs, WiMAX makes use of the concept of reference points. A reference point comprises the sum of interactions and actual protocols between two logical network domains, or entities. On the contrary, interface definitions typically only cover a single protocol used between two network elements. The following major reference points are introduced for WiMAX networks:

- *R1* comprises the 802.16 radio interface functionality between the WiMAX MS and the BS. To allow an MS capable of performing emergency services to indicate at the time of network entry that emergency services support is requested, the 802.16-2009 specifications offer emergency indication bits as part of the messages initially exchanged during network entry within the wireless MAC layer functionality. However, this is not supported by the Release 1.5 WiMAX network specifications and radio profiles based on 802.16. The support of emergency services in the WiMAX network that is available today does not impact the radio interface at the physical or MAC layer of the protocol stack.
- *R2* logically describes all end-to-end interaction between the MS and the CSN that transparently passes through the ASN. The main example here is the authentication procedure for secure network attachment based on the EAP protocol [140] where the actual authentication happens between the MS and the AAA server of the CSN. The indication of an MS for emergency network entry takes place within this EAP signaling and is therefore transparent to the underlying radio interface and to the BS.
- *R3* comprises the tunneling of user data and the AAA-related (RADIUS [141] or Diameter [142]) and IP mobility (MIP or Proxy-MIP) related signaling between ASN and CSN. Regarding emergency services, AAA signaling from the CSN to the ASN is used to install the appropriate QoS profiles and potential filtering rules for supporting emergency services and to provide corresponding MS-specific policy information to the ASN.
- *R4* between ASN-GWs allows network functions to be transferred between different ASN-GWs of the same or different ASNs due to MS mobility or network-internal load optimization.
- *R5* between two CSNs is required for roaming. The reference point comprises a subset of the R3 functionality, but in addition takes intermediate roaming exchange networks into account that might reside between a visited and a home CSN.
- *R6* between BS and ASN-GW supports tunneling of user data and allows the ASN-GW to provide central control for a potentially larger set of (up to a few hundred) base stations. Note that the BS does not directly interface with any network element outside the ASN.
- *R8* between two BSes optionally optimizes some ASN-internal signaling procedures that would otherwise involve and pass through the ASN-GW.

3.5.1.3 Relevant Network Enablers

WiMAX standards support application services with generic service enablers. These are kept generic, as no specific relationship between selected applications and the WiMAX network functions is assumed. One exception is the IP Multimedia Subsystem (IMS) developed by 3GPP as a service enabling platform itself and as a default system for providing VoIP over WiMAX. This does not, however, preclude or limit other VoIP technologies across WiMAX access.

Emergency-related considerations at the application service level are not part of the WiMAX network specifications due to the broad variety of different possible services. There is a clear separation between the network-level support for emergency services, and the application-level support. To use IMS as the default example, all emergency-call-related considerations and mechanisms are just those of the IMS specifications as developed by the 3GPP. The only relevant adaptations for IMS are described in [143] and include minor pieces like IMS service (P-CSCF) discovery support by the WiMAX network, and the transmission of the identity of the base station to which the MS is currently attached, which can be used as initial location information by the IMS system and the PSAP.

Important WiMAX-internal service enablers that are relevant within the scope of this section include the following:

- Built-in support in the AAA backbone signaling between ASN and CSN for network services like IP mobility and QoS management. AAA signaling across R3 and R5 allows the negotiation of network capabilities that are subsequently applied to individual subscriber sessions. User profile settings can be conveyed from the home CSN to the ASN entities.
- Accounting and charging that is able to support both simple tariff models and more advanced ones that account individually per service flow instead of just for the plain IP traffic. There are no measures that support emergency services accounting in particular. The home operator, however, has all related information available, so emergency-related accounting is subject to the operator's policy.
- Support for the dynamic control of the QoS applied to the individual service flows across the wireless interface. Besides the default, WiMAX native, QoS that is controlled by the AAA backbone of the WiMAX network, a slightly customized version of the 3GPP-specified policy and charging control (PCC) framework is available for WiMAX networks. All standardized measures for QoS control mainly target the service flows across the radio interface R1. Any end-to-end QoS that also covers the user traffic passing through the ASN, the R3 and R5 reference points toward the home CSN, and finally to the service provider is outside the scope of the WiMAX Release 1.5 network specifications. In the QoS-related control signaling on the WiMAX-defined reference points (especially on R6 and R3), appropriate parameters, for example, to prioritize flows related to an emergency service, are available. However, the actual use of such parameters across the WiMAX network elements remains deployment-specific.
- Location determination of an MS based on both the network and MS measurements, collection of the measured location information in the CSN, and means to support applications with such location information. Both possible approaches are possible, that is, the MS providing location information as part of the service-related signaling, as well as the WiMAX network providing the location of the MS to the service provider through a location server.

3.5.2 *Network Architecture for Emergency Services Support*

The above introduction to WiMAX already highlighted some key aspects of how emergency services are supported across WiMAX networks. This section elaborates on the overall approach from the perspective of the network reference model. WiMAX networks need to deal with emergency services and with regulatory requirements related to them in a different way compared to mobile cellular networks. The key differences are these:

- No legacy circuit-switched services like voice calls are available as fallback solution for emergency.
- No commonly available and standardized applications like a VoIP client supporting emergency calls will be installed on WiMAX devices. WiMAX network standards do not mandate any specific such application to be supported. This is because application aspects are largely out of scope for these standards.
- The only specific service supporting emergency services that the WiMAX specifications cover as default solution is the IMS. However, this is an optional solution and WiMAX users as well as operators can in general consider any type of VoIP solution or other service.
- WiMAX may be deployed by the NAP and NSP as a pure wireless broadband access service. The above aspects had to be considered at the time of designing a framework for emergency services support for WiMAX networks. This led to the network reference architecture shown in Figure 3.13 for WiMAX emergency services. It identifies the network domains and business entities involved.

In contrast to the basic WiMAX network reference model in Figure 3.12, two additional domains for the VoIP service providers (VSP) are introduced for emergency services support as shown in Figure 3.13. These are introduced for both visited and home networks to reflect

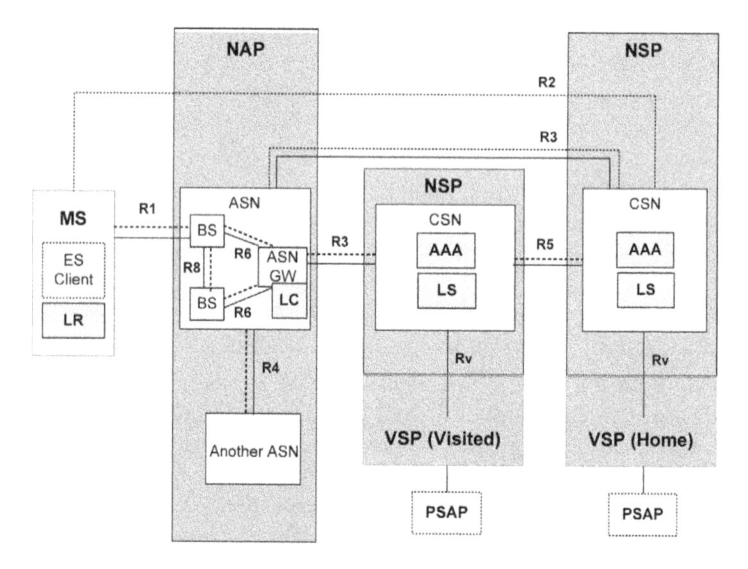

Figure 3.13 Network reference architecture for emergency services in WiMAX. (Adapted with permission from M. Riegel, D. Kroeselberg and A. Chindapol, Deploying Mobile WiMAX, John Wiley & Sons, Ltd., © 2009.)

roaming cases. The VSP's services can be accessed through either the home NSP or the visited NSP. The WiMAX network tunnels the user's IP traffic through the ASN and across R3 to the CSN. The IP traffic can "break out" of the WiMAX network either in the visited CSN or in the home CSN. This depends on where the tunnel for the user traffic terminates in the CSN. If Mobile IP is used to control macro (IP) mobility across R3, the home agent (HA) terminating the tunneling of user data can reside either in the home CSN or in the visited CSN. If assigned in the visited CSN, this is called "local breakout" within the context of this section. The impact of local breakout is that all user traffic can be routed to and from the Internet locally instead of passing through the home network. However, services specifically provided by the home CSN that cannot be reached directly through the Internet would not be reachable for the user. In contrast, if all IP traffic created by the user needs to pass through the home CSN, this may become challenging for emergency services. These are best provided locally, that is, in the visited domain.

For the feasibility of an emergency call, the business relationship between NSP and VSP is important, although the VSP domain is not considered to be part of the WiMAX network itself and as a result is outside the scope of WiMAX network specifications. However, Figure 3.13 captures the possible deployment options at a general level:

- WiMAX NSP and VSP are independent business entities with no specific relationship (making the WiMAX network a pure broadband access medium and ISP).
- WiMAX NSP and VSP are separate business entities with a dedicated business relationship. For example, the VSP can act as the default VoIP service provider to the WiMAX operator's subscribers. The requirement to fulfill regulatory requirements for emergency services support fall upon the VSP. However, the WiMAX NSP supports the VSP to match such requirements.
- WiMAX NSP and VSP are the same operator. Regulatory requirements for emergency services support directly impact the WiMAX operator.

Technically, the relationship between NSP and VSP is established by an interface between CSN and VSP as shown by the Rv reference point in Figure 3.13. WiMAX deployments that only target broadband data services are in principle not affected by regulatory requirements related to the support of emergency services. It is, however, the clear goal of the WiMAX network design to offer support to any VSP for enabling the VSP to match such regulatory requirements more efficiently. This includes, for example, support by the WiMAX CSN network regarding location information or special treatment and indication of an emergency call in the lower network layers. Figure 3.14 identifies the overall relationship between the stakeholders involved.

The WiMAX Release 1.5 network specifications for emergency services support [144] do not introduce any mandates regarding interfaces between the WiMAX CSN and the ASP infrastructure. This is because there are already many existing deployments and interface specifications that need to be considered. This also holds for existing deployment-specific or solution-specific choices for conveying location information. A generic interface in this area would need to cover many quite different scenarios and use cases spanning emergency services and location as well as identity management and correlation, QoS control by the application, and accounting or even payment for a very large range of services. Hence, any such effort would likely duplicate existing solutions. As a consequence, the

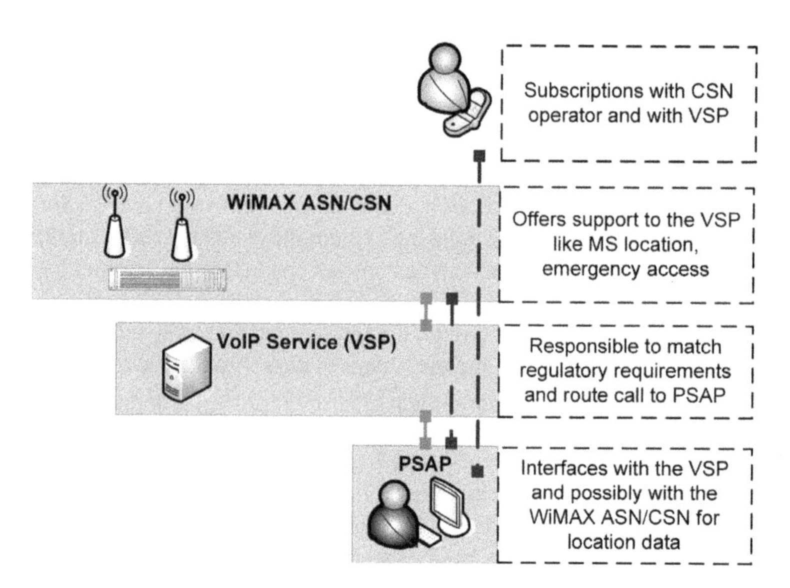

Figure 3.14 Relations between stackholders in WiMAX emergency calls. (Reproduced with permission from M. Riegel, D. Kroeselberg and A. Chindapol, Deploying Mobile WiMAX, John Wiley & Sons, Ltd., © 2009.)

approach for WiMAX networks is to not mandate any specific protocols for such interfaces for the support of emergency services. This is left to deployment-specific choices, or frameworks for VoIP emergency like the one developed by the National Emergency Number Association (NENA) [58].

3.5.3 The Fundamental Building Blocks

The WiMAX network design does not include mandatory VoIP functionality and targets wireless broadband IP connectivity in the first place. Based on this, for the support of emergency services, it can best offer a set of building blocks in a rather generic way. Operators know best how to meet their deployment-specific and regional needs and how to pick and utilize the right set of building blocks. WiMAX offers the following building blocks that are relevant for emergency services support:

1. The location of a WiMAX MS can be determined and provided to service providers via network-based methods or MS-based methods with assistance by the network.
2. Roaming support allows the WiMAX network to assign local breakout in the visited CSN and handle emergency services locally.
3. At the time of initial network entry, a WiMAX MS can indicate to the network that it requests emergency treatment.
4. AAA backend support can be adjusted dynamically on a per-MS basis for emergency. Examples are a dedicated QoS applied to the emergency sessions, or special authorization decisions to grant exceptional access in emergency cases. Hot-lining that can be triggered by the AAA server allows the MS to be guided (and limited) to a special server and webpage that may provide additional support in emergency cases.

5. A limited-access mode allows operators with such regulatory requirements in their region to enable emergency calls for WiMAX devices without valid subscription credentials. This may be relevant for roaming subscribers where the local WiMAX CSN operator does not have any roaming agreement with the subscriber's home operator. Another relevant case is an unprovisioned MS.

Most of the above building blocks are part of the overall WiMAX network architecture. Section 3.5.6 details WiMAX specifics for MS location. Some additional considerations specific to roaming, to service discovery and selection, and to local breakout can be found in section 3.5.4. The main purpose of the WiMAX Release 1.5 emergency services support document [144] regarding those general features like local breakout or hot-lining is to clarify how these are used and how they interact for emergency services.

In addition, one measure specifically added to support emergency services is an explicit indication from the MS to the WiMAX network that marks the network entry specifically for emergency services. This indication builds the foundation for addressing the above building blocks 3 to 5. Of course, this requires the MS in general to be aware of the user's desire to call for help.

Technically, the indication is designed into an early stage of the network entry procedure. It is realized with a WiMAX-specific extension (a so-called "decoration") of the subscriber's network access identifier (NAI) that is sent as part of the initial EAP phase. The EAP protocol performs the initial network entry authentication and carries the human-readable NAI identity mainly to ensure that the routing of the authentication to the home CSN works correctly. The MS will send a decorated NAI, that is, one that is extended by the string "{sm=2}." It indicates a service mode 2 for emergency network entry. The mechanism of decorating the NAI in WiMAX with a service mode is not only used in emergency cases. It can also indicate that an MS is entering the network to request over-the-air provisioning of a new subscription with a network operator.

It is in fact not the cleanest approach to "overload" the NAI string in an access-technology-specific way. The NAI format in general is defined by the IETF [145]. It consists of a username part and a realm part, like dirk@mywimaxoperator.com. This is mainly used for the purpose of AAA routing in the network, and any overloading of the NAI might potentially collide with proprietary NAI decorations chosen by different deployments and different access technologies. The main reason that this approach was chosen at the time of designing emergency services support in the Release 1.5 of WiMAX network specifications is that the standardized air interface communication at this time did not include support for a dedicated indication of emergency cases. Air interface communication is handled by chipsets in both devices and base stations. Obviously, these chipsets require stable specifications at a very early stage of the overall process of standards development. In summary, it was logical to avoid impacts to these chipsets and choose a solution for emergency indication that is fully transparent to the standardized R1 communication between MS and BS at the time of initial network entry.

The NAI that is carried within the EAP protocol identity messages is passing through the WiMAX network all the way toward the AAA server in the home CSN. With the MS sending an NAI including the WiMAX emergency decoration, this enables the AAA server to apply the appropriate policy to react on the emergency services request.

This does not directly impact the ASN network elements. However, the ASN-GW will be the first entity interpreting the NAI. By default, an ASN-GW just interprets the realm part of the NAI for routing the network entry request further on to the correct AAA server for this subscriber. However, the ASN-GW depending on the locally configured policy can already recognize the NAI emergency decoration and prioritize the network entry attempt and the resulting AAA message for network entry to the AAA server of the home network. Technically, even the BS could do this. However, the BS is in general not aware of the individual payloads or protocols carried by the wireless MAC layer (layer-2). EAP falls into this category, so the BS would not be able to identify and analyze EAP messages without performing expensive packet inspection.

3.5.4 Roaming Considerations and Network Entry

Although most initial deployments of Mobile WiMAX target the fixed-line replacement service offerings, roaming support is an integral part of the WiMAX Release 1.5 network architecture. For emergency services, however, roaming support is still limited.

This section will take a brief look at two different states that the MS can be in at the time of requesting an emergency call: either it is already connected to the network and to services provided by a VSP, or it is not connected to the network resources and has to perform initial network entry first.

The general problem to be solved is that emergency calls must be routed to the PSAP nearest to the emergency caller. Placing an emergency call to the emergency responder in the subscriber's home location clearly is of very limited value when roaming in a foreign country. As a consequence, local breakout to allow access to services offered through the visited CSN becomes a feature that impacts the emergency services support across WiMAX. Without such local breakout and with the subscriber connecting to a service provider for emergency that is reached through the home CSN, this "home" service provider, like a home VSP, needs to be able to discover the local PSAP and route the call accordingly.

Let us look first at the case where the MS has to perform initial network entry and can request emergency services support as part of the network entry signaling. Figure 3.15 shows the simplified steps for such network entry and emergency session establishment. When assuming a relatively homogeneous environment with interoperable IMS services being available in both the home and the visited domains, handling of emergency service requests from the WiMAX network perspective becomes relatively simple. IMS is used as example here as it is the only VoIP system directly considered as part of the WiMAX specifications.

Although the IMS system itself has no WiMAX specifics, the appropriate discovery and service selection procedures are in place and the usual steps to establish an emergency call across WiMAX ASN and CSN to the visited IMS system would include the following:

1. The MS performs initial network entry with indicating service mode 2 for emergency in the NAI.
2. Based on the NAI decoration, the AAA server as the network-side endpoint of the EAP exchange becomes aware of the emergency case. Depending on the local operator's policy and on the actual subscriber information, the MS is granted access to the WiMAX network resources.

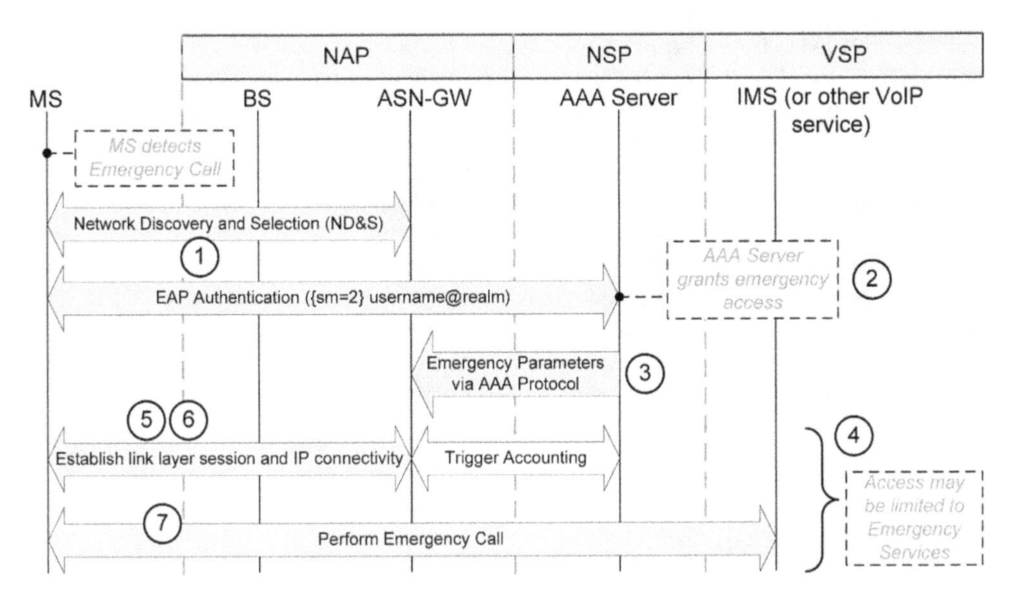

Figure 3.15 Emergency network entry.

3. The AAA server can pass required settings like QoS profiles or an indication about the actual authorization granted to the MS down to the ASN as part of the network entry signaling procedures.
4. The MS may be in limited service mode that, for example, limits the MS to access to the emergency IMS service only.
5. Furthermore, the AAA server selects and assigns the IMS system in the visited CSN and chooses the appropriate P-CSCF address that the ASN advertises to the MS. In line with the selection of the IMS system in the local CSN, the AAA server will also assign an HA or router in the visited CSN to establish local breakout. With this, the MS becomes able to directly reach the local IMS service.
6. The MS establishes IP connectivity with the visited CSN after successful authentication and authorization.
7. It uses the local IMS service to place an emergency call. The PSAP address must be known to the local IMS system, either by pre-configuration or by making use of a dynamic method like the LoST protocol [68].

Based on the actual roaming scenario, there may not be an appropriate VSP service like a local IMS, or the roaming agreement of the operators may not cover the AAA-based assignment of the local IMS service to the MS. In this case, a dynamic method like LoST would be required for discovering the correct PSAP. The home CSN will most likely not be aware of PSAPs in roaming locations based on pre-configured information.

The above considerations mainly hold for a rather closed scenario with tightly integrated networks. In any "non-IMS" case, the AAA server can still at least choose to select the visited CSN's HA and provide local breakout for the IP session. This would possibly make sense to support deployment-specific solutions for emergency services support in the visited operator's network, although in roaming environments these would often be of limited value because of the lack of support in generic roaming devices. When considering a service provider with no relation to the WiMAX operators for the emergency service, local breakout is in general less important. Such an independent service provider would be reachable through the Internet. It would be equally well accessible from either visited or home CSN.

For all the above cases, the assumption is that the MS performs initial network entry with emergency call indication. The WiMAX architecture fully covers non-roaming scenarios and comes with some limitations in roaming environment. On the contrary, many devices will already be connected to the WiMAX network at the time of emergency. Of course, the MS should not exit the network and perform a new entry to let the network know this is an emergency service request. It will try and stay in the active session with the WiMAX network. However, there are limitations to this case for a roaming MS:

- Based on the standards, the roaming MS is not necessarily aware of the information whether IP breakout and IMS services are the local ones coming from the visited CSN.
- If they are assigned with the home CSN, it is not possible to dynamically relocate the IP breakout to the visited CSN and discover local IMS services. So the two quite limited choices for the emergency caller are either to rely on the emergency call reaching a PSAP in the home location, or to exit the network, perform a new network entry with emergency indication by using the emergency decoration, and as a result, of course, delay the emergency call.

3.5.5 Limited Access

An MS may try to enter a WiMAX network with a subscription that lacks authorization for normal services, or with no subscription at all. Depending on the actual country and operator policy, access may still be granted to such devices for emergency services only. The MS would be restricted to using emergency services, but would not, for example, be able to access the Internet. Examples of such limited access to emergency services include cases where a subscriber has an empty prepaid account. Other scenarios are those where the MS would normally be rejected by the AAA server because it is trying to use an NAP that is not allowed by the home NSP's policy.

A rather technical classification of limited access cases distinguishes authorization and authentication. With a lack of authorization, the home NSP is still able to identify the subscriber based on the authentication procedure and the authenticated subscription identity. Authorization would normally fail, but the AAA server can be configured to ignore this and grant access in emergency cases. In contrast, if authentication fails, the AAA server cannot verify the identity of the access attempt and cannot relate it to any valid subscription. This may happen in roaming cases where the visited network is in no relation with the home NSP. Furthermore, the MS can be a blank one that does not have any subscription installed yet.

The following subsections detail these two cases and provide an overview of how WiMAX networks can support them.

3.5.5.1 Unauthorized Emergency Access

Technically, unauthorized access covers those cases where the MS can successfully perform EAP authentication with an existing subscription. For this, valid security keys are required as part of the subscription credentials installed in the MS. The home NSP AAA server can be discovered based on the NAI identity provided by the MS in the EAP procedure that contains the home realm information of the subscriber's NSP. Based on the authenticated NAI identity, the AAA server would deny a normal network entry due to lack of authorization. Examples are empty prepaid accounts, timed-out or barred subscriptions, or subscriptions not being authorized to use a certain NAP.

For the special handling of emergency calls, unauthorized access in WiMAX has a straightforward solution. This is based on the fact that the standard initial network entry procedures can technically be performed without problems. Any authorization decision and granting access or rejecting the network entry attempt is subject to the home NSP AAA server policy. Given this, there is a single point in the NSP's network where local policy regarding unauthorized access to emergency services can be implemented. The AAA server can also control the way in which access to IP services is actually limited by sending appropriate filtering rules to the ASN.

3.5.5.2 Unauthenticated Emergency Access

In several European countries and in North America, it is possible to place emergency calls with a mobile phone that does not contain any subscription. This may, for example, be a UMTS phone without a SIM card. In such cases, the mobile cellular network operator needs to skip any identification and authentication step during network entry. It is granting limited access to place an emergency call to a subscription-less, or unauthenticated, terminal. Although experience shows a dramatic level of misuse of this feature that negatively affects the whole infrastructure being operated for handling valid emergency calls, such features must be offered by any operator being subject to such local regulation.

General Considerations

When moving away from classic circuit-switched voice calls, unauthenticated access becomes even more of a challenge. Owing to the split between the VoIP application part and the network access (which may be provided by different and independent business entities), the WiMAX access network is in general not aware of the fact that a network entry procedure is performed specifically to access emergency services, or even a normal VoIP service. Granting unauthenticated emergency access for devices without a verifiable subscription would mean granting open access to the Internet. This would certainly be subject to substantial misuse.

WiMAX uses the emergency decoration for emergency network entries, so the network can learn that this entry is specifically meant for an emergency situation. However, this is still challenging for the network because the WiMAX operator can only control and limit the MS's access to well-known VoIP service providers. One example of such a configuration would be that the operator is running its own IMS VoIP system. Here, IP access can be limited to the required IP addresses of the emergency IMS system like a single P-CSCF address. In fact, the security considerations seem to rule out any emergency access to arbitrary service providers on the Internet, without any specific agreement between the VSP and the CSN operator.

It is hard to foresee how regulatory bodies will address such unauthenticated access for VoIP emergency in the future. Pure unauthenticated emergency calls with a blank device might be ruled out in general as unreasonable based on the known very high level of misuse. However, on the contrary, there are cases with a valid subscription being available in the MS, where the subscription just does not work for unclear reasons. One class of examples is roaming, where the visited WiMAX CSN just does not have any knowledge of the home network and cannot route the network entry signaling to any home AAA server for verification. Regulation in some countries might still require emergency services to be available in the case of missing roaming agreements between operators. This, of course, assumes that the visited network offers a VoIP technology compatible with the VoIP client on the MS.

One interesting thing to mention here is that the architectural considerations for a WiMAX deployment are different from those for plain Internet VoIP. One of the more prominent examples is that, on the Internet, roaming is typically not part of the architectural considerations behind available specifications, like in the IETF ECRIT effort [120]. There, the assumption is that there is always a "direct" connection between the end host and the service provider. However, issues are largely similar when considering limited access, and similar considerations for Internet VoIP emergency can be found in [124].

WiMAX Support for Unauthenticated Access

Wireless access to network resources comes with relatively different security and authentication methods per actual wireless technology. Technically, these methods are important factors for the design of unauthenticated access support. They need to support establishment of radio connectivity without the actual authentication step.

In WiMAX, it is a major benefit that the main authentication step is detached from the layer-2 protocol across the wireless link. It uses the EAP protocol as a pure container to carry the actual authentication method across the R1 reference point and through the ASN toward the AAA server in the CSN. The actual authentication method used, the so-called EAP method, is performed end-to-end between MS and AAA server. Only the resulting keys that the EAP method generates are securely distributed to and within the ASN. The BS, after receiving derived keys specific to the BS and MS pair, subsequently performs a second local security handshake. This is an integral part of the layer-2 wireless protocol across R1 but it is not affected by emergency network entry in any way. WiMAX networks limit unauthenticated emergency access support to the initial EAP protocol exchange. If the MS gets authorized by the AAA server for such limited access entry, all the following steps of the network entry procedure, including the handshake procedure across R1 to establish the wireless link protection, can be performed in the same way as for a regular entry with a valid WiMAX subscription. With this, the overall solution largely uses an appropriate configuration of the network access. It does not require any significant implementation effort in the MS and in the BS.

The standardized solution to allow unauthenticated emergency access to WiMAX networks as per [136] is based on two commonly available features across all WiMAX equipment:

- The EAP-TLS method [146] that must be supported by all WiMAX devices and that is already used to secure dynamic over-the-air provisioning of the MS. In brief, EAP-TLS makes the well-known TLS protocol with certificate-based authentication and establishment of a secured tunnel between the TLS endpoints available as an EAP authentication method. EAP methods based on TLS are also available for access authentication in WiFi environments like enterprise scenarios.

- Device manufacturers install WiMAX-Forum-compliant X.509 device certificates [147] and keys in all WiMAX devices. Such certificates can be cryptographically verified and allow device (MAC address) authentication to the network operator at the time of initial network entry. Furthermore, the device can authenticate the network to which it is attaching based on the corresponding certificate provided by the NSP's AAA server. Systems that base their security and trust on cryptographic certificates, with X.509 being the most common certificate format, require a public key infrastructure (PKI) including root certificate authorities. These entities are commonly trusted root entities issuing certificates directly for devices or NSP servers, or for subordinate certificate authorities. The roots of trust for the WiMAX PKI are hosted by the WiMAX Forum (see [148]).

Based on these prerequisites, any WiMAX device is prepared to perform a network entry to WiMAX networks based on EAP-TLS device authentication. The pre-installed device certificate is used as security credential for MS authentication in the TLS handshake.

This is in fact authenticated access. However, in cases where the involved NSP has not issued a subscription for this specific device, technically authenticating such an unknown device can only verify the device's MAC address. This has no meaning to the operator as there is no subscription. The WiMAX NSP authenticates an identity that is unknown to the NSP's AAA server, so the authentication step does not provide any value besides generating the keys to protect the wireless link. It still needs to be considered an unauthenticated network entry.

Given this, it is up to the AAA server's local policy whether to grant access in such cases and allow the network entry to continue. A reasonable step for any such MS requesting emergency access is to install pre-configured QoS profiles for emergency services that can be dynamically provided by the AAA server to the ASN at the time of network entry. Again, appropriate filtering of the limited IP connectivity provided to the MS is a very important part of the solution to prevent any MS from misusing the unauthenticated access and gain access to resources that are otherwise charged for.

As mentioned above, device authentication can at least provide the authenticated MAC address of the WiMAX device to the network operator. The value of such an MAC address in the case of misuse is obviously limited. However, it could at least be used to address a number of misuse cases like cloned devices that may be blocked by the operator to reduce the negative impact on the emergency infrastructure resources.

As a final remark, it is obvious that this solution only addresses the network access part. With this, the issues for a concrete deployment are reduced to the actual application layer technology aspects. This, unfortunately, remains a major issue for cases where the actual VoIP systems on the terminal and in a visited NSP's network are not compatible.

3.5.6 Location Support in WiMAX

Location information does not offer a service to the subscriber on its own. It enhances numerous applications that usually have both time and accuracy requirements on the location information. These have to be taken into account when looking at the different methods for realizing location that are possible in WiMAX networks.

Network measurements can be collected from the WiMAX ASN and be used by the CSN or partnering operators, including a PSAP. They provide a good source of information for determining location. The network operator may also provide services in cooperation with the subscriber's device, by collecting location measurements from the MS. Alternatively, the network can provide additional information based on network measurements to the device to support location determination in the device, where assisted GPS (A-GPS) serves as an example. Especially for the support of emergency services with appropriate location, autonomous GPS is clearly not acceptable as the time to acquire a location is far too long. Finally, as a simple measure for providing initial location information to an emergency service like a PSAP, the MS can include the cell or base station ID in the application signaling to the server. IMS-based emergency services over WiMAX use this method. However, the scalability of such an approach between different operators is limited, due to the fact that the application server has to be able to resolve the cell ID to the corresponding geographical location. The emergency service application server may even be agnostic to the actual access technology used by the device.

Summarizing the above, it makes a lot of sense that a WiMAX NSP can offer location information to authorized parties like VoIP application providers on request. This is supported through a standardized location interface to the NAP that is able to perform the actual location measurements in the access network. The above features are especially relevant to NSPs integrating a commercial VoIP offering in their network that is subject to the regional emergency requirements.

3.5.6.1 Location Architecture

The individual functions and network entities that extend the standard WiMAX network reference architecture based on [149] and that are highlighted in ref. 3.16 include the following:

- A Location Server (LS) in the CSN. It is responsible for determining the actual MS location based on measurement data received from the LC in the ASN or from the MS itself.
- A Location Controller (LC) function in the ASN. This requests measurements from the different LAs of the base stations controlled by the LC. The LC typically resides in the ASN-GW.
- Location Agents (LA). These serve as the function measuring the actual location data in the base station.

In roaming, the location server can reside in the home CSN as the home LS (HLS) or in the visited CSN as the visited LS (VLS). Either or both may be involved in providing location information for a single request. Furthermore, standardized location support is available and can be implemented as part of the different reference points of the WiMAX network:

- *R4 and R6*, as the intra- and inter-ASN reference points, include control messages and TLVs to enable location with mobility support.

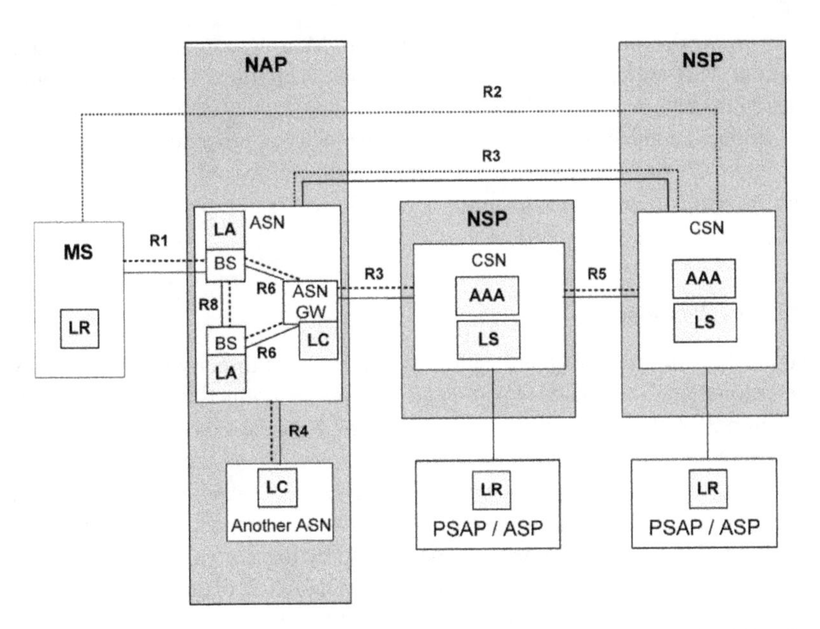

Figure 3.16 WiMAX Location Support. (Adapted with permission from M. Riegel, D. Kroeselberg and A. Chindapol, Deploying Mobile WiMAX, John Wiley & Sons, Ltd., © 2009.)

- *R3* allows the LS in the CSN to request measurement information related to a subscriber from the serving ASN. This measurement data is used by the LS to determine the location of the subscriber's MS.
- *R2* between the MS and the CSN covers all direct communication between MS and LS via two possible application protocols. The location client in the MS is called a location requester (LR).

External entities requesting location information are also referred to as LRs. These roles would apply to the servers of the emergency services provider. Such LRs connect with the operator's LS to request location information. These are not considered to be a functional part of the WiMAX network itself. Such LRs logically connect with the WiMAX CSN across a reference point denoted U1 in the reference model as shown by Figure 3.16. WiMAX does not provide a standardized protocol solution for this reference point (besides an informative Universal Services Interface (USI) framework [150]). The reason for this is not to duplicate the already existing deployments and protocol choices like the HELD protocol [35] of IETF for the Internet space, or the Mobile Location Protocol (MLP) protocol [151] of the Open Mobile Alliance (OMA) for the mobile cellular world.

The network location procedures based on R4/R6 and R3 are referred to as "control-plane location." R2 procedures that directly involve the MS are referred to as "user-plane location." WiMAX allows these two different levels of location support to be combined. Some abstract examples are as follows:

- Device-initiated user-plane location. The MS requests assistance data from the LS across R2 and performs the location calculation.

- Device- or network-initiated mixed-plane location. Measurements from both the MS and the LC in the ASN are used as input to the location calculation. The actual calculation can take place in the MS or in the CSN LS.
- Network-initiated control-plane-only location. The location procedure is triggered either by the CSN internally or by an external LR. The LS will retrieve MS measurements from the LS in the ASN and calculate the location. The MS itself is not involved.

3.5.6.2 MS Involvement

User-plane location summarizes the location procedures that directly involve the MS for the location determination. The related signaling procedures are defined for the R2 reference point shown in Figure 3.16 that connects the MS with the LS in the CSN. Different standardized solutions for such location communication between a device and the network are defined by standards organizations, including the Open Mobile Alliance (OMA) and the Internet Engineering Task Force (IETF). WiMAX adopts and provides explicit support for two application-layer location configuration protocol frameworks developed by these organizations.

The frameworks that are directly considered in WiMAX location are the OMA Secure User Plane Location (SUPL) protocol and the WiMAX Location Protocol (WLP), which is largely based on the location framework developed by the IETF Geopriv effort and the HELD protocol. This section will provide a brief overview about the main conceptual differences and WiMAX-specific profiling.

Owing to their different conceptual approaches, the two frameworks target different ecosystems. SUPL is driven by the mobile cellular industry and specifically targets the requirements of mobile cellular operators. In contrast, WLP mirrors the fact that WiMAX in a simple deployment mainly serves as broadband access to the Internet and is based on IETF location technology. Hence, it best reflects requirements and use cases from the Internet community.

A major difference in the architectural design considerations of the two protocols is that the IETF HELD protocol [35] used in WLP is independent of the underlying access technology, so there are no WiMAX-specific adoptions that would impact the basic protocol functions. As the actual access technology is generally viewed as being independent of IP-based services and applications requiring location, HELD does not consider roaming but assumes the location server to be a local one in the respective access network.

On the contrary, SUPL comes with a full-blown roaming protocol architecture with coverage for intermediate visited networks and location servers to best enable mobile roamers. The SUPL architecture takes specifics of the underlying access technology into account and requires some building blocks to be adapted to best meet the specific access requirements. This is also the case for SUPL over WiMAX, where SUPL uses a WiMAX-specific security framework.

The SUPL architecture defined by OMA [152] smoothly integrates into the WiMAX roaming network reference model (see also Figure 3.12). It includes an SUPL-enabled terminal (SET) mapping to the WiMAX MS and the SUPL location platform (SLP) mapping to the LS as the main functions. Application Servers making use of SLP-provided location information logically implement an SUPL Location Agent (SLA) that interfaces with the SLP, which is comparable to the WiMAX-defined LR function for applications.

Regarding the actual protocol stack across R2 in WiMAX SUPL-based deployments, the exchanges between MS and LS are based on the Userplane Location Protocol (ULP) [153].

The communication is TCP/IP-based and is secured by the PSK-TLS [154] protocol that is a variant of the commonly used Transport Layer Security (TLS) protocol based on pre-shared key authentication. The root session key used by WiMAX systems to enable PSK-TLS is derived from the EAP-based access authentication generated EMSK key.

WLP is based on IETF standards for formatting and transfer of geographic location information. It is based on the HELD protocol that in its basic version provides a simple mechanism to allow a device to request location from a location server in the access network. The matching communication model is based on the assumption that it is the responsibility of the device itself to provide location information to application service providers as part of the application signaling itself. However, protocol extensions like [82] to allow an LS to request location information measured by the device are used in the WiMAX domain. This, for example, enables the use cases in WiMAX where location information is provided by the WiMAX LS directly to the emergency service provider. Conceptually the LS when mapped into the WiMAX network resides in the CSN instead of the access part of the network, so the terminology used in IETF is different from the WiMAX view, where roaming is explicitly taken into account and where the network side splits into ASN and CSN.

Considering security, HELD is conceptually different from the access-based keying used to secure SUPL exchanges. It assumes certificate-based authentication of the LS to allow the MS to verify that the location server is not a rogue one. Client verification is simply based on the device's IP address, or other network identifier that is verified by the network, with return-routability checks (meaning a rogue device would not get the HELD response carrying the actual location information). Protection of exchanged data is based on a TLS tunnel with server-only authentication. Within the WLP solution for WiMAX, the approach of aligning the security solutions is supported and recommended. This results in WLP security between WiMAX MS and LS that is also based on PSK-TLS with shared keys derived from the EAP authentication phase during initial network entry. This is in fact the same solution that is used for SUPL over WiMAX. The IETF HELD security model, however, is not precluded within WiMAX, allowing it to be used in solutions where pre-arranged keying is less easy, for example, prepaid or temporary access situations or wireless broadband scenarios where residential gateways are employed.

3.5.6.3 AAA Support of Network-Based Location

Control-plane signaling in WiMAX makes substantial use of AAA protocols like RADIUS and Diameter. This also affects location procedures across the WiMAX internal reference points that are based on these protocols. Location support is partially integrated with the AAA infrastructure. Examples include the following:

- Authorization of location information requests by third parties against the subscriber's profile and settings, meaning that the CSN AAA server has to be consulted for any location requests. The LS is responsible to serve only authorized requests for location information of a specific subscriber, including those sent by verified location requesters like PSAPs in the case of an emergency call.
- MS IP address identification and security key distribution to the LS for the protection of user-plane location signaling.

- LC identification in the ASN (also in the case of mobility where the LC moves between different ASN-GWs).
- Support for accounting to allow network operators to apply proper charging for the use of location information, either to subscribers, or to other parties like ASPs.

The message flows specified in [149] include an AAA-based exchange between the LS and the AAA server, where the AAA server is consulted for deciding whether to grant authorization prior to the LS collecting the requested location information from the ASN.

Communication between the LS in the CSN and the LC in the ASN uses the R3/5 reference point and is based on either RADIUS or Diameter, extended by a set of WiMAX-specific attributes or AVPs for exchanging the location-related information.

Conceptually, the AAA communication between LS and LC is designed to be separate from the basic AAA signaling path for network access between an ASN-GW's AAA client and the AAA server. However, at least for RADIUS, the standard AAA signaling for a specific subscriber and MS between ASN and CSN can also include the signaling required for location support.

Figure 3.17 shows an example message flow where the LS retrieves measurements and calculates location information when receiving a location request by an external LR like a PSAP.

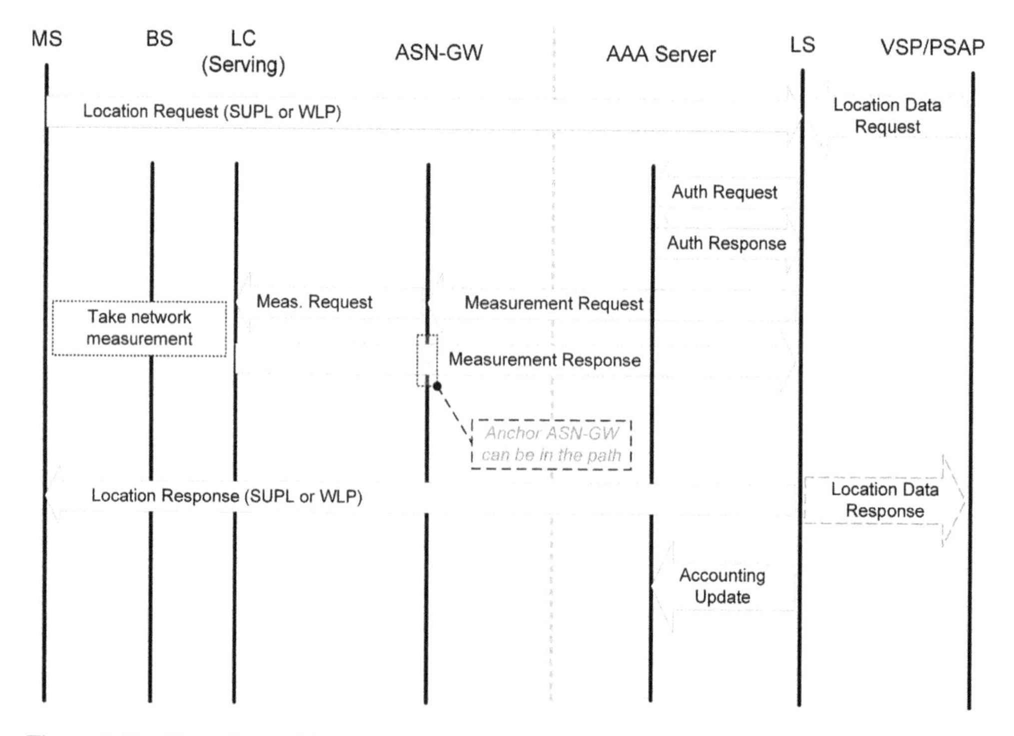

Figure 3.17 Network-based location retrieval after handoff via anchor AAA client. (Adapted with permission from M. Riegel, D. Kroeselberg and A. Chindapol, Deploying Mobile WiMAX, John Wiley & Sons, Ltd., © 2009.)

The ASN supports MS mobility and the ASN-GW controlling the MS and providing the LC function can change for an MS moving between base stations. In such cases, the AAA path carrying the location signaling can continue to follow the standard WiMAX AAA path that keeps the AAA client and the (anchor) authenticator in the ASN-GW where the MS originally entered the WiMAX ASN. In this case measurement responses from LC and LS are relayed across R4 through the original (anchor) ASN-GW. This increases the signaling load across R4, but otherwise simplifies AAA deployment because for the MS there is still a single R3 AAA client for the CSN. Alternatively, the LC in the new serving ASN-GW may directly act as a second AAA client just for location support and send all response messages directly to the AAA server in the CSN (indicated by the dotted box in Figure 3.17).

3.5.7 Conclusion

In summary, the WiMAX Forum network standards provide a framework for the support of emergency services [144] that describe how existing core network functionality is used for supporting emergency services across WiMAX access. Furthermore, specific technical building blocks like the NAI-based emergency indication during network entry or support for unauthenticated and unauthorized emergency network entry are available and can be enabled by operators based on regional requirements or based on their preference.

Based on the Release 1.5 core network functionality of WiMAX, support for emergency calls in WiMAX networks is covered in a reasonable way for non-roaming cases. This especially holds in cases where the WiMAX operator runs the VoIP service itself, or relies on partnering service providers within a controlled environment. Owing to its complex nature, roaming support for emergency services between WiMAX NSPs is only available with limitations. As a general observation, especially the area of generic VoIP emergency services support in combination with nomadic or fully mobile IP broadband access still has a lot of challenges, although the IETF ECRIT effort made substantial progress in this area.

Finally, it becomes clear that full emergency services support requires a sound integration of the different enablers within the WiMAX network, like QoS, location support, AAA messaging, and the VoIP service. Emergency services are an "application" that serves as a very good example for how important it is to integrate these diverse enablers across an operator's network and across different networks working together.

3.6 3GPP

Hannu Hietalahti
Nokia

3.6.1 Introduction

From the landline telephone user's perspective, an emergency call looks like just another phone call between the caller and the Public Safety Answering Point (PSAP) where the PSAP call-taker answers the call. In mobile phones there are some differences that are also visible for the user, such as being able to make calls with limited service when registration for normal calls is not allowed, without opening the key lock, the use of local country-specific emergency numbers when roaming abroad, and so on.

That simplified view is all the user needs to know, but the technical implementation is more complicated. The regulators determine the minimum set of requirements the system must fulfill, whichever technology is used to route the emergency calls from the user to the PSAP. These form an important part of the external requirements to the system.

The technical view on the design and implementation of emergency services for mobile phones can be split into two main types, which are the Circuit-Switched (CS) and Packet-Switched (PS) domains. Different call control and routing protocols are used in CS and PS domains, so it is better to look at each of them separately, even though both need to respond to common external requirements.

Mobile systems face tougher security, location, and authentication challenges than fixed systems, which know where the wireline connection leads to and thus the subscriber's location is always known. Fixed phones do not roam, but a mobile system faces very challenging requirements related to authorizing emergency calls for roaming subscribers whose subscription does not allow them to make normal calls in their present location. Also certain very common fraud cases hitting mobile systems are non-existent in fixed telephony.

Here the focus is on 3GPP-specific requirements, and a procedural signaling flow approach is taken on CS and PS emergency calls, assuming that most readers consider that easier to follow than a specification-oriented analysis of the tasks of each affected system entity.

3.6.2 Requirements

3.6.2.1 External Requirements of the System

Regulatory and legal requirements apply on any emergency telephony system. The technical implementations that provide emergency call capability in the User Equipment (UE) and in the network must comply with the mandatory external requirements. Typically such requirements placed on the system are non-technical but they will have an impact on the technical implementation.

The regulatory requirements are typically common to all licensed band mobile telephony systems, and they include items such as:

- positioning of the user;
- limited service emergency calls;
- prioritization, both in the UE and in the network;
- emergency numbers, global and also local national ones;
- different service numbers for each service versus single number for multiple services;
- local configuration of PSAPs; and
- routing of calls to the PSAP.

3.6.2.2 3GPP Specifications Requirements

The 3GPP system specifications reflect the external requirements to the mobile telephony system. The majority of emergency call service requirements are common to CS and PS domain calls, and they are specified in 3GPP TS 22.101 [155]. This section deals with the 3GPP service requirements. Some of them apply on one of the domains only.

While the service requirements for emergency calls in this section are mostly common to CS and PS domains, the architectures and protocols covered in sections 3.6.3 and 3.6.4 are independent and specified separately.

Support of Emergency Calls

3GPP networks must support emergency calls in the CS domain, but the UE support of an emergency call is only required on the condition that the UE supports speech calls. Hence, a data card UE with no speech call support does not need to support emergency calls.

In the PS domain, 3GPP specifies emergency calls only via IP Multimedia Subsystem (IMS). IMS must support emergency calls, but so far only voice media are supported.[2] Multimedia emergency calls are being considered; however, not only is the introduction of multimedia service an engineering challenge but also substantial parts of the PSAP equipment and processes need to be upgraded as well.

Subject to local regulation, the emergency calls may need to receive higher priority than normal calls. Emergency calls must be routed to the nearest PSAP according to national regulation, and that often requires mapping of the user's location to the PSAP that is responsible for that area.

It is also required that an emergency call must be possible without a valid subscription, and in that case it is the only service that is allowed. This requirement splits into two main cases: one where no subscription of any kind is available due to completely lacking or invalid SIM or USIM card; and the case when a valid (U)SIM is available but no roaming agreement exists between the local operator and the user's home operator. The former case means that the subscriber cannot be identified at all. In the second case, the user is known, but is not authorized for normal services in the serving network, and consequently there is no network interface to authenticate the user.

3GPP systems must also provide configurability for the operator to either allow or not allow emergency calls without (U)SIM. Evolved Packet System (EPS) in Release 8 introduces yet another restriction, as the support of SIM is discontinued and no normal calls are allowed without USIM.

Emergency Numbers

Technically, the simplest way to configure emergency numbers is to use only the specified well-known emergency numbers 112 and 911. However, this is not only a technical issue but requires education of users as well. Hence, also other emergency numbers need to be supported as well. Many countries also use multiple emergency numbers based on their tradition to distinguish between different emergency services.

The 3GPP system provides several mechanisms for the operators to configure local emergency numbers. The home operator can configure additional emergency numbers of the home country on the (U)SIM and the serving operator can download additional local emergency numbers to mobile phones over the radio interface. In that way a roaming user's phone can be primed with the local emergency numbers. The advantage of pre-configured local emergency

[2] At the time of writing, standardization for non-voice emergency services (NOVES) for IMS is work in progress in the 3GPP. A study has been conducted and can be found in TR 22.871 [156].

Table 3.1 3GPP emergency services categories

Police
Ambulance
Fire brigade
Marine (coast) guard
Mountain rescue
Manually initiated eCall
Automatically initiated eCall

numbers is that the emergency calls using those numbers can be identified as emergency calls immediately at the mobile phone when requesting network resources to establish a call. To use those local emergency numbers, the roaming user must of course be aware of the local numbers in each country. Local emergency numbers are often valid for an entire country, but small areas are not uncommon either. Since the configuration parameters received from the serving network are valid within that network, it is necessary to memorize the downloaded emergency numbers as long as the UE stays in the same country. The country is identified by the Mobile Country Code (MCC), and the local emergency numbers remain valid even if the UE roams to another operator's network in the same country.

In some countries, different emergency services are identified by different emergency numbers. In this case, different emergency requests are routed to different PSAPs, each responding to their specific emergency service requests. In the single PSAP approach, it is up to the PSAP operator to find out from the caller which emergency services need to be dispatched, but the service categories allow the telephony system to route known categories to their dedicated PSAPs.

The emergency services categories specified in the 3GPP system are listed in Table 3.1. It is also possible that new categories will be added later, but the list is already enumerated in Release 8 (Rel-8) of the 3GPP specifications. Also, combinations of these categories are possible. Usually the different categories can be identified by different emergency numbers, such as, in Japan, 110 for police, 118 for coast guard, and 119 for fire or ambulance.

When establishing an emergency call, it must be possible for the user to hide his or her identity,[3] but only if that is allowed by the local regulation. Even if identity hiding is allowed, the serving operator may override the user's request.

However, in all countries and irrespective of whether any local emergency numbers are supported or not and whether multiple emergency service categories are used or not, both 112 and 911 are always considered as emergency numbers.

Even in the case when the UE was not aware that the number dialed by the user is a local emergency number, the serving network has the option of processing the call as an emergency call. For 3GPP system-specific reasons, such a call will differ from a normal emergency call case, since the UE will set up the call as a normal call without any priority, without any

[3] In the context of telecommunication, the term "identity" refers to the calling party identity, such as an E.164 number (a phone number) or an SIP URI.

specific indication of an emergency call, and thus subject to any restrictions that might apply on normal calls.

Some local emergency numbers have quite a long tradition and they are recognized as emergency calls if no other information on local emergency calls is available, that is, when there is no valid (U)SIM available. Those numbers are 000, 08, 110, 118, 119, and 999.

3GPP specifications define four different methods of configuring emergency numbers to phones:

- default numbers 112 and 911;
- default numbers when no (U)SIM is present;
- additional local emergency numbers configured by HPLMN on the (U)SIM; and
- additional local emergency numbers downloaded over the radio interface.

Location of Emergency Calls

Some countries have introduced a regulatory requirement to indicate the caller's location with given accuracy to the PSAP in emergency calls. 3GPP specifications allow the networks to support this requirement, where necessary.

The user may have requested location privacy, but for emergency calls location information is an essential part of the service and therefore there is no regulation that allows the caller to obtain emergency services support without disclosing his or her location. As such, a location request from a PSAP will override the user's preferences to dispatch first responders.

The detailed service requirements on emergency call location are specified in 3GPP TS 22.071 [111].

Emergency Callback

Some countries require the capability for the PSAP operator to call back to the emergency caller after the initial emergency call has been disconnected. The initial emergency call may have been disconnected prematurely before sufficient information was provided. Missing information may be related to the type of incident or regarding location information in the case that automatic location information was not available or sufficiently accurate.

The callback is based on the caller's contact information, such as a phone number, that was received during the initial emergency call. Therefore, the system must be able to supply the calling party identity to the PSAP for possible callback. While the emergency call processing in a 3GPP system is different to normal call processing in many ways, the callback handling is treated like a normal call. Consequently, the callback in limited service state is not specified.

Supplementary services that might be registered and activated are handled differently in the CS and the PS domains. 3GPP TS 22.004 [157] specifies Teleservices, such as normal call, emergency call, and the supplementary services that are related with Teleservices. Only multicall and operator-controlled enhanced multi-level precedence and preemption are applicable to the CS domain emergency call, which is Teleservice 12. Most call-related supplementary services are only defined for a speech call, which is Teleservice 11. Callback is processed as a normal call, and no exceptions on supplementary services have been defined for the CS domain.

Some PS domain emergency services require exceptional processing of supplementary services, as specified in 3GPP TS 22.173 [158]:

- call diversion;
- call transfer;
- call barring (of incoming calls);
- hold; and
- conference call.

If a call can be identified as a callback, then any diversions, call transfer, and barring of incoming calls must be suppressed. Furthermore, the user receiving a callback cannot put the callback on hold or join a conference call. To fulfill this requirement, it must be possible to distinguish a callback from normal calls. The protocol solution for the Session Initiation Protocol (SIP) using an additional callback indicator is still a work in progress [159]. When finalized, this callback indicator in an SIP message will tell SIP entities, such as SIP proxies and the SIP user agent, to suppress undesired supplementary services and to give the callback higher priority treatment.

Other Requirements

Both the network and the mobile phone can support either CS domain, PS domain or both, and, depending on the combination of supported and subscribed services, either CS or PS calls might not be possible. In such cases the emergency calls will obviously be made in the available domain. But if both domains are supported and available, then according to 3GPP Rel-8 it is strongly recommended that the UE should initiate its emergency call attempt in CS domain. It is foreseen that cellular IMS and VoIP calls become more commonplace and this CS domain preference will disappear in the future.

The above requirement has been changed from 3GPP Rel-9 onwards to make CS and PS domains equal. This means that the phone may start its initial emergency call attempt in either domain, but if that attempt fails, there is a strong recommendation to make a re-attempt in the other domain, if that is supported by the phone.

The capability to transfer auxiliary data during an emergency call is required by 3GPP TS 22.101 [155]. That can be information on the position of the user or related with the automated emergency calling device embedded in a vehicle. This requirement does not yet determine how the data transfer takes place, but it can happen before, during or after the voice part of the call, and it may offer a possibility for the PSAP to request more information. Both the voice and data components must be combined in routing to hit the same PSAP that handles the call.

This automated emergency calling support in vehicles is known as eCall. eCall phones that are permanently configured to support only eCalls can be be embedded in vehicles as part of the car passenger safety system. These vehicles may not issue an emergency call for years or never. The benefit for the user is the increased safety via automated emergency call capability if it is needed some day. For a network operator it would be very inefficient to offer a service where millions of vehicle-based eCall-only devices would consume signaling resources with their repeated location registrations without creating chargeable traffic. Therefore, an eCall phone is required to register only when it is attempting to make an emergency call. In such a case the eCall phone will also need to stay registered for a while after the call in case the PSAP requests more data or needs to perform a callback (in case there is a voice communication capability).

3.6.3 *Emergency Calls in the CS Domain*

3.6.3.1 CS Domain System Elements

The Mobile Equipment (ME) is defined as a phone without (U)SIM and a UE is defined as an ME with a (U)SIM, that is, a phone with a subscription. Here, the notation (U)SIM is used to mean either SIM or USIM in the cases when they are interchangeable.

Only GERAN and UTRAN access networks are considered in the CS domain since E-UTRAN is a PS-only access technology and CS emergency calls are thus not applicable. The main tasks of an access network are:

- radio connectivity;
- paging;
- security procedures;
- broadcast information; and
- assignment of dedicated channels.

The (V)MSC is the mobile switching center that has the main responsibility for routing calls inside the network and between them. The main procedures that run between the MSC and the UEs are:

- registration;
- authentication; and
- call processing.

The GMSC is a Gateway MSC that interconnects other networks to the PLMN:

- gateway to other networks; and
- routing of mobile terminating calls to the (V)MSC of the called subscriber.

The HLR is Home Location Register that stores subscriber information and security-related data:

- subscriber data; and
- authentication data and security keys.

3.6.3.2 CS Domain Architecture

Both when in the home network and when roaming the call control processing is done by the (V)MSC in the local serving network. In Figure 3.18, the VMSC/VLR B is serving the roaming user.

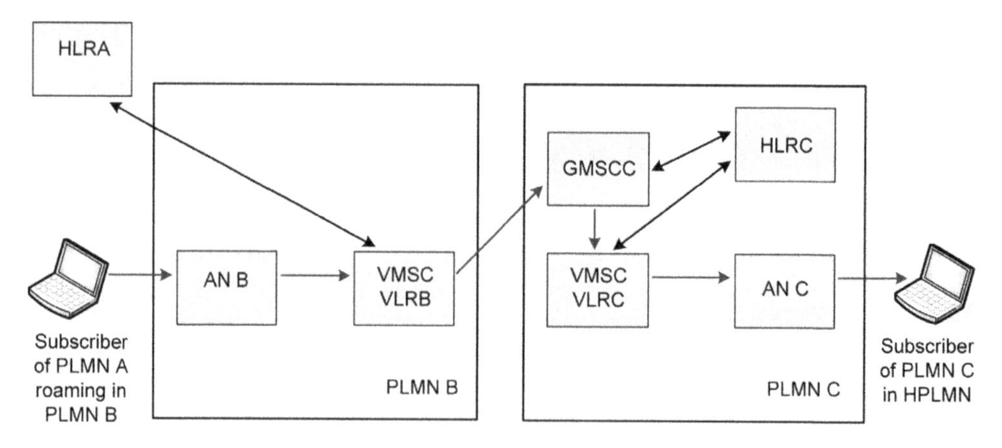

Figure 3.18 User A calls User C while roaming to VPLMN B. (© 2009. 3GPP™ TSs and TRs are the property of ARIB, ATIS, CCSA, TTA and TTC who jointly own the copyright in them. They are subject to further modifications and are therefore provided to you "as is" for information purposes only. Further use is strictly prohibited.)

An emergency PSAP is always a local service. As the emergency numbers vary between different countries, the home network could not reliably detect emergency numbers in all cases. The geographical or service-based work split of PSAPs is part of the local configuration. Therefore, it is up to the local serving network to detect an emergency call attempt and to route the emergency call locally to the appropriate PSAP.

Basic call handling is defined in 3GPP TS 23.018 [160]. At registration time the user is authenticated by the VMSC using the security parameters obtained from the HLR at the HPLMN of the subscriber. When subsequent calls are being set up, it is the task of the serving VPLMN to determine where to offer the Mobile Originating (MO) calls that the user initiates. Normal mobile-to-mobile calls outside of the serving PLMN are routed to the destination PLMN via its Gateway MSC. The HLR of the called user is queried for subscriber data and the VMSC at the destination network then pages the target UE to set up the call, as shown in Figure 3.18.

The VMSC serving the roaming user must be able to detect the attempt to initiate an emergency call, and instead of routing the call based on the target phone number, as is the case for normal calls, it needs to be routed locally to the appropriate PSAP, as shown in Figure 3.19.

Since the UE may not always be aware of the local emergency numbers, it is possible for either the UE or the serving (V)MSC to detect an emergency call attempt. If the UE is aware that the call being established is an emergency call, then it will indicate it in its request to the (V)MSC.

If the user has dialed a local emergency number and the UE did not recognize the emergency call, then the call is initiated as a normal call. The serving (V)MSC can still consider it as an emergency call based on the dialed number and route it as an emergency call. This case only applies to emergency calls while the UE is normally registered for speech services.

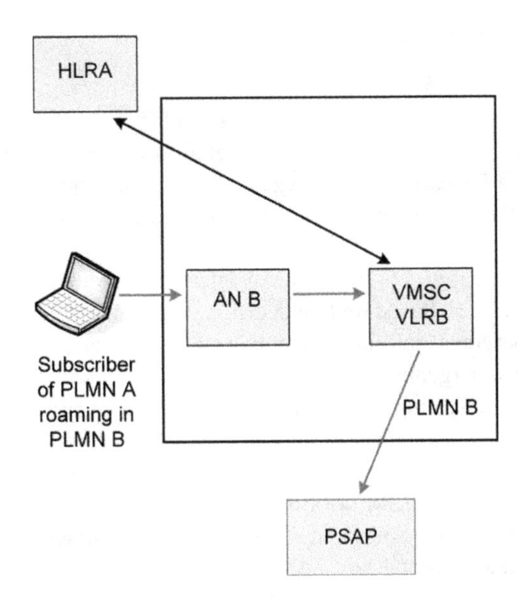

Figure 3.19 CS emergency call while roaming to VPLMN. (© 2009. 3GP™ TSs and TRs are the property of ARIB, ATIS, CCSA, TTA and TTC who jointly own the copyright in them. They are subject to further modifications and are therefore provided to you "as is" for information purposes only. Further use is strictly prohibited.)

Normal calls are not allowed in limited service state, but emergency calls can be allowed. Emergency calling in limited service is an exceptional case. The phone can end up in limited service state for several reasons. Either there is no (U)SIM card or the user is based on terms of subscription not allowed to access that network or part of the network. A third case also exists when there is no network coverage of any network, but from an emergency call viewpoint that is not relevant since no calls are possible when there is no connection to a network.

When there is no (U)SIM card inserted in the phone, the user cannot be authenticated or even identified reliably. In some countries emergency calls are not allowed in this case based on regulatory requirements, and the operators need to implement that regulatory mandate by rejecting such emergency call attempts.

Several network access control mechanisms exist to protect the network against overload, for example, when a major disaster occurs and a high number of users start initiating calls at the same time. In such cases, the network can either block calls by allowing only part of the phone population in the affected area to make calls. That can be achieved via an Access Control mechanism where each 10% of the users can be granted or denied permission to make calls. Domain Specific Access Class Barring (DSAC) can be used to bar access to either CS or PS domain or both during peak load. An even more dramatic way to block specifically emergency calls is to indicate that emergency calls are not allowed in the cell.

The phone will only attempt to initiate an emergency call when it is allowed by the network to do so, but there is still no guarantee that it will succeed, as the network may still reject the attempt.

3.6.3.3 3GPP Protocol Solution

Detection of Emergency Numbers

The emergency numbers defined in 3GPP TS 22.101 [155] are common to the CS and the PS domains, so the same numbers must be considered as emergency numbers by the phone irrespective of whether a CS call or an IMS call is being established. Some of this analysis is hard coded in the Mobile Equipment (ME) and some of it can be performed only by the UE with a (U)SIM card.

Primarily it is the responsibility of the phone to determine whether the number dialed by the user is an emergency number. Based on that decision, the phone will have to branch to either normal call setup or emergency call setup. In both a CS domain and a PS domain IMS call, the procedures to initiate an emergency call differ from those for normal call setup. In this way the regulatory requirement of allowing emergency calls also without subscription can be fulfilled, as the serving network gets an indication of the attempted emergency call. This allows the network to process emergency call exceptions in suppression of subscription checking, and to apply special call processing rules for prioritization and routing.

If the phone does not know at the time of call initiation that the dialed number is an emergency number, then it will initiate a normal call. This is the case if a non-standard emergency number is used and the number has not been distributed to the phone to be included in the default emergency numbers list. The network can still detect the emergency number during the number analysis and route it toward the PSAP. The analysis of emergency numbers takes place before taking any other routing decisions or before checking whether any supplementary services might apply. This method is obviously restricted to the phone's normal service mode, as in limited service the phone can only initiate calls toward the numbers it knows to be emergency numbers.

Provisioning of Emergency Numbers

The numbers 112 and 911 are well-known emergency numbers, but several countries still use other numbers as well, to distinguish different emergency calls to different emergency services or based on tradition. A simple way to handle the multitude of emergency numbers that exist globally would be just to collect a full list of all numbers that are used as emergency numbers in all countries. Unfortunately, the "collision" between the local emergency numbers of one country with local service numbers of some other country does not make this approach feasible. In fixed telephony, this local reuse of three-digit short codes was not a problem, but it become a problem with mobile phones that roam freely. As an example, in Japan, 118 is the emergency number for sea rescue, but in Finland, 118 is used for phone directory inquiries. Where should the serving Finnish network route a 118 call dialed by a roaming Japanese subscriber? Another solution idea would be to harmonize the emergency numbers globally. This approach would require a lot of negotiations between national authorities, and educating the users would be a slow process. Therefore, the mobile phone system must adapt to the needs of society and provide configurability for local needs.

The solution is to define some numbers to be always considered as emergency numbers and to allow the operators to add on to the UE's emergency numbers list any numbers that are required for local needs. 3GPP TS 22.101 [155] specifies that, when initiating a call, the phone must consider the following numbers as emergency numbers:

1. 112 and 911 are always emergency numbers;
2. any additional emergency numbers that are specified as such on a (U)SIM;
3. 000, 08, 110, 118, 119, and 999 are considered as emergency numbers by ME without a (U)SIM; and
4. any additional emergency numbers that have been downloaded by the serving PLMN.

Several countries still do not use 112 or 911 as emergency numbers, but the UE must always consider those as emergency numbers, and thus they cannot be used for any other purpose anywhere. Even if the national emergency number is, say 999, both 112 and 911 will also work as the emergency number. Those inbound roamers who are used to well-known numbers 112 and 911 at home can in this case use the numbers they already know.

In countries where additional emergency numbers are used, the operators can configure the additional numbers on their subscribers' (U)SIM. When no (U)SIM is available, then the ME is in limited service state and can only initiate emergency calls. Thus, there is no conflict with any possible local service number and also additional well-known emergency numbers in item 3 above trigger the establishment of an emergency call.

The inbound roamers to a country that uses additional emergency numbers can be covered by the serving operator downloading the additional emergency numbers to any phone during dedicated CS or PS mobility management procedures. Ideally, this can be done when a new roaming phone is attaching to VPLMN in a foreign country.

The UE handling of multiple emergency numbers is not strictly specified, as 3GPP specifications only define the external protocol requirements that can be observed from the outside in black-box testing but leave the actual implementation up to each phone manufacturer. Nevertheless, a logical model of the emergency number handling is a list of emergency numbers that always at the very minimum has at least the well-known numbers 112 and 911 included. If additional default numbers for (U)SIM-less ME, pre-configured on (U)SIM or received from serving network apply, they are appended to the list. The resulting emergency numbers list is the set of numbers that trigger emergency call establishment.

Indication of Emergency Calls and Routing in the Network

Once the phone has determined that an emergency call needs to be initiated (based on the user-entered dial string), the call establishment signaling will differ from the normal call. The serving network is given the necessary information to process the call in a special way.

Prior to sending the initial SETUP or EMERGENCY SETUP message, the Call Control of the UE will indicate establishment cause to the lower layers. The specific establishment cause "Emergency call" is used for the EMERGENCY SETUP message to distinguish the emergency call case from others for prioritization and suppression of access rights checking.

The CS emergency call flow diagram in Figure 3.20 gives an example for an emergency call setup. Note that this is not the only possible call flow.

When setting up an emergency call, the UE will use the EMERGENCY SETUP message to distinguish the request from normal call setup. It is not possible to encode a target number in the EMERGENCY SETUP message. In the absence of the target number, the serving MSC has to route the call based on the context, and with the help of location information or a locally configured route to a PSAP. The advantage of this approach is the elimination of the obvious fraud case via a limited service emergency call to a non-PSAP target number to avoid charging.

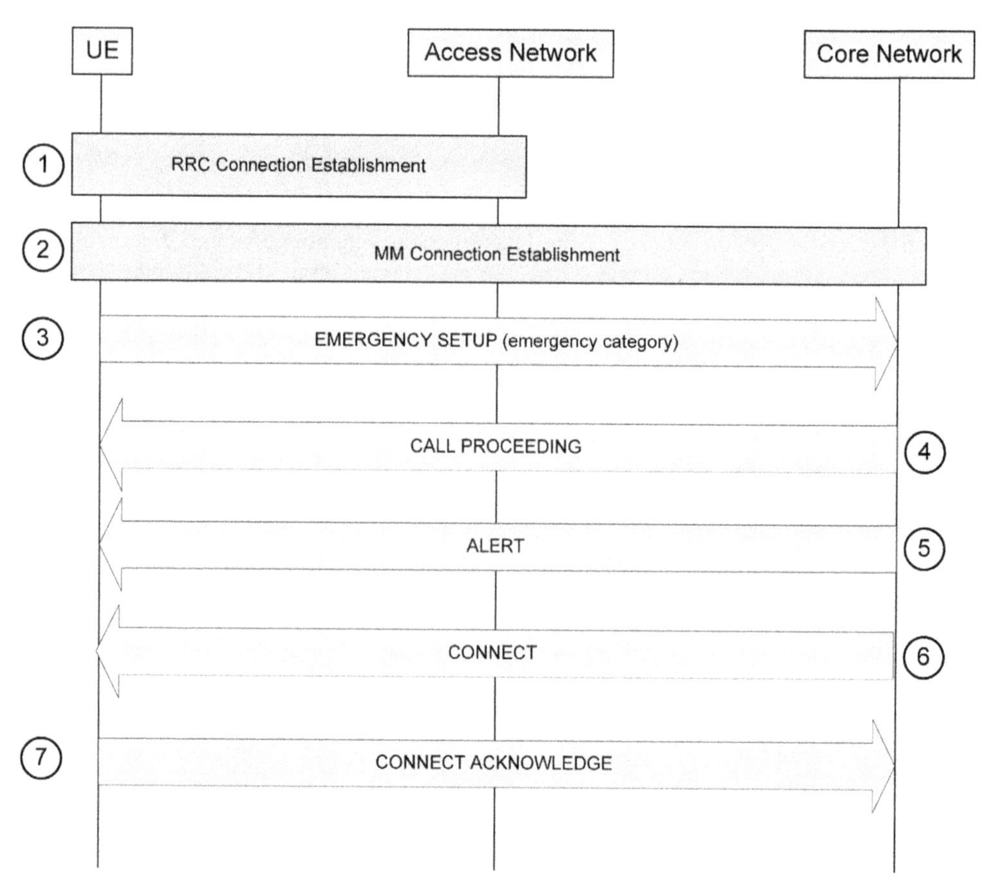

Figure 3.20 CS emergency call flow example (see notes in Table 3.2). (© 2009. 3GPP™ TSs and TRs are the property of ARIB, ATIS, CCSA, TTA and TTC who jointly own the copyright in them. They are subject to further modifications and are therefore provided to you "as is" for information purposes only. Further use is strictly prohibited.)

Compared to a normal SETUP, an EMERGENCY SETUP message can optionally contain emergency category information, which, as specified in 3GPP TS 24.008 [161], can be any of the categories listed in Table 3.1. Since the categories are represented as a bitmap in the EMERGENCY SETUP message, a combination of more than one category is also possible. If the MSC cannot resolve the required emergency category, then it will route the call to the default PSAP. This would be the case when a legacy MSC does not support the emergency category indication or a supporting MSC cannot resolve the indicated combination of multiple emergency categories.

eCall

The specification work to enable auxiliary data transfer during emergency calling had a strong requirement to make the eCall transparent to existing networks so that it can be deployed

Table 3.2 Notes to Figure 3.20

Step	Description
(1) – RR	Radio channel establishment using establishment cause "emergency call" to indicate emergency call. In limited service case "emergency call" is the only allowed establishment cause
(2) – MM	Mobility Management request for MM connection with CM service type "emergency call." Ciphering and integrity protection are not possible without a (U)SIM
(3) – CC	EMERGENCY SETUP contains no target information but it may contain emergency category
(4) – CC	CALL PROCEEDING acknowledges that the serving MSC successfully decoded the requested EMERGENCY SETUP and is processing the call
(5) – CC	ALERTing the remote user (at PSAP)
(6) – CC	The remote user (at PSAP) answers to the call (CONNECT)
(7) – CC	Originating UE acknowledges the call connection (CONNECT ACKNOWLEDGE)

rapidly. This narrowed down the variety of technical options that would fulfill the requirement, and in-band information transfer during emergency call was chosen [162]. An in-band modem is either integrated into the phone or acoustically coupled to it, and it preprocesses the auxiliary information before the speech codec. The call information remains transparent for the network, and the PSAP needs to decode the received in-band information to benefit from the auxiliary data. The MSC will be affected though, since it must be able to detect the initiation of an eCall based on the indication given by the UE in the service category of the EMERGENCY SETUP message.

Separate emergency categories have been specified for automatic and manually initiated eCalls. This is done to fulfill the requirement that it must be possible for the operator to route all eCalls to a dedicated PSAP that handles automatically initiated calls where the emergency caller might not be able to speak or provide any assistance.

As it is foreseen that a very high number of eCall devices will be deployed in vehicles in the near future, the location update requests from the eCall devices would become a significant burden on the network resources if normal mobility management procedures were used. An eCall device that only supports eCall is not expected to make or receive any normal calls, and therefore a so-called eCall inactivity procedure has been introduced. This inactivity procedure means camping on cells of the selected PLMN with location updating disabled to maintain radio silence unless an emergency call needs to be made.

Limited Service Case

When the phone is normally registered to the serving PLMN, it will keep constantly re-selecting the best serving cell among so-called suitable cells that are defined for each 3GPP radio technology. The suitable cell criteria are defined for GERAN in 3GPP TS 43.022 [163], for UTRAN in 3GPP TS 25.304 [164], and for E-UTRAN in 3GPP TS 36.304 [165]. Some of the non-3GPP access technologies also define their corresponding criteria for cell (re-)selection.

Even though the radio-specific parts of the suitable cell criteria may differ, the service-state-related logic is in principle common, and, rather than quoting each of them, they can be summarized as one. To qualify as a suitable cell, a cell must fulfill the following criteria:

- It is part of the registered PLMN (or if registration has not yet taken place, the selected PLMN), or a PLMN that is equivalent to the registered (selected) PLMN.
- The cell is not barred.
- The cell is not part of a forbidden area.
- Cell selection criteria for cell quality are fulfilled.
- Additional access-specific restrictions do not apply.

"Equivalent" PLMN above means another PLMN that a PLMN has indicated to UE as equivalent to itself for PLMN selection and cell selection purposes.

When the phone has no (U)SIM card inserted or the card has been deemed not valid, then it cannot have any registered network either, and thus all cells of any network are unsuitable. Lack of valid (U)SIM always leads the phone to limited service, but this is not the only limited service case that exists. Other cases occur when the phone has a (U)SIM card, but it is either not valid or has no subscription in the area where the phone resides. The difference between the two cases is that a valid (U)SIM means the user can be identified and even authenticated if the network has a connection to the HLR or HSS in the user's HPLMN. When an ME without a valid (U)SIM has to identify itself, it will use its IMEI as the user identity and that cannot be authenticated.

In limited service state the ME will camp on what we call an acceptable cell. "Acceptable" is again defined for each 3GPP radio technology. It can be considered a technically valid and available cell for which the UE has no subscription to use it. The acceptable cell criteria are also specified in the same 3GPP TSs as the suitable cell criteria, but the requirements list is much shorter:

- The cell is not barred.
- Cell selection criteria for cell quality are fulfilled.

In limited service the phone will keep re-selecting the best acceptable cell while maintaining radio silence. If it has to initiate an emergency call, it will always come as a "surprise" for the network without prior registration or knowledge of the whereabouts of that phone. Thus emergency calling must be indicated as the purpose for the connection at the initial radio resource layer request for the signaling channel, to override in the network the policing of the subscription-based access right to (this part of) the network.

Responding to unauthenticated malicious emergency calls and unintended "pocket calls" are a substantial part of the PSAP workload and therefore the regulators in several countries have taken steps to block either all limited service emergency calls or at least (U)SIM-less emergency calls. When necessary, the network can fulfill such regulatory requirement by rejecting a call attempt that is not allowed.

3.6.4 Emergency Calls in PS Domain

Before we start the description of the functionality, the involved entities are explained.

3.6.4.1 IMS System Elements

- *UE*. The UE is a mobile IMS phone with a (U)SIM.
- *ME*. The ME is a phone without a (U)SIM and thus with no subscription.
- *P-CSCF*. The Proxy-CSCF is the first point of contact in the IMS network, that handles the security association, checks the emergency calls, and maintains the path to the S-CSCF. The following functions are performed:
  - security context,
  - checking of emergency numbers in emergency calls that were not detected by UE,
  - policing allowed versus not allowed emergency requests,
  - prioritization of emergency calls,
  - routing emergency calls to the E-CSCF.
- *I-CSCF*. The Interrogating CSCF provides a gateway to the IMS part of the PLMN for incoming requests and selects a S-CSCF to process the request further. The following functions are performed:
  - gateway toward other networks,
  - selects the S-CSCF.
- *S-CSCF*. The Serving CSCF handles the call processing and routing in the IMS networks serving the initiating and destination user. Normally, the UE first needs to register with the S-CSCF, which works as its registrar. The following functions are performed:
  - registrar for both normal and emergency registrations,
  - call handling.
- *E-CSCF*. The Emergency CSCF is a specific CSCF that takes the call handling role from the S-CSCF for emergency calls, since the emergency calls require processing rules that are specific to emergency calls only. The following functions are performed:
  - determine the appropriate PSAP based on location information,
  - route emergency calls toward the PSAP,
- *LRF*. The Location Retrieval Function retrieves and stores the location of the UE that the CSCFs can query to obtain information about where to route the emergency calls.

3.6.4.2 IMS Architecture

Contrary to the local control model of CS calls, the IMS architecture is based on a home control model. Figure 3.21 shows the architecture for a normal outgoing call from UE A to UE B. Both are roaming to a visited network but have registered to the S-CSCF in their home network. Also in the roaming case the S-CSCF is located in the HPLMN and therefore a local breakout is needed for local services.

The security context between the UE and the P-CSCF as well as the path from the P-CSCF to the S-CSCF residing in the home network is established at registration time. When establishing a call, the UE sends an INVITE toward the P-CSCF, which routes the session invitation to the S-CSCF. Based on the home network of the destination, the S-CSCF then needs to route the call toward the home network of the target user, where it is up to the I-CSCF to assign an S-CSCF for call routing. The S-CSCF serving User B needs to determine whether to route the call internally in the home network or further on to the visited network toward B.

Owing to possibly different emergency numbers, the home network could not reliably detect emergency calls in all cases, and it has no information on the PSAPs in the visited country

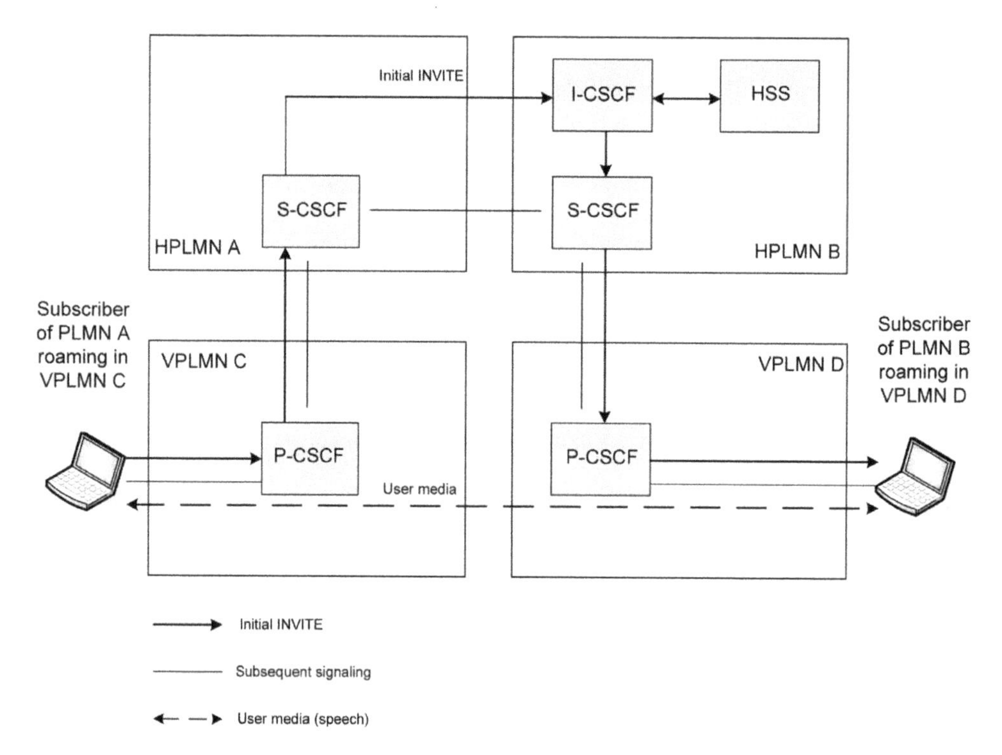

Figure 3.21 IMS: call between roaming subscribers. (© 2009. 3GPP™ TSs and TRs are the property of ARIB, ATIS, CCSA, TTA and TTC who jointly own the copyright in them. They are subject to further modifications and are therefore provided to you "as is" for information purposes only. Further use is strictly prohibited.)

either. Therefore, in roaming cases, a local breakout to the PSAP needs to be done at the serving network, to route the call to the local PSAP.

IMS registration is required before the UE can initiate a normal session, but, depending on the situation, an emergency session can be established with or without prior emergency registration. If the UE is already registered in its home network, then it does not need to perform an emergency registration prior to initiating an emergency session and can use the existing registration to set up the emergency call. If the UE is not IMS registered or is roaming to a visited network, then it shall execute an emergency registration. As an exception to this rule, an emergency session without registration has been specified. A UE without UICC and thus with no credentials cannot pass the authentication; it can attempt an emergency call without registration. It is up to the regulator in each country to determine whether or not such anonymous access is allowed. Emergency registration is not a subscription-based service, and thus barring and roaming restrictions are ignored. A Rel-9 UE has to perform emergency registration in all cases when it knows that the outgoing call is an emergency call.

Emergency registration, as shown in Figure 3.22, is done to register the UE with the S-CSCF in the home network even though the emergency call routing and resolving of emergency categorization will be up to E-CSCF to handle in the serving network.

The home control model where the serving S-CSCF resides in the user's HPLMN is challenging for any local services, including emergency calls. Emergency calls are not routed

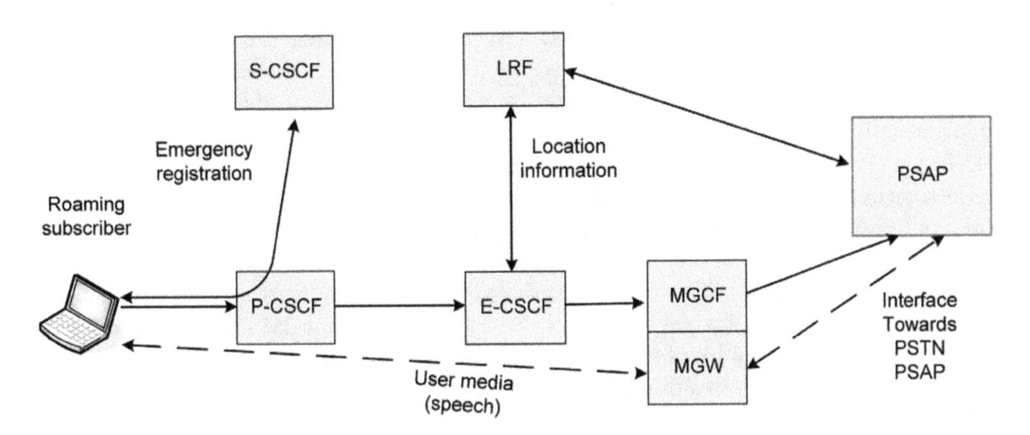

Figure 3.22 IMS emergency registration and call path. (© 2009. 3GPP™ TSs and TRs are the property of ARIB, ATIS, CCSA, TTA and TTC who jointly own the copyright in them. They are subject to further modifications and are therefore provided to you "as is" for information purposes only. Further use is strictly prohibited.)

internationally and the S-CSCF of the HPLMN would not have enough information to route the call to the appropriate PSAP in the visited country. The local breakout bypasses the S-CSCF by utilizing an E-CSCF in the serving network to provide some of the S-CSCF functionality for normal calls with the emergency call-related enhancements. The E-CSCF will resolve the routing of emergency calls. It is also responsible for obtaining the user's location with reasonable accuracy for call routing and then offering the call toward the PSAP that handles the emergency calls in the calling user's present location. The PSAP can be reachable via PSTN or IP telephony. A PSTN PSAP is reachable via the CS telephony network and therefore an MGCF is included in the signaling path to interwork between the IMS SIP and the CS call control protocols. An IP-based PSAP is directly reachable from the IMS without any protocol interworking, and the emergency call signaling is routed from the E-CSCF directly to the PSAP.

The first use of the location information is to identify the right PSAP. This may be followed by a second step at the PSAP, where the caller's location information can be used to dispatch emergency services to the right street address.

3.6.4.3 IMS Protocol Solution

Detection and Provisioning of Emergency Numbers

3GPP IMS can run on multiple mobile and fixed access technologies that can be used in different business ecosystems. Each ecosystem and access technology comes with its specific requirements that the upper layers must take into account. The 3GPP IMS protocol specification clearly identifies a common IMS part that applies irrespective of the access technology, and access-specific adaptation layers for each supported access technology are defined. Technically, it would be more correct to call IMS access adaptive rather than access-independent.

Both detection and provisioning of emergency numbers are mechanisms that are shared between CS and PS domain emergency calls. Hence, if a number is identified as an emergency number for a CS domain call, then an IMS UE must consider that number an emergency number also for IMS calls.

Call establishment procedures are obviously completely different, due to the use of two fundamentally different call control protocols. The CS call control is specified in 3GPP TS 24.008 [161] and the IMS call control is defined in 3GPP TS 24.229 [114] and uses IETF SIP.

Emergency calls were not supported at all in early IMS versions before Rel-7. An early Rel-5 or Rel-6 IMS phone shall not attempt to establish an IMS emergency call if it detects that the dialed number is one of the local emergency numbers but must initiate a CS emergency call instead. If the UE is not aware that the dialed number is an emergency number but the P-CSCF detects it as an emergency number, then a non-supporting IMS network will reject the emergency call with a hint for the UE to try a CS domain emergency call instead. This also saves the HPLMN from a difficult task of discovering the most appropriate PSAP in the visited country when the user is roaming.

Emergency Registration

Normally, emergency registration is a prerequisite for the initiation of an emergency call, unless the UE knows that it is already IMS registered in the HPLMN or when it is lacking the credentials to pass the authentication. The HPLMN matching is done by comparing the Mobile Country Code (MCC) and the Mobile Network Code (MNC) of the registered network with the MCC and MNC of International Mobile Subscriber Identity (IMSI) on the (U)SIM.

Emergency calls are supported in IMS from Rel-7 onwards, but the bearer level services are not specified for all radio access technologies. One of the main goals of 3GPP in Rel-9 has been to add the regulatory emergency call requirements also to 3GPP access technologies UTRAN and E-UTRAN PS domain procedures to allow IMS emergency calls in limited service state. In Rel-7 an IMS UE is able to initiate an emergency call in 3GPP networks. Such an emergency call can proceed, assuming network configuration allows, but will be processed as a normal call and only in normal service mode.

When a UE has detected an IMS emergency call attempt, it must first determine whether it is registered to the HPLMN or the VPLMN, based on the access technology registration status. Unless the UE is already registered to the HPLMN both in access and at the IMS level, it will have to either perform an emergency registration before attempting to initiate an IMS emergency call or attempt an emergency call without registration. The latter is not allowed in all networks. In Rel-9 this condition was changed to require the UE to skip PLMN matching and to perform emergency registration even while in HPLMN, if the UE is registering over a 3GPP access technology. This, of course, only applies when the UE is aware of all local emergency numbers and knows it is establishing an emergency call.

The IMS emergency session without IMS registration bypasses the security procedures and requires exception handling of many connectivity layer procedures. This case only applies on emergency sessions that are initiated from UEs without sufficient security credentials. Based on regulatory requirements, this limited service emergency call case is not allowed in all countries.

As shown in Figure 3.23, the emergency registration procedure uses the same messages as the normal IMS registration, but there are differences in the encoding of some parameters. The UE must include an "sos" URI parameter in the Contact header field to indicate to the network that it is requesting emergency registration. A supporting registrar (S-CSCF) that accepts the emergency registration request must echo the "sos" URI parameter back in its 200 OK

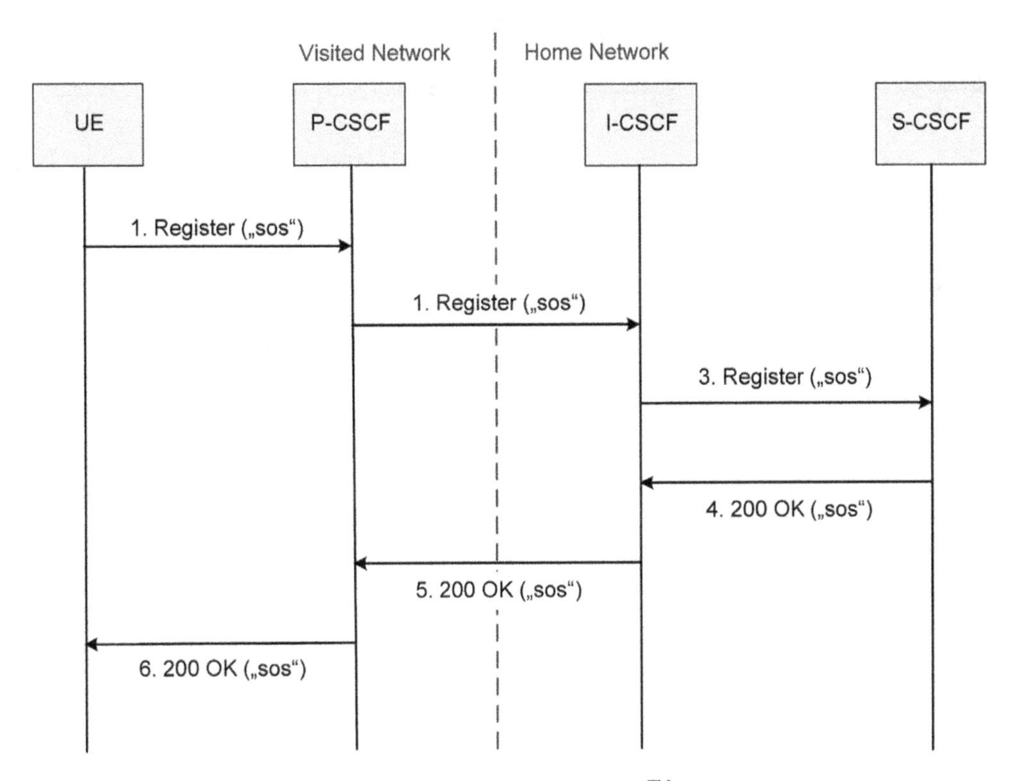

Figure 3.23 IMS emergency registration flow. (© 2009. 3GPP™ TSs and TRs are the property of ARIB, ATIS, CCSA, TTA and TTC who jointly own the copyright in them. They are subject to further modifications and are therefore provided to you "as is" for information purposes only. Further use is strictly prohibited.)

message. The registered contact address is therefore restricted for emergency use only. The UE must primarily include in the To and From header fields either the first pre-provisioned public user identity or the default public identity of the registered set obtained in preceding normal registration. If neither is available, then the UE will fall back to using a derived temporary user identity. Emergency registration must be independent of any other registration and any SIP signaling, and user media related with the emergency registration must be restricted to emergency calls only.

Once the emergency registration is completed with reception of a 200 OK message, the UE shall not subscribe to the reg-event package. Normally the UE performs re-registration when approaching the end of the registration lifetime, but in the case of emergency registration no re-registration is done, unless the emergency call is ongoing or the user initiates a new emergency call before the expiry of the emergency registration. No explicit de-registration is signaled from either side, but the emergency registration is left to expire.

Emergency Call

Contrary to the CS emergency call initiation with the EMERGENCY SETUP message, the IMS has not defined a specific SIP message for emergency session setup. Instead, the INVITE

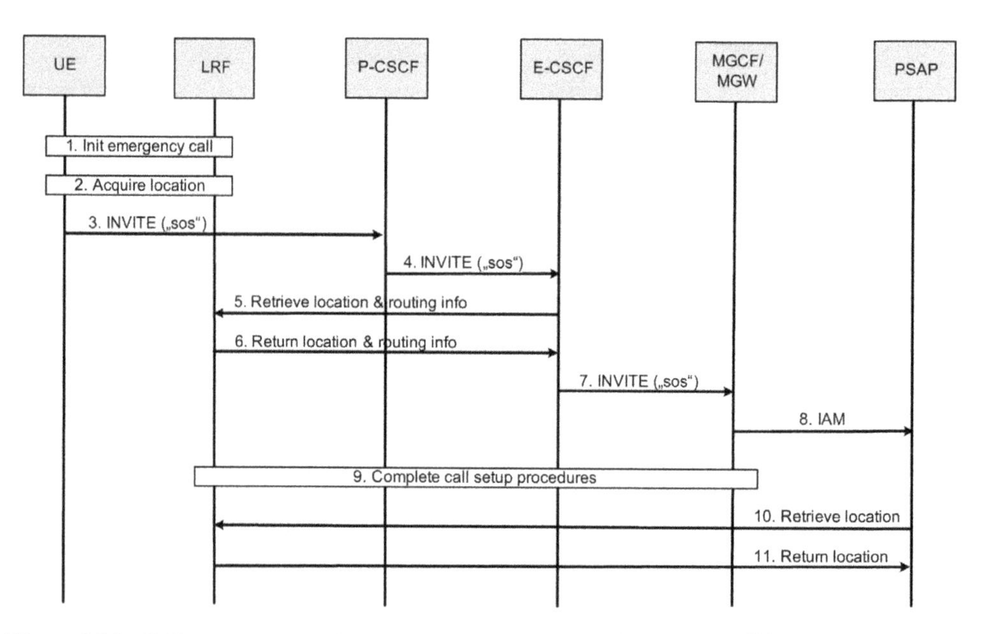

Figure 3.24 IMS emergency call flow (see notes in Table 3.3). (© 2009. 3GPP™ TSs and TRs are the property of ARIB, ATIS, CCSA, TTA and TTC who jointly own the copyright in them. They are subject to further modifications and are therefore provided to you "as is" for information purposes only. Further use is strictly prohibited.)

method is reused also in emergency session establishment, and the emergency use is indicated by using an emergency service URN with service type "sos" as request URI. The IMS emergency call flow shown in Figure 3.24 is based on 3GPP TS 23.167. In addition, the possibility of sub-services can be included. The variety of sub-service types is wider than in CS domain emergency call categorization, with RFC 5031 [11] already initially defining the following values, with the possibility to add more later if needed:

- Ambulance
- Animal control
- Fire
- Gas
- Marine rescue
- Mountain rescue
- Physician
- Poison
- Police

In addition to normal originating call establishment, some specific requirements apply to IMS emergency call initiation. The UE must include in the From and the P-Preferred-Identity header fields of the INVITE message a user identity that was registered in the emergency registration. The Request-URI and the To header fields must contain a service URN with the top-level service type set to "sos" and optionally a sub-service type indicating the emergency

Table 3.3 Notes to Figure 3.24

Step	Description
1.	UE detects the emergency call attempt
2.	UE obtains its location
3.	Session initiation with INVITE including "sos" indication and location information
4.	P-CSCF detects emergency session and forwards the INVITE to the E-CSCF
5.	E-CSCF queries the LRF to determine the appropriate PSAP
6.	LRF returns location and routing information
7.	E-CSCF routes the call toward the nearest PSAP
8.	PSTN PSAP requires protocol conversion from SIP to ISUP
9.	Call establishment procedures as in normal interworking case
10.	PSAP requests user location to dispatch emergency services
11.	LRF returns stored user location

category. The Contact header field must include the same "sos" URI parameter that was used in emergency registration. Also location information must be included if available to the UE.

If the UE was already registered with the home network and did not need to perform emergency registration, then it uses the already registered public user identities from the normal registration.

Even after registration, the UE can receive a 380 (Alternative service), in which case it attempts IMS emergency registration followed by an IMS emergency call in another network or a CS emergency call.

For an unregistered IMS emergency call, the UE must discover a local P-CSCF, and unprotected signaling related with emergency call establishment must be allowed since no authentication can take place. An unprotected anonymous INVITE for an unregistered IMS emergency call must still indicate the "sos" URI parameter and the service URN with a service type "sos" to identify it as an emergency call. The missing public user identity is replaced with the access-specific equipment identity in the P-Preferred-Identity header field. For 3GPP access technologies, the equipment identity is the International Mobile Equipment Identity (IMEI), which the UE is only allowed to use as its mobile identity if it has not got any subscription-based identity.

The P-CSCF on the network side also keeps a list of emergency numbers, so that it can detect emergency calls from those UEs that are not aware of local emergency numbers. It will also need to detect the emergency service URN from those UEs that have detected the emergency call attempt. The P-CSCF selects an E-CSCF, a network element serving specifically emergency calls, and routes the emergency call toward it. The analysis of emergency numbers takes place first before taking any other routing decisions.

When the E-CSCF receives an INVITE for an IMS emergency call, it first needs to determine the user's location that has been provided. This location determination is for routing to the appropriate PSAP and not for dispatch of first responders. This routing decision can be made by the E-CSCF (based on a local policy database) or the E-CSCF can query a LRF. If the next hop routing decision points to an IP-based PSAP, then routing the emergency call toward such PSAP would be just normal SIP routing. If the call needs to be routed to a PSTN-based PSAP, then the E-CSCF must be prepared involving gateway functions.

Several enhancements to IMS were made for Rel-9 when full emergency call support in the limited service state was added to the EPS and the GPRS connectivity layers. GPRS refers to the 2G and 3G General Packet Radio Services. In this case, however, the Rel-9 enhancements are more applicable to the 3G GPRS.

3.6.4.4 PS Domain Connectivity Layer

Emergency Support over Different Bearers

3GPP Rel-5 did not offer any emergency support at the GPRS bearer level. A GPRS network would not reject attempted emergency calls if the IMS layer can handle them, but there is no emergency-specific support for priority or limited service calls.

Emergency call support was added to IMS in Rel-7, but in practice it was possible to support IMS emergency calls over WLAN access only.

The major revolution in 3GPP Rel-8 was the introduction of EPS, which consists of the new E-UTRAN radio and the PS-only Evolved Packet Core (EPC). The initial EPS release did not support the regulatory exceptions for limited service emergency calls; those were added in 3GPP Rel-9.

EPS supports emergency bearers to distinguish the emergency session signaling from the normal session signaling, both for prioritization and for overriding of subscription-based access rights checking. In the case when the UE knows that the session being initiated is for an emergency, then emergency EPS bearer context must be used. There can be cases when the UE is not aware of a local emergency number, and in that case the emergency session can proceed, but then neither prioritization nor limited service state calls are possible.

One of the main enhancements in 3GPP Rel-9 was the introduction of limited service access to the 3GPP PS domain, and the IMS adaptation layer for GPRS and E-UTRAN was enhanced to specify the usage of emergency bearers for emergency services. When the UE detects an emergency call attempt, it will use emergency bearers for both IMS signaling and media plane. Rel-9 allows the support of also limited service access, where required by the regulator.

If the UE does not have sufficient security credentials to pass the connectivity layer authentication, then it is not likely that IMS authentication would succeed either. In that case unregistered emergency session initiation procedures will be used. Lack of security context then becomes an issue and, in addition to the anonymous access in bearer level, the IMS design will need to make an exception for policing signaling protection. Normally, only protected signaling via the established security context is allowed, but unprotected signaling needs to be allowed for an emergency call. This is controlled by allowing emergency session-related signaling using the same IP address and unprotected port number as was used for session initiation but no other signaling will be allowed. Since IMS registration is omitted in this case, the UE needs to obtain an IP address, establish connectivity, and discover a P-CSCF to convey the session setup signaling to IMS core.

PS Domain Emergency Calls in GPRS and EPS

In the early days of EPS work it was discussed whether the EPS Mobility Management (EMM) signaling layer should be specified and designed as a subset of the existing GPRS GMM or as a standalone protocol entity. The evaluation of the alternatives in Rel-8 led to standardization of the EMM and the EPS Session Management (ESM) protocols as standalone protocol entities. Both are identified with their own protocol discriminator. Subsequently, it was logical also to start a new specification, 3GPP TS 24.301 [166], to contain the EMM requirements alongside 3GPP TS 24.008 [161], which contains the GPRS GMM specification. Extension of the 3GPP packet connectivity protocols to support IMS emergency calls meant mirroring many of the changes to both GMM and MME.

The emergency attach procedure when the UE is in limited service state is one of the requirements that applies to GMM and to EMM. Anonymous access was initially considered as part of the early GPRS specification versions, but that work was not completed until the PS-only EPS system in Rel-8 added more requirements for limited service access to the PS domain. Emergency registration was then added in Rel-9 to allow emergency calls when normal registration is not possible.

The scope of the attach procedure is broadened to cover emergency attach. The UE must indicate an emergency as the reason in the attach type. The network can use this indication to suppress subscription checking, security procedures, and prioritization.

The UE must identify itself for any attach, regardless whether it is a normal or an emergency attach procedure. The mobile identity can be a temporary identity assigned by the network, a subscription-based IMSI read from the (U)SIM, or the identity of the ME without a (U)SIM. A temporary identity, a TMSI in GPRS and a GUTI in EPS, is used primarily, if one has been assigned to the UE. The IMSI is used only if no temporary identity exists.

When a phone (ME) requesting emergency attach has no (U)SIM and therefore none of the above identities, then as the last resort it can use IMEI as its identity in emergency attach. For this case the GMM and the MME protocol entities of the network can be configured to skip authentication of those phones that do not have the security credentials to pass it.

If the network does not support (U)SIM-less emergency calls, then, based on seeing the IMEI as the mobile identity, it can reject the attach and indicate in the cause value that an IMEI-based attach is not accepted in this network.

For emergency use the network can accept attach procedure or a GPRS Routing Area Update (RAU) or a EPS Tracking Area Update (TAU) from an area that is forbidden for the UE, but in that case the corresponding forbidden list in the UE is not affected. This is logical as, even though the UE can exceptionally gain access to the restricted area, it would still not be allowed normal service in that area.

Normally, a successful registration causes an update of the forbidden list. For example, the PLMN or the area within a PLMN where registration was successful would get removed from the forbidden list. As an exception to this, a successful emergency attach, RAU or TAU to a forbidden area does not affect the corresponding forbidden lists. This makes sense, as the emergency attach can bypass security procedures and thus it cannot guarantee subscription-based access rights to the same PLMN or the area within PLMN.

If the UE has successfully registered to the network and has emergency bearer resources when moving to an area where the UE has, as part of the subscription, no normal services, then during the RAU or the TAU the network must tear down all normal bearer resources and must only maintain the emergency bearer resources. This case is related to regional roaming restrictions and the UE roaming outside of its regionally provisioned area in an allowed PLMN. The logic behind it is that an emergency call does not allow the user to extend to other services or to areas that are not allowed otherwise.

The EPS attach procedure can be combined with the assignment of a default bearer by bundling EMM and ESM messaging together. GPRS attach and PDP context establishment on the other hand are separate procedures. The network can detect the emergency case based on request type "emergency" that is included as part of EPS and has been added to the GPRS procedures.

Bearer resource modification is not allowed for any PDN connection that has been established for emergency services. If the network receives such a request from the UE, it must reject it.

Emergency calls are protected against network-initiated detach. If the UE in GPRS has an emergency PDP context active, then a network-initiated detach request by the Home Subscriber Server (HSS) causes all normal PDP contexts to be deactivated. The same principle holds also for EPS contexts.

The policy for selection of an access domain, if CS and PS service is available, varies over different 3GPP releases. Up to Rel-9 the choice of the emergency call domain is very straightforward: use the CS domain when possible. But the full IMS and PS emergency call capability changes this policy and now the phone will have two domains to choose from.

Femto-cell operation is covered in 3GPP specifications via home cells and Closed Subscriber Group (CSG). The access to home cells is defined in Rel-8 and enhanced in Rel-9. Emergency calls are possible via a CSG cell. A UE that is emergency attached and has established an emergency EPS bearer via the CSG cell shall inhibit manual CSG selection to another CSG cell as long as the UE is attached for emergency services. If anonymous access to the EPS is not allowed, then the MME uses the reject cause #5 "IMEI not accepted" to prohibit an attach request with an IMEI as the mobile identity.

The reject cause #11 indicates that the emergency attach is not allowed in this network.

EPS-Specific Emergency Call Topics

In addition to the common PS domain requirements, additional requirements apply to EPS. The main principles are aligned between GPRS and EPS, but some of the fundamental design differences force also corresponding special case handling for EPS emergency calls.

The EPS security context requires a USIM, and SIM security is not supported. Emergency calls with a SIM inserted in an EPS capable UE must still be provided, which creates a new exception case that never existed before. Improved security was also introduced when adding 3G services, but the difference there is that UTRAN access is allowed also with a SIM.

Emergency calls are treated with higher priority and processed separately from the normal call cases due to different security requirements. A UE must not request multiple parallel PDN connections for an emergency, and a UE in limited service is not allowed to request another parallel PDN connection if it already has one PDN connection for emergency purposes.

EPS does not have a parallel CS architecture. Unlike in UTRAN, in E-UTRAN there is no support for CS and PS domain services on the same cell. This difference also has relevance for emergency calls. If a CS domain emergency call is needed, then it will require a switch from the E-UTRAN to the GERAN or to an UTRAN cell. As such, it is no longer only a higher-layer application call control decision.

When the network accepts an attach or a TAU from a UE, it can include an EPS network feature support indication in the accept message. This is the way for the serving network to indicate whether IMS voice service is supported and whether emergency bearer services are available. The IMS voice services indication is used by the UE to determine the default domain for outgoing emergency calls, as specified in 3GPP TS 23.221 [167].

Domain selection is based on the current registration status information and UE preferences. The following issues affect the UEs selection of a domain for emergency call purposes:

- CS and IMS registration state;
- possibly ongoing calls in either domain;
- pre-configured user and operator preference; and
- UE and network voice emergency call capabilities.

When any of the capabilities required for an IMS emergency call are missing, then the CS domain must be used instead. When both domains are possible based on the registration state, the known UE capabilities and the received indications of network support come into play. For this case it is possible for the operator and the user to pre-configure preferences to influence the decision between the CS and the PS domain.

Limited Service Case

Basically the same limited service cases as in CS domain exist, namely one with a valid (U)SIM and one without a valid (U)SIM. The latter case cannot be authenticated. If there is no roaming agreement between the VPLMN and HPLMN, then even the former case cannot be authenticated.

Additionally to these alternatives, E-UTRAN creates another exceptional case, since the 3GPP Rel-8 and later specifications for EPS specify that E-UTRAN access is only allowed with valid USIM. Thus, a valid subscription with a SIM that allows access to other access technologies of the selected PLMN is not sufficient for normal services in EPS. Emergency calls without a valid subscription are allowed in some countries, and so the system design must also consider this case.

Allowing emergency calls in limited service forces bypassing of security mechanisms. To prevent fraud, usage is restricted to valid emergency cases only.

In GERAN and UTRAN the solution for emergency calls from ME without sufficient credentials was simply to omit authentication and ciphering. In UTRAN, integrity protection is skipped. In E-UTRAN there can be no authentication either, but the handling of this case differs from the GERAN and the UTRAN procedures. After skipping the authentication, the security procedures continue as normal but a null algorithm is used.

If the UE has no security credentials and authentication has been omitted, the algorithms and keys are not available for security mode control. However, security mode control is not skipped; rather, a null algorithm with keys that are generated locally by the EMM entities of both the UE and network is employed. A null algorithm is used for both ciphering and integrity protection. Integrity protection is not specified for GERAN and even ciphering is optional, even though usually supported and activated in live networks when possible. After the establishment of a security context, ciphering and integrity protection are normally mandatory in UTRAN and E-UTRAN. The UE will then require both ciphering and integrity protection, but for the emergency case the UE must accept the null algorithm in SECURITY MODE COMMAND.

If an authentication failure occurs during an emergency attach, then the MME is allowed to make an exception to processing of the failure and allow the emergency attach procedure to continue. Similarly, the EMM entity on the UE side must be prepared for the attach procedure to continue even after the UE has detected an authentication failure. If the network is configured to skip the authentication, then it can respond with a SECURITY MODE COMMAND, which the UE must consider as implicit acceptance of authentication. The command will indicate a null algorithm, since in this case USIM-based security is not available.

Mobility Between the PS and the CS Domain

If IMS support is not available either in the phone or in the network, or if CS service is preferred, then a UE that is camping on an E-UTRAN cell can request CS fallback for an emergency call. To prepare for a possible CS fallback, the UE must perform a combined attach to the PS and the CS domain. If the need for a CS fallback arises later on, then the UE will send a service request to the MME. Once on a GERAN or an UTRAN cell, the UE will initiate the MM connection establishment for the CS emergency call. The CS fallback is intended as an interim mechanism for speech service support in E-UTRAN networks until IMS speech services are supported.

Handovers of calls between GERAN and UTRAN are commonplace, also for emergency calls. PS domain emergency calls in EPS add further complexity. The main issues are transfer of the call between the PS IMS and the CS, as well as the security context management, which normally requires a USIM in the EPS.

If an IMS emergency call has already been started in the PS domain, then radio level handover is not sufficient to ensure mobility between the CS and the PS domain. At the same time it is necessary to start the call control protocol. This transfer is called Single Radio Voice Call Continuity (SRVCC). SRVCC from the E-UTRAN to the UTRAN requires the handover command also to carry sufficient information to reconstruct the CS mobility management and call control protocol state.

The session data of all IMS emergency calls is stored by the application server that is responsible for the VCC functionality until the end of the emergency session. This anchoring of the call is done to prepare for the possibility that the call might need to be transferred to the CS domain. When session transfer is needed, the MME initiates a SRVCC toward the MSC server, which prepares the CS part of the transfer and the handover command is sent to the UE. Based on this radio layer signaling, the CS MM entity moves locally to MM CONNECTION ACTIVE state and the CS CC entity moves to the Call Active state.

EPS security is different from the security offered by UTRAN and E-UTRAN. Therefore, the security context is not transferable as such between the 3GPP access technologies. It is necessary to re-generate security parameters during SRVCC between E-UTRAN and GERAN/UTRAN. In handover from E-UTRAN to UTRAN, syntactically correct UTRAN ciphering and integrity protection keys are derived from the EPS security key so that none of the confidential security parameters need to be sent over the radio interface as part of the session transfer procedure. If the authentication has been skipped in E-UTRAN due to lack of a valid USIM, then instead of seeding the key derivation with the locally created keys for a null algorithm, the already existing GERAN and UTRAN practice to omit the security procedures can be used.

The Voice Call Continuity procedures are also defined for other access technologies, such as GERAN and 3GPP2 access.

3.6.5 *Identified Overload Problems*

3.6.5.1 Introduction

If a major disaster, like an earthquake, a tsunami or a terrorist attack, hits a densely populated area, then the call volume is so high that not all calls can be served. Congestion can occur at any point in the system, from the radio interface to the PSAP.

Without any congestion control mechanisms, the telephony system would be vulnerable to uncontrolled overload, where hardly any requests get served. Without prioritization between requests, even the calls that complete may not be the most important calls.

Therefore, both the telephony system and the PSAP operators support a congestion handling mechanism. In principle, the congestion control mechanisms are general and apply on any overload situation. Since major disasters are one foreseeable source of congestion, the overload control mechanisms are considered here from the emergency services perspective.

3.6.5.2 Network Overload

Multiple mechanisms are specified to control the phone's access to the network:

- cell barring;
- access class barring;
- barring of emergency calls in cells;
- domain-specific access control;
- paging permission access control; and
- service-specific access control.

The procedures listed are means to control and restrict the permission for a phone to attempt a normal call as well as an emergency call. Since the phone is unaware of the network load situation, congestion control via barring procedures is controlled by the serving network. Whenever any of the above restrictions apply, it is an indication by the network to the phone that emergency calls are not possible.

Passing these hurdles and being allowed to make an emergency call attempt still cannot guarantee success. The networks obviously will do all they can, including prioritizing emergency calls higher than normal calls in resource allocation. Major disasters can lead to such a high number of emergency calls that they alone can congest parts of the system. Radio propagation problems that also cause normal calls to fail can impact emergency calls. If the phone is camping on a weak distant cell, the radio link might not be good enough to provide successful radio connection. Being able to initiate an emergency call is only permission to attempt the call. The only way to know whether it will succeed or not is to try.

The most coarse level cell-based network access control is directly related not to emergency calls but to network maintenance. It allows the operator to deny any normal phones from camping on a cell that is being (re-)configured and thus is not in operational use. Only special test phones that are designed to support cell barring override are able to camp on a barred cell so that the operator staff can use it for their trials. Normal phones do not consider such a cell as a candidate for cell selection or cell re-selection procedures. Therefore, neither normal nor emergency calls will be processed in barred cells.

Access class barring allows or denies the access to a cell for the nominated groups of users. The user's access class is stored on the (U)SIM data file "Access Control Class" in two octets as specified in Section 4.2.15 of 3GPP TS 31.102 [168]:

- Bits 0–9: Normal access classes
- Bit 10: Not stored on (U)SIM, see access class 10
- Bit 11: For PLMN use
- Bit 12: Security Services
- Bit 13: Public Utilities (e.g., water/gas suppliers)
- Bit 14: Emergency services
- Bit 15: PLMN staff

All users should have one of the normal access classes marked in their (U)SIM. Ideally, an operator should distribute users evenly to the available normal access classes. That way it is possible to control the access of the entire subscriber population by 10% groups and it is also possible to alternate the blocked access classes over the time until the congestion disappears.

Most users have got only one access class in the bit range of the normal access classes, namely from 0 to 9. Some users may have been granted a special access class in the bit range from 11 to 15. Normal access classes apply in any PLMN but the special access classes 11 to 15 are only valid in the HPLMN and the EPLMN. Special access classes 12 to 14 apply in any PLMN of the home country. Emergency services staff, people maintaining the communication network, and public authorities can also have one or more of the special access classes in the bit range from 11 to 15 in addition to their normal access class.

The serving network signals over the radio interface those access classes that are allowed to perform uplink access. Uplink signaling for any purpose including emergency calls is permitted only to those UEs that have got at least one allowed access class.

In normal conditions, all access classes are allowed uplink access to the network, but when congestion is detected, the network administrator has the option to start turning off one or more normal access classes. This is a very powerful method to protect the network against overload by granting access to the network only to those users belonging to certain user groups. In an extreme case, all normal access classes can be barred to allow calls for only those UEs that belong to one of the special access classes.

In Rel-8 the access class barring is enhanced for E-UTRAN to allow a certain percentage of phones to access the network. This is achieved by the network indicating the probability for barring as a percentage. Before attempting uplink access, each affected phone must draw a random number. If the random number is higher than the network-given threshold, then access is allowed. If the random number is lower than the given threshold, then the phone has to start a back-off timer to wait for the next opportunity to draw another random number. It is expected that the random number lottery will be subject to phone conformance testing.

The (U)SIM data file definition, defined in 3GPP TS 31.102 [168], defines access class (AC) 10 as a special case. In addition to the access class information, the network can send an indication whether emergency calls are allowed or not by using AC 10. It is not possible to store AC 10 on a (U)SIM and technically it is not even an access class at all. Instead, it should rather be seen as an attribute telling how each access class must behave in terms of emergency calls. If AC 10 barring is enabled, then emergency calls are not allowed for any normal access classes. In this case MEs without a (U)SIM and thus with no access class at all are also barred.

If the UE is configured with special access classes, then emergency calls are allowed despite the AC 10 barring.

Domain Specific Access Control (DSAC) was added in Rel-6 with the intention to distinguish between CS and PS domain uplink access. The logic behind the service is that, when a CS domain is becoming congested due to some major disaster resulting in a high number of calls, it may still be possible to communicate in the PS domain. Technical implementation is straightforward: the serving network broadcasts the domain-specific barring information as part of the system information telling the barring status for both CS and PS domains independently. Each one of the normal and special access classes can be barred on either the CS or the PS domain or both, independently of each other. Uplink access is forbidden for a UE with all access classes barred in the domain that would be used for the connection.

Paging Permission Access Control (PPAC) was added in Rel-8 with the intention to give the UE the right to respond to paging requests for terminating calls when a mobile originating calls is barred. The barring status is indicated to the UE by the serving network via cell broadcast messages. PPAC can apply on either the CS or the PS domain or both and it also indicates the barred and allowed access classes.

Service Specific Access Control (SSAC) is a Rel-8 feature and it affects PS domain IMS calls only. It reuses the proportional Rel-8 access class barring giving an indication of the relative percentage of the barred population of phones that must be affected by the barring. In SSAC the restriction does not occur at the radio layers irrespective of the service, but it only affects the IMS MMTel service. The radio layers receive an indication of the barring probability that must affect the MMTel services, but it is up to the IMS layer to apply the indicated restriction for attempted outgoing MMTel calls.

As specified in 3GPP TS 25.304 [164], none of these access control mechanisms affect the cell selection or cell re-selection procedures. This means that the phone is required to camp on the best cell according to the cell (re-)selection criteria irrespective of the access control information, except for cell barring, which removes the cell from the cell selection candidates list. Any possible barring information is checked before requesting radio connection, or in the case of PPAC, before responding to paging request. The UE is not allowed to hunt for a cell that would allow access to it, and for good reason, as that would cause serious problems in radio network planning and load control. SSAC does not apply to IMS emergency calls but it can be used to adjust the available services to the network capacity and to ensure sufficient network capacity for emergency calls.

3.6.5.3 PSAP Overload

Major emergency situations can push PSAP and the emergency rescue resources beyond their limits. Another common source of substantial amount of workload is unintended and inappropriate emergency calls.

Theoretically the emergency services including PSAPs could be scaled to cope with any foreseeable workload. In practice, it is possible to organize the work of both PSAPs and dispatched emergency units for seamless cooperation and load sharing, both between the different services and between different areas where the PSAPs are organized regionally. Completely bulletproof configurations that can handle any foreseeable workload are rather expensive and quite rare in real life.

Inappropriate and unintended emergency calls can take a very high proportion of the PSAP capacity. As an example, the Finnish emergency services authorities have reported that the average rate of non-emergency calls has exceeded 20% [169] since the reorganization into 15 regional PSAPs in Finland in 2006. This is just one example case, and even higher figures have been reported in other countries. The received non-emergency calls are split further into unintended calls that were made by mistake, silent calls, and inappropriate or malicious calls. Owing to the nature of emergency services, all of these take time, as the PSAP staff have to process every call.

In many countries the majority of emergency calls are nowadays mobile calls and therefore these calls also dominate the PSAP workload. PSAPs are monitoring their own performance based on predetermined targets for the average time to respond to calls, the length of waiting times for callers who have given up waiting, and so on. The high number of non-emergency calls has led some national administrations to ban the emergency calls from phones that are in limited service state, as explained earlier in this section. Limited service state means usually that the phone has no (U)SIM at all but it can also be due to the user's HPLMN having no roaming agreement with any of the available PLMNs. In either case, the phone has no subscription to operate in normal service mode in that area.

The logic behind blocking emergency calls in limited service is that, while it theoretically seems reasonable to allow any emergency calls in order to avoid leaving anyone outside of the service, in practice the misuse of limited service emergency calls, for example, for test calls by stolen phones, seems to put more lives at risk than they save.

If the regulator requests blocking of limited service emergency calls, the operators can enforce this policy by rejecting emergency call attempts from phones that have not been able to register to the network normally. In CS domain emergency calls there is the option of blocking the service in the VMSC if the requested service is not allowed. In IMS the same thing still applies in most practical scenarios and the request will be rejected in the SGSN or the MME.

4

Deployment Examples

Chapters 2 and 3 illustrated international standardization efforts with regard to location, location protocols, and architectures. Normally, one would assume that these standardization efforts are then implemented and deployed in various countries. In this chapter we want to illustrate that the road to deployment is, unfortunately, not so simple. Various countries have started their own efforts to develop transition architectures.

With four examples, we have tried to illustrate developments to move toward VoIP emergency calling and an IP-based emergency services infrastructure. Trying to capture different national and state-level deployments turned out to be quite challenging. First, there are not many who know the technical details or who are interested in writing about them. Second, the status (in terms of technical details, and deployment progress) of the project changes frequently, therefore making them less ideal for a book.

We nevertheless decided to include a chapter about deployment examples in this book to illustrate what various countries have planned to do (from a technical point of view), and what happened to these efforts. We will see similar developments in other regions in the near future as well, and there are, in our opinion, lessons to be learned.

In section 4.1 Jakob Schlyter illustrates plans for interconnecting VoIP providers with the Swedish emergency services network. For this interconnection, a few questions arise:

Q1 How does the VoIP service provider know which PSAP to route the call to?
Q2 What protocols are used for interconnection between the VSP and the emergency services network?
Q3 Where does location information for dispatch come from?

In the planned deployment in Sweden, these three questions have been answered as follows.

A1 A VoIP provider gets the address of the PSAP pre-configured (out of band). This is considered to be an interim solution before LoST deployments catch up.
A2 For the interconnection, a detailed set of technical requirements using SIP are defined, as described in detail by Jakob.

Internet Protocol-Based Emergency Services, First Edition. Hannes Tschofenig and Henning Schulzrinne.

A3 For detailed location information, the expectation is that the location protocols developed in the IETF Geopriv Working Group, as described in Chapter 2, will close this gap and either allow the end device to provide high-quality location information or to allow a PSAP to retrieve location information from a location server in the access network.

Today, the regulatory situation in Sweden only places obligations for providing emergency services support onto those who offer voice communication using E.164 numbers. Even then there are ways to escape that obligation by indicating that no emergency services support is offered to end customers. While there are no objections against the described architecture in Sweden, the deployment is still work in progress due to a lack of funds.

The architecture developed in the United Kingdom by the NICC, as described by Ray Bellis in section 4.2, starts with the assumption that only UK-based VSPs will be covered. It is also interesting to note that the interactions between the end device and the VSP are outside the scope of the standardization work. A different way of seeing this requirement is that the solution needs to work even with devices that do not contribute functionality to the successful emergency calling other than the voice call itself. A commonly used term for this design assumption is "interworking with legacy devices." Furthermore, the UK emergency services model operates stage 1 and stage 2 PSAPs. Stage 2 PSAPs shall be left unmodified with the presented solution, and even for the stage 1 solution no changes to the PSTN call handling infrastructure are assumed.

These are quite a number of constraints to work with and, as Ray explains, various components from the NENA i2 architecture have been reused for this purpose. Owing to the nature of the UK emergency services infrastructure, routing emergency calls from a VSP to one of the stage 1 PSAPs is fairly easy. As long as you know that you are in the United Kingdom, routing to any of the stage 1 PSAPs will get you to the stage 2 PSAPs for help. The protocol for interconnection of the voice signaling is still using legacy PSTN technology; changes are focused on the control plane protocols required for the location lookup only. The location information for dispatch is then provided by the location servers in the access provider networks.

At the time of writing, the UK architecture had not yet been deployed but all specifications are completed. The main obstacle for deployment had been lack of appropriate regulation that puts incentives on various stakeholders, such as ISPs and VSPs, to implement and deploy the described infrastructure components.

In section 4.3 Guy Caron describes the efforts in Canada for developing a IP-based emergency services solution, which builds on the NENA i2 design. The communication interaction between the end device and the VSP is again considered outside the scope, and interworking with legacy PSAPs is also assumed. The UK and proposed Canadian models show differences in the procedure for discovering the location servers: the UK model relies on the DNS, and the proposed design uses the WHOIS database by the American Registry for Internet Numbers (ARIN).

The developed Canadian i2 architecture was rejected after a long consultation process in 2010. Guy explains the history of the emergency services work in Canada and offers an analysis of the decisions. We believe that the lessons that can be learned from the activities in Canada are also applicable to other countries. In the meantime, over-the-top VoIP providers are left with either no solution (if they are located outside Canada; similar to the situation in the United Kingdom) or with an interim solution.

Finally, in section 4.4, Byron Smith describes an effort to build a new emergency communications infrastructure that has a scope different from the previously discussed deployment examples, which focused on providing access to emergency services infrastructure by VoIP emergency callers. Byron focuses on a development that is now quite common in the emergency services field: legacy emergency services networks get replaced with IP-based networks for cost-saving reasons.

With regard to the lessons that can be learned, many of the national transition architectures required a lot of work in developing the architecture, and despite their tailored nature they progressed to deployment extremely slowly. Typically, it is very difficult to get all relevant stakeholders involved in the development of these transition architectures; consequently, these architectures suffer from various limitations. Often, the end device manufacturers are absent from the discussion since they care more about getting involved in a standardized solution at an international level, instead of country- or region-specific efforts. Since the regulatory framework does not require many application and VoIP providers to offer emergency services support, the amount of VoIP emergency calls that need to be supported is often rather minor. Requirements are created by those who do not have to fund the resulting deployments. Typically, there are substantial challenges in obtaining the necessary funding, especially for access network operators to deploy location servers to serve IP location. Owing to the custom-made design, off-the-shelf equipment is often not available, which increases the costs even further.

Meanwhile we have also seen various successful deployments of NENA i3 in different parts of the United States, and also IP-based emergency services network deployments in Europe. We hope to share the experience with those latest developments online via our webpage.

4.1 Emergency Calling in Sweden

Jakob Schlyter
Kirei AB

4.1.1 Introduction

This section describes an architecture for providing Voice over IP (VoIP) connectivity to the Swedish emergency services PSAP. The primary purpose of this service is to allow a Voice Service Provider (VSP) to be able to connect to 112 services over a native VoIP interface. A VSP may be any company, service provider, organization or residential user that runs a VoIP service. The Swedish emergency services model therefore follows the general directions of emergency calling using the Internet, as described by IETF ECRIT WG [121] in RFC 5012 [7]. The SOS Network Termination Point (SOS-NTP) is the IP-enabled interface of the PSAP that will receive VoIP emergency calls. The SOS-NTP supports the receipt of emergency calls using SIP (Session Initiation Protocol).

This section includes a description of the available interfaces, procedures for routing of emergency calls in different call cases, information to be transferred in the emergency call,

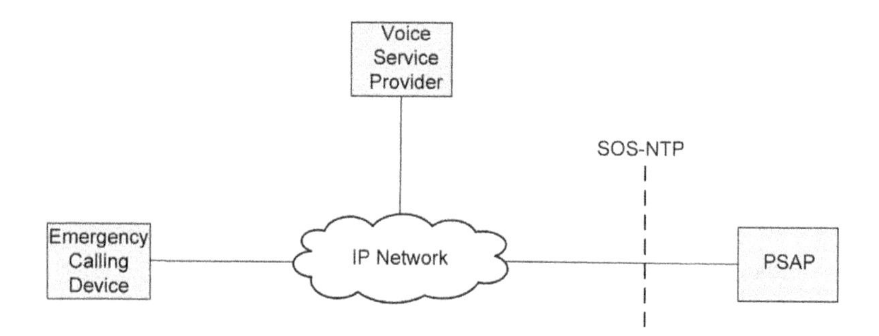

Figure 4.1 SOS-NTP interconnection point to the PSAP. (© Jakob Schlyter, 2012.)

and testing of the emergency call service. Information about the VSP's internal network operation is out of scope. How IP packets are transported between the emergency caller, the VSP and the PSAP are out of scope as well.

This section is concerned with technical issues and is assumed to be used by VSPs in their interconnection to the PSAP, as shown in Figure 4.1 (with the SOS-NTP marker).

Even though different types of traffic flows and call scenarios exist, they basically fit the generic case shown in Figure 4.2.

1. An emergency caller uses an SIP User Agent (UA/Phone) to initiate an SIP call to reach the PSAP (e.g., using sip:112@vsp).
2. The VSP forwards the emergency call to a predefined PSAP SIP URI (sip:112nnn@112.se in our example).
3. The SOS-NTP accepts the incoming SIP call and forwards it to the PSAP.

Note: nnn in the SIP URI equals municipality identity codes and 112.se is the domain used for the emergency services in Sweden.

4.1.2 Overview

The operation of efficient emergency services requires that necessary information concerning the emergency caller is made available to the PSAP. The mandatory information components are Calling Party Identity and Caller Location. Furthermore, routing information and address

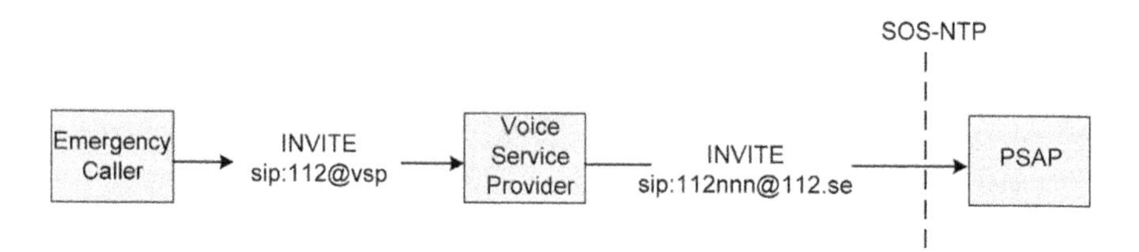

Figure 4.2 Generic emergency call flow. (© Jakob Schlyter, 2012.)

information of the subscriber caller may be made available to the emergency service authorities in the incoming call or by subsequent requests.

4.1.2.1 Calling Party Identity

The Calling Party Identity is used by the PSAP for two purposes:

1. to make it possible for the emergency call-taker to call back; and
2. as one of several methods to find the address and location of the emergency caller.

4.1.2.2 Caller Location

The location of the caller is used for two purposes:

1. to facilitate routing of an emergency call to the appropriate PSAP/emergency call-taker (e.g., using the municipality identity code); and
2. to enable geographical location of the emergency caller, enabling dispatching of rescue resources to the right place (municipality identity code and supplementary information derived from, for example, calling party identity).

4.1.2.3 Routing Information

Routing information is used for multiple purposes:

1. to enable routing to the appropriate PSAP/emergency call-taker (e.g., based on municipality identity code);
2. to convey information on the access type the call was made from to the emergency call-taker; and
3. to convey information on the area the emergency call was made from to the emergency call-taker (e.g., municipality identity code).

The routing information is assigned to the emergency call by the emergency caller or the originating VSP. In the case when it is assigned by the emergency caller, for example, a corporate network, the VSP cannot guarantee routing to the appropriate PSAP. In the case when the assignment of municipality identity codes is done by the VSP, the municipality identity code shall represent the NTP.

The routing procedure shown in Figure 4.3 can be described as follows:

- 1a, 1b The routing information can be assigned by either the emergency caller or the VSP.
- 2 An IP network transfers the routing information by transparently forwarding the SIP signaling to the SOS-NTP.
- 3 The PSAP uses the routing information.
- 4 The emergency call-taker at the PSAP might use the routing information.

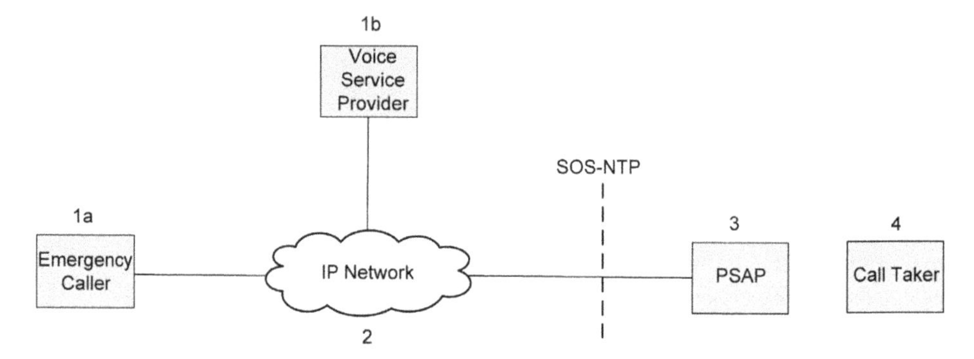

Figure 4.3 Routing overview. (© Jakob Schlyter, 2012.)

4.1.2.4 Address and Location Information

Location information is used by the emergency call-taker to locate the emergency caller. It can be a geographical address, for example, street name and number, or a position expressed in coordinates. The location can be retrieved using two methods:

1. The Calling Party Identity is used as an identifier in a request to a database or location server where the mapping of Calling Party Identity into geographical address or position is made available.
2. The address information is received in or derived from the incoming emergency call.

The location information can be of three types:

1. NTP of a fixed telephone line (e.g., geographical address of subscriber/emergency caller);
2. NTP of a mobile telephone (e.g., location of base station); or
3. address related to subscriptions (e.g., home or billing address of subscriber).

Note that in the case of mobile telephones, both address (usually billing address) and location can be available. These need not give the same information.

4.1.3 Protocols for PSAP Interconnection

A VSP connects directly to the PSAP at the session layer. The signaling protocol used at the session layer is SIP. IP networks are used to connect the VSP to the PSAP at the IP layer. How this is done is out of the scope of this section.

To make it possible for the PSAP to get the necessary information, the following information has to be transferred in the initial SIP INVITE request:

- Calling Party Identity for the identification of the emergency caller in tel URI or SIP URI format.
- Called Party Identity for the identification of PSAP and routing information in SIP URI format.

4.1.3.1 Format of Calling Party Identity

The originating VSP can use either of the following formats:

1. Telephone Uniform Resource Identifier, tel URI
   - Format of tel URI defined in RFC 3966 [80] and RFC 3261 [124]
   - Generic example: tel:+[International E.164 number]
   - Illustrative example: tel:+4686785500
2. SIP Uniform Resource Identifier, SIP URI
   - Format of SIP URI defined in RFC 3261 [124]
   - Generic example: sip:+[International E.164 number]@[vsp-domain]
   - Illustrative example: sip:+4686785500@example.com

   The following SIP headers are used, in preferred order, to identify the calling party identity:
   - P-Asserted-Identity header (RFC 3325 [130])
   - P-Preferred-Identity header (RFC 3325 [130])
   - From header (RFC 3261 [124])

The originating VSP should either:

- insert a P-Asserted-Identity header providing a tel URI or an SIP URI with a numerical user part containing Calling-Party-Identity as an international E.164 number prefixed by "+"; or
- insert a P-Preferred-Identity header providing a tel URI or an SIP URI with a numerical user part containing Calling-Party-Identity as an international E.164 number prefixed by "+"; or
- construct the From header to include an SIP URI with a numeric user part that can be used for dial-back purposes – this means that the user part of the SIP URI shall be an international E.164 number prefixed by "+".

The VSP shall assure the Calling Party Identification.

4.1.3.2 Format of Called Party Information

The Called Party Identity shall contain two main pieces of information:

1. the Called Party Identity;
2. identification of originating area.

The originating VSP shall use the SIP Uniform Resource Identifier (SIP URI) format:

- Format of SIP URI defined in RFC 3986 [171] and RFC 3261 [124]
- Generic example 1: sip:112[municipality-identity-code]@112.se
- Illustrative example 1: sip:112274@112.se (call originating from a fixed line in Mellerud)
- Example 2: sip:sos@112.se

Note: Only the SIP URI used in the example 1 will provide call routing to the correct emergency call-taker. In example 2, routing to the correct emergency call-taker is not possible due to lack of municipality identity code in the Request URI. The emergency call might be routed to an emergency call-taker without local knowledge (e.g., an emergency call from Mellerud using example 2 (sip:sos@112.se) might be answered by an emergency call-taker anywhere in Sweden). The string 112.se is the domain for the emergency service in Sweden.

The use of format in example 2 is primarily for future use when the LoST function, as defined by IETF ECRIT WG, has been introduced. LoST is used to translate a dialed emergency number, such as sip:112@example.com or urn:services:sos, to a called URI acceptable by SOS-NTP, for example, sip:112274@112.se, as described in RFC 5031 [11] and RFC 5222 [68].

4.1.3.3 Identification of Originating Area

The originating VSP shall assign a municipality identity code. The code shall be transferred in the initial SIP INVITE request as part of the Request URI.

4.1.4 Protocol Standards

The SOS-NTP supports the VoIP signaling protocol SIP and media transport protocol RTP. These protocols must be used by the VSP when connecting to the SOS-NTP for emergency calls.

At least the following SIP and RTP related RFCs are supported by SOS-NTP:

- RFC 3261 [124]: Session Initiation Protocol
- RFC 3264 [172]: An Offer/Answer model with SDP
- RFC 3311 [173]: The SIP UPDATE Method
- RFC 3325 [130]: Private Extensions to SIP for Asserted Identity within Trusted Networks
- RFC 3550 [174]: A Transport Protocol for Real-Time Applications

In addition, it is recommended that the following SIP-related RFCs are supported by the SOS-NTP:

- RFC 3312 [175]: Integration of Resource Management and SIP
- RFC 4032 [176]: Update to SIP Preconditions Framework

At least the following SIP and RTP related RFCs must be supported by connecting VSPs and emergency callers:

- RFC 3261 [124]: Session Initiation Protocol
- RFC 3263 [177]: SIP: Locating SIP servers
- RFC 3264 [172]: An Offer/Answer model with SDP
- RFC 3550 [174]: A Transport Protocol for Real-Time Applications

4.1.5 Media

The SOS-NTP supports at least the following audio codec standards, variants, and packetization times. Additional codecs and packetization times may be supported:

- G.711 (RFC 3551 [178], packetization time 0–200 ms)
- G.729 (RFC 3551 [178], packetization time 0–200 ms)
- AMR (RFC 4867 [179], packetization time 0–200 ms)

4.1.6 Emergency Call Routing

The routing of an emergency call to the SOS-NTP will not be based on municipality identity codes. The DNS will be used to find the appropriate SOS-NTP. The municipality identity code has to be transferred to the SOS-NTP to enable the PSAP to deliver the emergency call to the correct emergency call-taker. The municipality identity code has to be included, if delivery to the correct emergency call-taker is to be achieved, in the Request URI of the initial SIP INVITE request at the last hop (VSP to SOS-NTP or emergency caller to SOS-NTP). The municipality identity code will be displayed (if available) to the emergency call-taker.

4.1.7 Testing

The SOS-NTP will provide an SIP URI for testing purposes. This URI can be used to test functionality and reachability of the SOS-NTP. There are two SIP URIs that are defined for testing purposes:

- sip:sos.test@112.se
- sip:112493@112.se

It is recommended that the VSP continuously perform connectivity tests to the SOS-NTP. This can be achieved using, for instance, the SIP OPTION message as a keep-alive function. The SOS-NTP might use rate-limiting or any other necessary technology to suppress traffic overload.

4.1.8 Examples

This section includes a couple of examples and use cases that show how different types of VSPs can connect to the SOS-NTP. In these examples, only fixed subscribers are described. Nomadic IP subscribers are only handled correctly once the full SIP-based PSAP solution is in place (according to IETF ECRIT standards [121]) or if the nomadic subscriber uses some sort of offline system to update the current location.

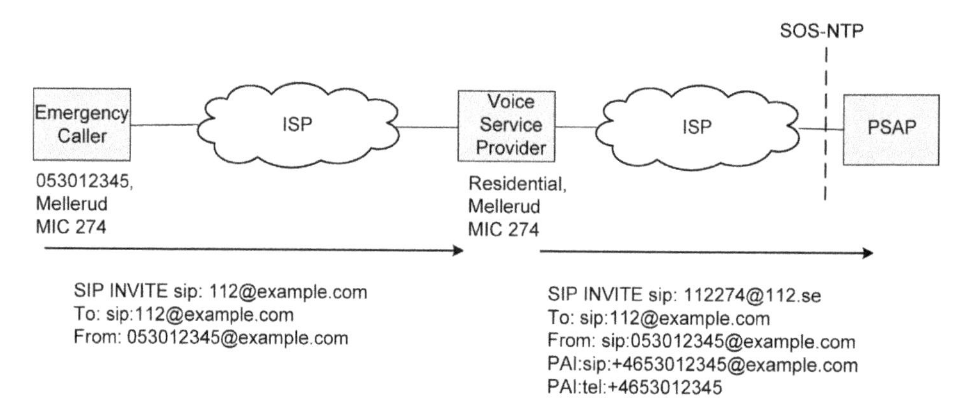

Figure 4.4 Residential user emergency call scenario. (© Jakob Schlyter, 2012.)

4.1.8.1 Residential User Running a VoIP Service

Since basically anyone with an Internet connection, a DNS domain, and an SIP server can provide their own VoIP service, it is a good example to start with. These users must be able to connect to the SOS-NTP for emergency calls. As these systems usually only have a few subscribers, it is easier to keep track of the subscribers. If the system is used for fixed line replacement, the SIP server used by the VSP (that means the residential user himself in this scenario) is configured to add the appropriate additional information on the SIP trunk used for the SOS-NTP.

As seen in Figure 4.4, the following actions take place:

- The Emergency Caller dials the dial string for emergency services, in this case 112. The Emergency Caller is located in Mellerud with the MIC of 274. The Caller has the Swedish national phone number of 053012345, a Mellerud number.
- The UA used by the Emergency Caller sends an SIP INVITE to the VSP (the SIP server of the residential user, responsible for the DNS/SIP domain example.com).
- The VSP authenticates the incoming INVITE as it would any outbound call. Then, it detects the dialed emergency dial string and uses any internal logic (e.g., dial plan) to construct the SIP INVITE that is sent to the SOS-NTP. This includes adding the MIC associated with the current subscriber and adding a PAI with an authenticated Calling Party ID in full E.164 format prefixed by "+".
- The VSP forwards the SIP INVITE to the SOS-NTP.

4.1.8.2 Company Running a VoIP Service

It is becoming more and more common that companies use VoIP internally in their corporate networks. VoIP is used both within the company at one location as well as between different geographical locations. If multiple geographical locations are served by the same VSP, it is important that the VSP, based on the location of each subscriber, inserts the required additional information appropriate for that specific subscriber.

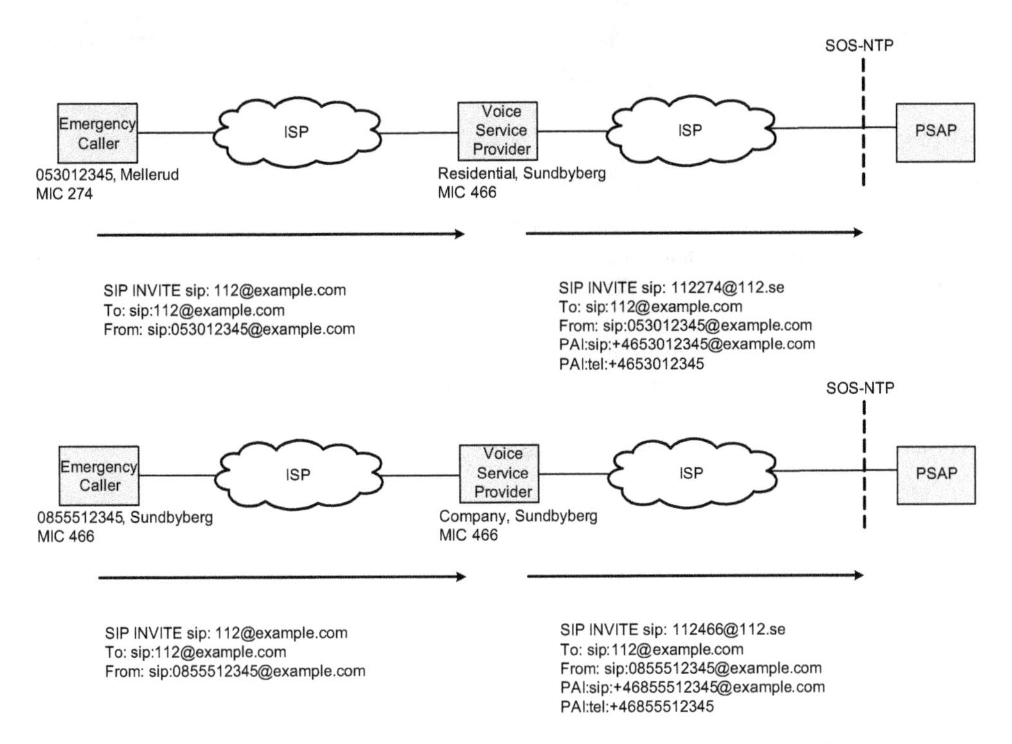

Figure 4.5 Emergency call scenario from a company running its own voice service. (© Jakob Schlyter, 2012.)

For instance, if the VSP server is located in Stockholm, and has a local office in Mellerud, emergency calls originating from Mellerud must be sent to the SOS-NTP with the correct SIP Request URI containing the MIC of Mellerud, not Stockholm, and correct Calling-Party-Identity.

As seen in Figure 4.5, the following actions take place:

- The Emergency Caller dials the dial string for emergency services, in this case 112. The Emergency Caller in the first example is located in Mellerud with the MIC of 274. The Caller has the Swedish national phone number of 053012345, a Mellerud number. The Emergency Caller in the second example is located in Sundbyberg with the MIC of 466. The second caller has the Swedish national phone number of 0855512345, a Stockholm/Sundbyberg number.
- The UA used by the Emergency Caller sends an SIP INVITE to the VSP (the SIP server of the company, responsible for the DNS/SIP domain example.com).
- The VSP authenticates the incoming INVITE as it would any outbound call. Then, it detects the dialed emergency dial string and uses any internal logic (e.g., dial plan) to construct the SIP INVITE that is sent to the SOS-NTP. This includes adding the appropriate MIC associated with the specific subscriber (MIC 274 or MIC 466) and adding a PAI with an authenticated Calling Party ID in full E.164 format prefixed by "+".

- The VSP forwards the SIP INVITE to the SOS-NTP. If the VSP uses local PSTN breakout in Sweden to deliver the emergency call, this is out of scope for this section.

4.1.8.3 Multinational Company Running a VoIP Service

This scenario is basically the same as the previous one, with the addition that the VSP is actually situated outside the national border of Sweden. From a technical point of view, this makes no difference. As long as the VSP is able to set the correct additional information appropriate for the location of each subscriber, the call will be routed to the appropriate PSAP. The SIP connection between the VSP and the SOS-NTP has no limitations based on origin. However, depending in IP routing agreements and policies, different (or not all) SOS-NTPs may be visible from different ISPs.

As seen in Figure 4.6, the following actions take place:

- The Emergency Caller dials the dial string for emergency services, in this case 112. The Emergency Caller in the first example is located in Mellerud with the MIC of 274. The Caller has the Swedish national phone number of 053012345, a Mellerud number. The Emergency Caller in the second example is located in Sundbyberg with the MIC of 466. The second caller has the Swedish national phone number of 0855512345, a Stockholm/Sundbyberg number.

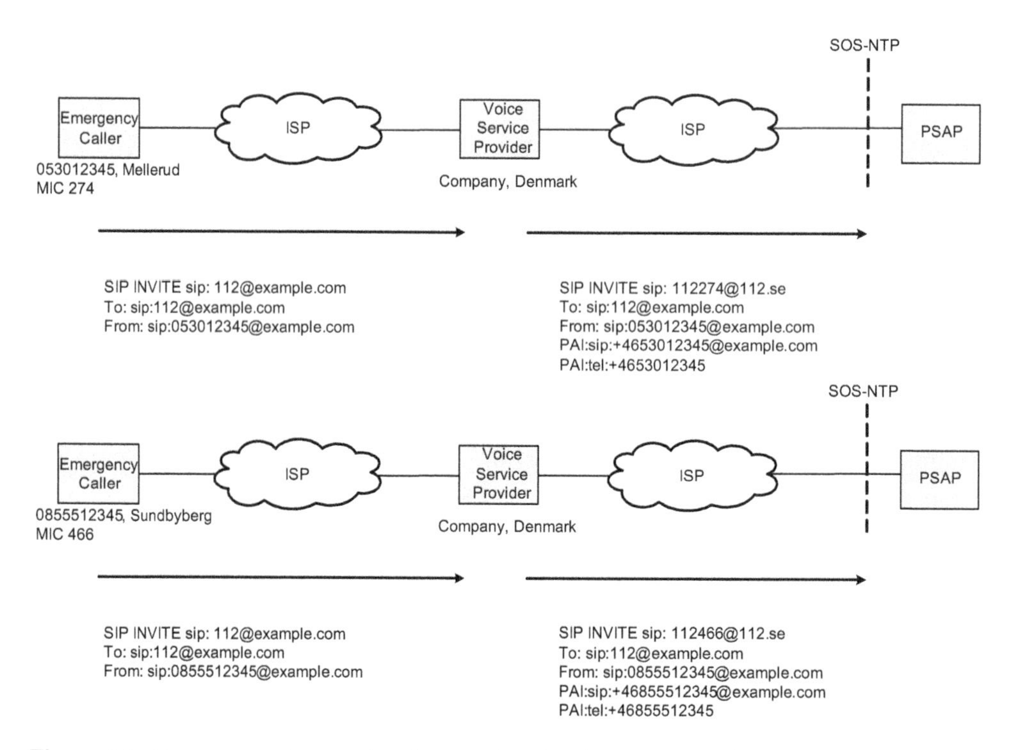

Figure 4.6 Emergency call scenarios for a multinational company running its own voice service. (© Jakob Schlyter, 2012.)

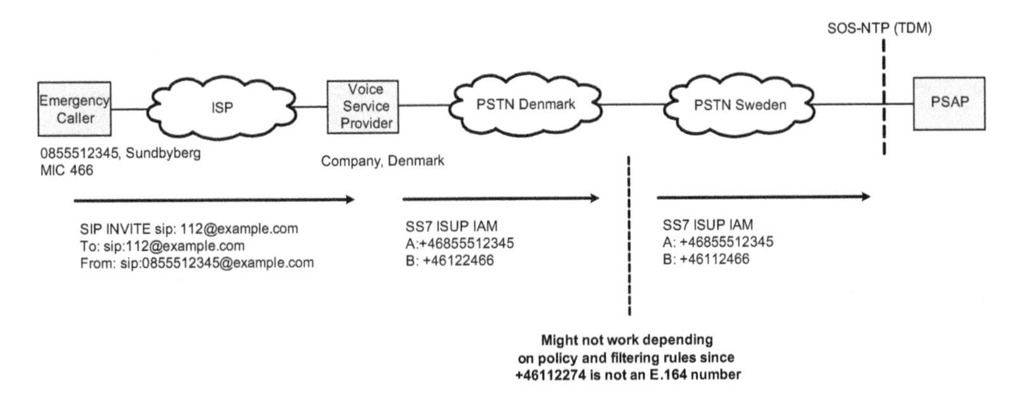

Figure 4.7 Example of PSTN breakout in Denmark, forwarding the emergency call over the PSTN to Sweden. (© Jakob Schlyter, 2012.)

- The UA used by the Emergency Caller sends an SIP INVITE to the VSP (the SIP server of the company, responsible for the DNS/SIP domain example.com) in this case located in Denmark, outside the national borders of Sweden.
- The VSP authenticates the incoming INVITE as it would any outbound call. Then, it detects the dialed emergency dial string and uses any internal logic (e.g., dial plan) to construct the SIP INVITE that is sent to the SOS-NTP. This includes adding the appropriate MIC associated with the specific subscriber (MIC 274 or MIC 466) and adding a PAI with an authenticated Calling Party ID in full E.164 format prefixed by "+".
- The VSP forwards the SIP INVITE to the SOS-NTP. A description for cases where the VSP uses a PSTN breakout inside or outside Sweden is outside the scope of this section, but Figure 4.7 illustrates the case.

Note that some PSTN providers do not allow routing of non-E.164 numbers between the national borders of the PSTN network. This means that it might not be possible for the VSP to send +46112274 (emergency call from a subscriber in Mellerud, Sweden) from a PSTN connection outside Sweden to the Swedish PSTN. Therefore, the emergency call might fail.

4.1.8.4 Telecommunication Provider With Residential Subscribers

The telecommunication provider acts as VSP and residential subscribers connect. The VSP most likely will service subscribers from many different geographical locations and, on a per-subscriber basis, the VSP must be able to add the required additional information to the SIP Request URI and Calling Party ID.

As seen in Figure 4.8, the following actions take place:

- The Emergency Caller dials the dial string for emergency services, in this case 112. The Emergency Caller in the first example is located in Mellerud with the MIC of 274. The Caller has the Swedish national phone number of 053012345, a Mellerud number. The Emergency

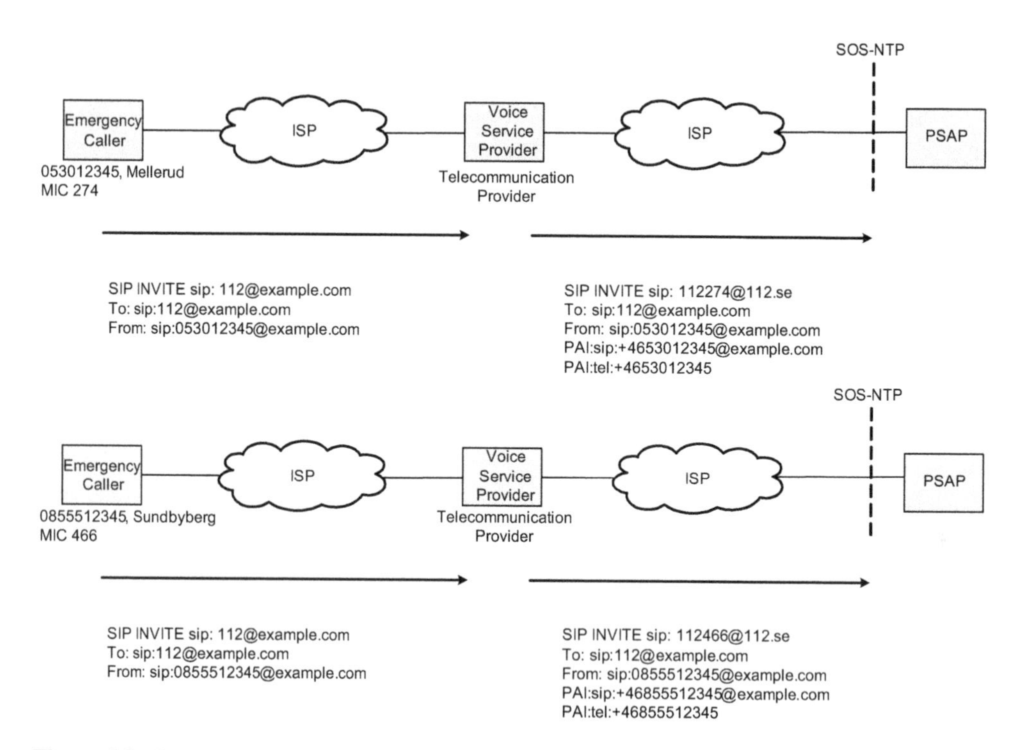

Figure 4.8 Two emergency call scenarios for a telecommunication provider running a voice service for residential Users. (© Jakob Schlyter, 2012.)

Caller in the second example is located in Sundbyberg with the MIC of 466. The second caller has the Swedish national phone number of 0855512345, a Stockholm/Sundbyberg number.

- The UA used by the Emergency Caller sends an SIP INVITE to the VSP (the SIP server of the telecommunication provider, responsible for the DNS/SIP domain example.com).
- The VSP authenticates the incoming INVITE as it would any outbound call. Then, it detects the dialed emergency dial string and uses any internal logic (e.g., dial plan) to construct the SIP INVITE that is sent to the SOS-NTP. This includes adding the appropriate MIC associated with the specific subscriber (MIC 274 or MIC 466) and adding a PAI with an authenticated Calling Party ID in full E.164 format prefixed by "+".
- The VSP forwards the SIP INVITE to the SOS-NTP.

4.1.8.5 Telecommunication Provider With SIP Trunk to Companies

In this scenario a company is running its own voice service but uses an SIP trunk to a telecommunication provider for PSTN connectivity. Both of the entities are acting as VSPs. Emergency calls from the company (acting as local VSP) are sent to the SIP trunk provider (acting as telecommunication VSP), which will forward the call to the SOS-NTP.

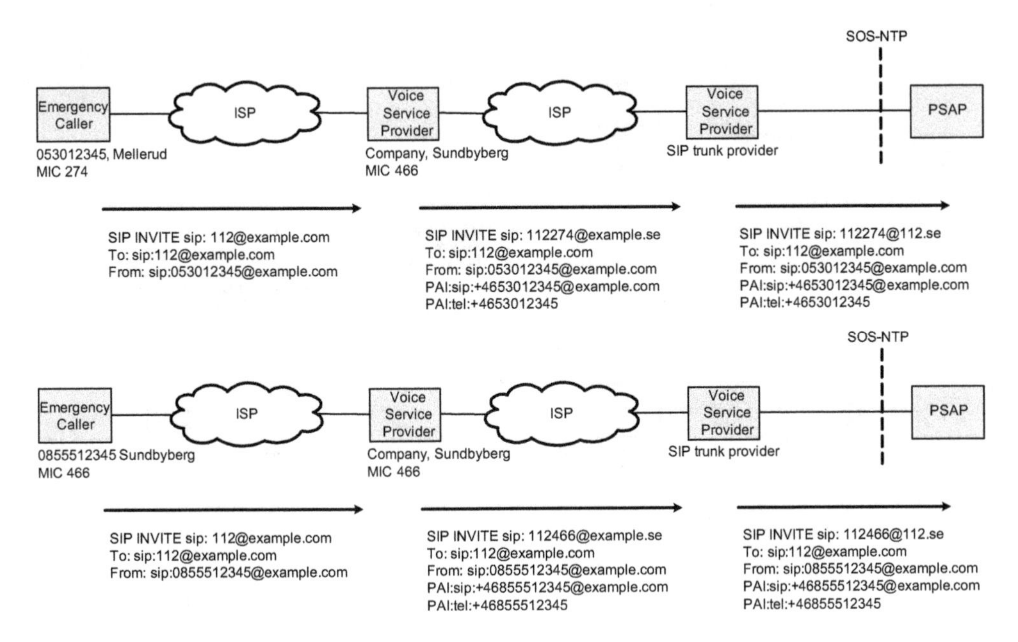

Figure 4.9 Emergency call scenario where SIP trunk provider only forwards the call. (© Jakob Schlyter, 2012.)

Depending on agreements between the company and the SIP trunk provider, the company or the SIP trunk provider can be responsible for adding the required additional information to the SIP INVITE sent to the SOS-NTP. If the agreement is that the company is solely responsible for this, the SIP trunk provider only acts as a forwarder and forwards the SIP INVITE to the SOS-NTP.

As seen in Figure 4.9, the following actions take place:

- The Emergency Caller dials the dial string for emergency services, in this case 112. The Emergency Caller in the first example is located in Mellerud with the MIC of 274. The Caller has the Swedish national phone number of 053012345, a Mellerud number. The Emergency Caller in the second example is located in Sundbyberg with the MIC of 466. The second caller has the Swedish national phone number of 0855512345, a Stockholm/Sundbyberg number.
- The UA used by the Emergency Caller sends an SIP INVITE to the VSP (the SIP server of the company, responsible for the DNS/SIP domain example.com).
- The VSP authenticates the incoming INVITE as it would any outbound call. Then, it detects the dialed emergency dial string and uses any internal logic (e.g., dial plan) to construct the SIP INVITE that is sent to the upstream SIP trunk provider. This includes adding the appropriate MIC associated with the specific subscriber (MIC 274 or MIC 466) and adding a PAI with an authenticated Calling Party ID in full E.164 format prefixed by "+".
- The SIP trunk provider may authenticate the incoming INVITE as it would any inbound call on the SIP trunk. Then, the SIP trunk provider forwards the SIP INVITE to the SOS-NTP.

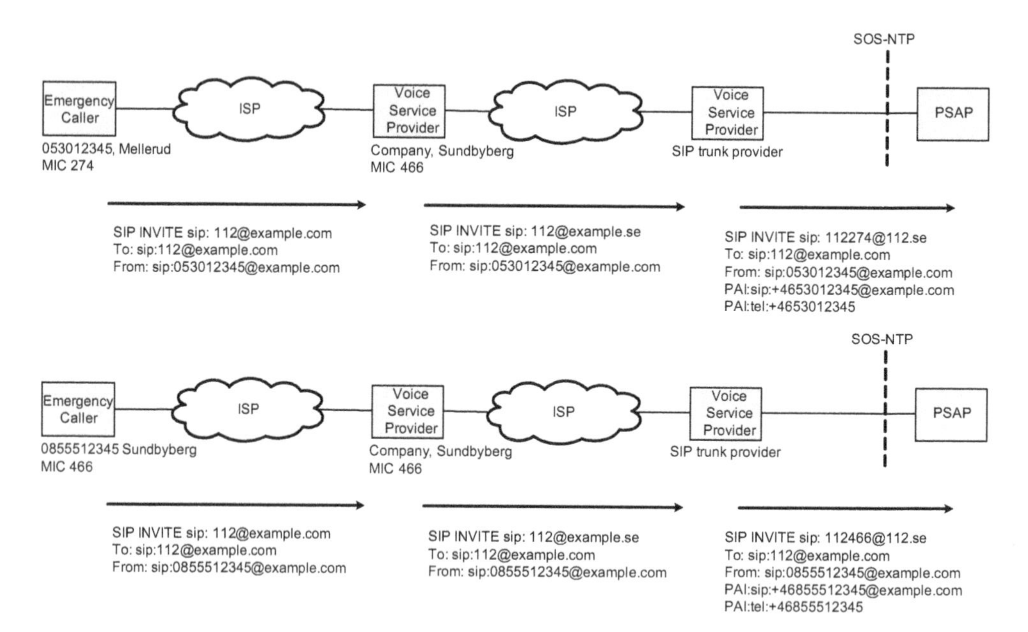

Figure 4.10 Emergency call scenario where SIP trunk provider adds information. (© Jakob Schlyter, 2012.)

If the agreement is that the SIP trunk provider is responsible for adding the additional information, the SIP trunk provider must add the required additional information to the SIP INVITE sent from the SIP trunk provider to the SOS-NTP.

As seen in Figure 4.10, the following actions take place:

- The Emergency Caller dials the dial string for emergency services, in this case 112. The Emergency Caller in the first example is located in Mellerud with the MIC of 274. The Caller has the Swedish national phone number of 053012345, a Mellerud number. The Emergency Caller in the second example is located in Sundbyberg with the MIC of 466. The second caller has the Swedish national phone number of 0855512345, a Stockholm/Sundbyberg number.
- The UA used by the Emergency Caller sends an SIP INVITE to the VSP (the SIP server of the company, responsible for the DNS/SIP domain example.com and acting as local VSP).
- The VSP authenticates the incoming INVITE as it would any outbound call. The VSP forwards the call to its upstream SIP trunk provider (the VSP responsible for the SIP/DNS domain example.se) as it would any other external call.
- The SIP trunk provider, acting as upstream VSP, authenticates the incoming SIP INVITE. Then, it detects the dialed emergency dial string and uses any internal logic (e.g., dial plan) to construct the SIP INVITE that is sent to the SOS-NTP. This includes adding the appropriate MIC associated with the specific subscriber (MIC 274 or MIC 466) and adding a PAI with an authenticated Calling Party ID in full E.164 format prefixed by "+".
- The SIP trunk provider forward the SIP INVITE to the SOS-NTP.

4.2 UK Specification for Locating VoIP Callers

Ray Bellis
Nominet

4.2.1 Introduction

NICC Standards Limited (formerly the Network Interoperability Consultative Committee) [180] is the UK body responsible for telecommunications standardization, wherever possible drawing on the work of international standards development organizations, such as ITU-T, ETSI, and the IETF. This section describes the efforts documented in NICC Document ND 1638 [181].

4.2.2 The Regulatory Environment

As the United Kingdom is an EU Member State, telecommunications in the United Kingdom is regulated according to the Communications Act 2003 (CA2003), which implements the requirements of the EU Framework Directive (2002/21/EC). CA2003 dissolved the previous licensing regime for telecommunications providers, hence removing the concept of a "Licensed Operator." Anybody "providing an electronic communications network or electronic communications service" is instead now termed a "Communications Provider" and is subject to the Act.

It is worth noting that, while the previous licensing regime typically only covered the traditional PSTN telecoms operators, the new definition also covers all Internet Service Providers (ISPs) and VoIP Service Providers (VSPs).

More information about the regulatory situation in Europe can be found in section 7.2.

4.2.2.1 General Conditions of Entitlement

Specific operational and technical requirements are not specified in CA2003, but may be imposed by OFCOM, the UK communications regulator, as General Conditions of Entitlement as described in §45 of the Act. Quoting the Act, such conditions must be:

1. objectively justifiable in relation to the networks, services, facilities, apparatus or directories to which it relates;
2. not such as to discriminate unduly against particular persons or against a particular description of persons;
3. proportionate to what the condition or modification is intended to achieve; and
4. in relation to what it is intended to achieve, transparent.

General Condition 4 (Emergency Call Numbers)
At the time of writing, the latest General Conditions include the following:

> The Communications Provider shall, to the extent technically feasible, make Caller Location Information for all calls to the emergency call numbers "112" and "999" available to the Emergency Organisations handling those calls.

A number of additional General Conditions (relating to call quality, reliability, etc.) only apply to those telephone services that meet the four "gating criteria" that define whether that service is a "Publicly Available Telephone Service" ("PATS"):

- a service available to the public;
- for making and receiving national and international calls;
- for accessing emergency services; and
- through a national or international phone number on a numbering plan.

Since many VSPs did not provide access to the emergency services, they did not satisfy all four criteria and were therefore not subject to those additional General Conditions.

However, in 2007 General Condition 4 was updated so that it applied to any service "enabling origination of calls to numbers in the National Telephone Numbering Plan." This effectively required all VSPs to offer emergency services access, in the process forcing them to satisfy the four PATS criteria, in turn requiring those VSPs to meet the full PATS obligations. General Condition 4 also applies to any ISP whose data network is used for the transit of VoIP calls to the emergency services.

4.2.3 Standards Development

Currently, VoIP providers are required to regularly supply data to the Emergency Operators that maps from telephone number to customer address for all fixed-location customers.

For customers with nomadic equipment the data indicate only that the number belongs to a nomadic user, prompting the emergency operator to specifically request the caller's location. NICC created a VoIP Location Working Group as a subgroup of NICC's Technical Steering Group in 2005 in anticipation of General Condition 4. The specification aims to provide a technically feasible solution to the problem of nomadic users.

4.2.3.1 Scope

NICC's current work in this area is limited in scope to the case of a UK customer calling the emergency services using VoIP technology from within the United Kingdom. Future work will consider the following:

- foreign visitors requiring assistance within the United Kingdom; and
- UK travelers requiring assistance abroad.

4.2.4 The Current UK Emergency Services Structure

4.2.4.1 Emergency Handling Authorities

Figure 4.11 shows the current emergency services architecture of the United Kingdom. There are two Emergency Handling Authorities (EHA) in the United Kingdom, those being operated by British Telecommunications plc (BT) and by Cable and Wireless (CW). All communication

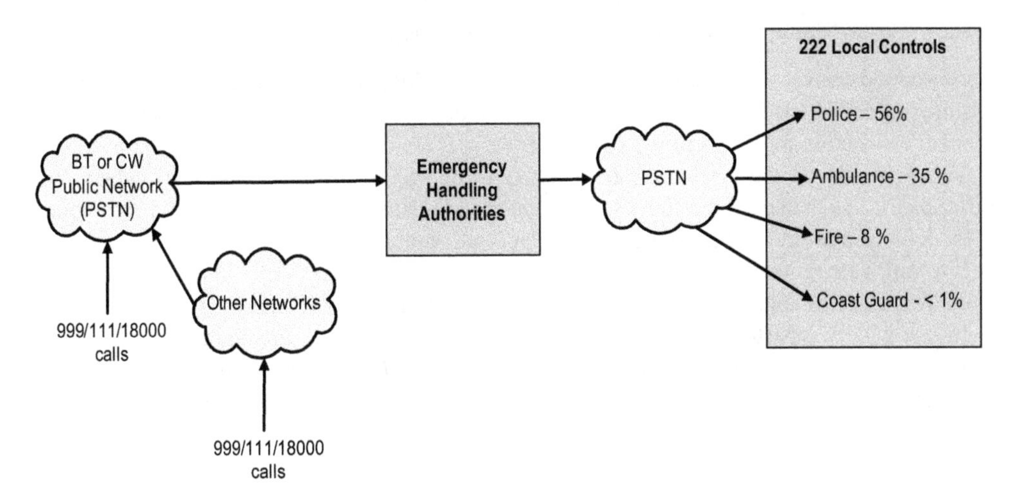

Figure 4.11 Today's emergency services architecture in the United Kingdom.

providers (CPs) with interconnect to the PSTN are required to contract with one (or both) of those.

All emergency calls are delivered to an EHA, which then selects the relevant Emergency Authority and routes the call onward, including location information.

4.2.5 Principles Driving the Specification

The following assumptions guided the authors of the UK specification.

1. User-managed versus network-derived location:
   End users cannot be required (nor relied upon) to keep their VSP up to date with their location information, even in the case of "fixed" services.
2. Smart device versus smart network:
   End user devices shall in the long term remain unable to obtain their own location or supply it in-band to the emergency services.
3. Separation of PSTN access and Internet access:
   End users will in many cases have separate contracts for telephony services and Internet access services.
4. Scalability:
   It would be impracticable to establish a fully transitive trust relationship for exchange of location information between every VSP and ISP within the United Kingdom.
5. EHA architecture:
   EHAs shall in the medium term remain reachable only via the PSTN.

The implications on the design of each of these assumptions are explored below.

4.2.5.1 User-Managed Versus Network-Derived Location

As described above, the existing system for providing location information for VoIP callers requires each VSP to regularly send the EHAs a complete database of mappings from telephone number to location for their entire customer base.

This is similar to the system for regular fixed-line PSTN users. However, a fundamental difference is that, when a fixed-line user changes their location, their telephony service provider is by definition aware of that change – the telephone number is tied to a specific location.

If a VoIP user moves house, there is no technical means for their VSP to become automatically aware of that change – the VSP must rely on their customer telling them.

In one widely reported case [182] in Canada in 2008, failure to maintain current address information allegedly contributed to the death of a toddler when an ambulance was dispatched to the family's previous address, some 2500 miles away from where it was required.

The specification therefore relies on network-derived location information, similar to the systems used by the mobile telephone networks.

4.2.5.2 Smart Device Versus Smart Network

Standards to allow smart VoIP telephones to determine their location and subsequently convey that information to called parties are still under development.

Furthermore, even when those standards are completed, it will be many years before they are universally deployed. The specification therefore assumes that the end device plays no part whatsoever in supplying its location, further reinforcing the requirement for network-derived location information.

4.2.5.3 Separation of PSTN Access and Internet Access

Many (if not most) VoIP users obtain their telephony service and Internet access services from different suppliers. Furthermore, nomadic users might use VoIP services from many and varied wireless hotspots where little or no personal identifying information is supplied.

Essentially, the VSP simply cannot automatically determine the physical location of a nomadic user. However, every Internet access service is tied to a physical location at some level. ADSL lines are associated with fixed PSTN lines. Leased lines and LAN extension services are also delivered to fixed locations.

WiFi hotspots are always back-hauled via some form of fixed Internet access technology, although the coverage area of each hotspot does mean that there will always be a certain amount of uncertainty in the location information. Even "mobile" (or cellular) IP access services such as 3G and LTE are ultimately fixed, given that the cellular operator can identify which cell is currently serving any given subscriber and better yet radio triangulation may be used to further improve the accuracy of the location information.

The specification therefore works on the basis that the ISP should be the primary source of location information – every ISP should always be able to identify which subscriber is using any particular IP address and the physical location to which the service has been delivered.

In cases where the Layer 2 transport service is provided by a third party (e.g., via local loop unbundling), the ISP may need to further rely on information supplied by the physical line access provider.

4.2.5.4 Scalability

For this section, we assume that there are m VSPs and n ISPs, where both figures are believed to be at least several hundred.

Distribution of location information has many privacy and security implications, and such information may not be freely given to anyone that asks for it. An ISP should only supply the location information to authorized and trusted third parties.

If the VSP had to obtain location information from the ISP and then forward that to the EHA, a very large trust network would have to be established – every ISP would have to have a trust relationship with every possible VSP. This requires $m \times n$ relationships to be maintained.

Instead, the architecture specifies that the VSPs will supply location-keying information (the caller's public IP address) directly to the EHAs, which the EHAs will then supply to the right ISP.[1] The ISP will then return the actual location information to the EHA.

In this way, only two limited trust networks are required – one between the EHAs and VSPs, and the other between the EHAs and ISPs. No direct exchange of information between VSP and ISP is required. Only $m + n$ relationships need to be maintained.

4.2.5.5 EHA Architecture

There are no declared plans from either EHA to upgrade their systems to receive VoIP calls directly, so the expectation is that in the medium term calls will continue to be delivered over the PSTN.

However, there is only very limited capacity for additional data within the PSTN's SS7 signaling protocol. Therefore, the location-keying information allowing a caller's location to be determined must be supplied out of band.

4.2.6 Putting It All Together

Figure 4.12 demonstrates the UK transition architecture for emergency services, and the individual functions are described below.

4.2.6.1 VoIP Positioning Center (VPC)

The VPC acts as the gateway between VoIP callers and the Emergency Operators. Whenever an emergency call is placed via a VSP, the VSP recognizes that the call is for the emergency services and sends the additional location-keying information to the VPC over an out-of-band IP message[2] at the same time as the call itself is passed to the EHA over SS7. Both the in-band SS7 signaling and the out-of-band IP message contain the caller's CLI and the VSP's unique identifier. Using these two pieces of information, the additional location-keying information can be associated with the incoming PSTN call.

---

[1] See the description of the IAIC in section 4.2.6.2 for more information about exactly how the EHA determines which is the right ISP.

[2] The NICC specification for these out-of-band messages between the VSP and the VPC is based on the V2 interface from the US National Emergency Number Association's i2 specification [183].

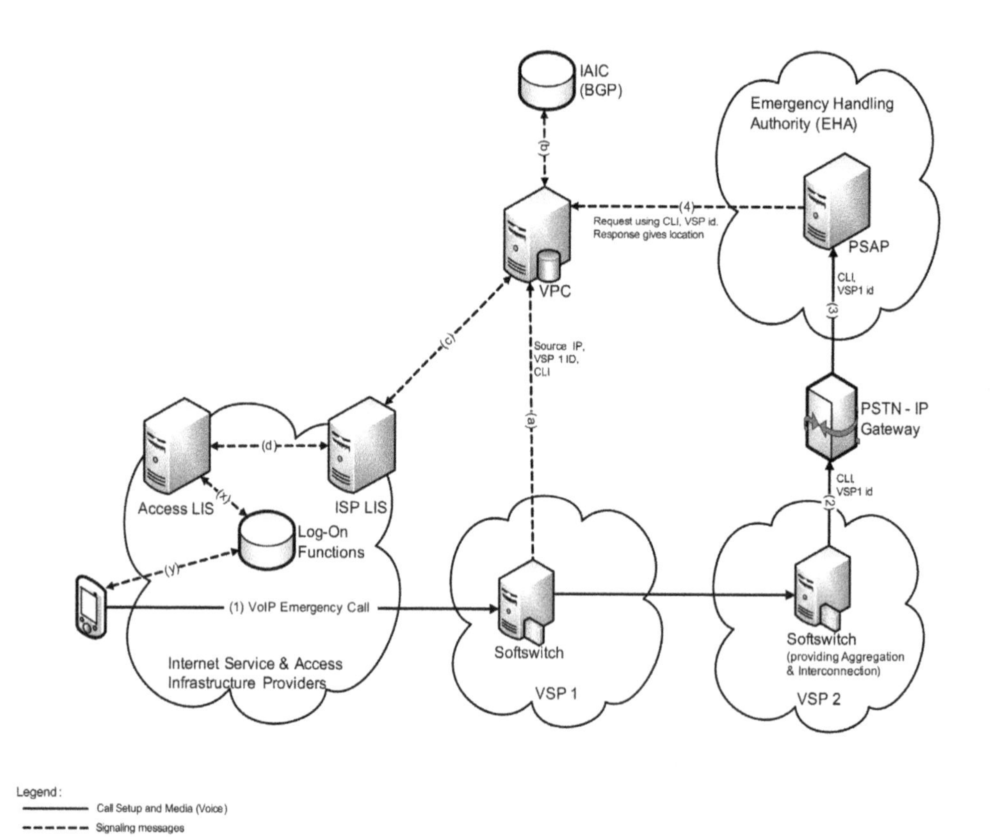

Figure 4.12 Proposed UK VoIP transition emergency services architecture.

Having determined the public IP address of the caller, the VPC may use the HELD [31] protocol to query a Location Information Server (LIS) to obtain a "PIDF-LO" XML data structure that contains the civic address of the site to which the internet access service is provided.

Once the civic address has been determined, the VPC may then route the call onwards to the appropriate call-handling center.

4.2.6.2 IP to ISP Address Converter (IAIC)

To use the HELD protocol, the VPC must first determine (or "discover," in IETF parlance) which LIS actually has the required information for the IP address in question. Two discovery mechanisms are described in more detail in the subsequent text below.

The VPC will initiate simultaneous LIS discovery queries using both mechanisms. While the BGP4 mechanism is likely to return its response soonest, it is considered that the DNS mechanism (where implemented) will be more accurate. Hence the VPC will wait for a period of time of the order of 200 ms for a DNS response before failing over to BGP4-derived data.

BGP4-Derived LIS Discovery

Each IP address on the Internet is part of an "Autonomous System," where each such Autonomous System (or "AS") belongs to an individual ISP, or in some cases to larger corporate enterprises.

Each AS connects to one or more other ASs – this global mesh of autonomous systems is what we think of as "the Internet." The BGP4 routing protocol is used by the routers at the edge of each AS to exchange IP routing information with the adjacent routers at each of the neighboring ASes.

Any router that has a "full" copy of the global BGP4 routing database can immediately identify to which AS any particular IP address belongs. The mechanism is therefore for the IAIC to operate such a router, and continuously monitor for changes to the global BGP4 routing database so as to maintain a dynamic database that maps from IP address ranges to AS numbers.

To subsequently determine the appropriate LIS, the IAIC would also need to build a (relatively) static database to map from AS numbers to LIS URIs. Ongoing maintenance of this database by mutual cooperation between the EHAs and the ISPs would be required.

Note that a significant disadvantage of this mechanism is that it does not allow the ISP to specifying alternate LIS URIs for different parts of their network.

DNS-Derived LIS Discovery

There is already a global distributed database based on IP addresses – the "in-addr.arpa." subtree of the Domain Name System (DNS).

To use IP addresses as the "key" field for a DNS database lookup, the IP address is converted into reverse octet format, hence the IP address "192.0.2.100" maps to the domain name "100.2.0.192.in-addr.arpa." A similar method maps IPv6 addresses under the "ip6.arpa" tree.

Under each such mapped domain name, one can usually find "PTR" records that convert the IP address back into a host name. The process of taking an IP address and finding a host name from it is known as "reverse DNS," since it is the opposite of the more common use of DNS (which is converting from host names to IP addresses).

For LIS discovery purposes, an additional "NAPTR" record that contains the URI of the relevant LIS is stored in the appropriate part of the reverse DNS tree for each IP address or network. This method is described in detail in [74].

4.2.7 Implications for Access Network Providers

4.2.7.1 Access to Downstream (Layer 2/Layer 1) Information

If the ISP provides Layer 3 Internet access via a lower-layer wholesale product (such as BT's IPstream xDSL products), they may not be able to control whether end user credentials are tied to a specific physical access line.

In this case the ISP must rely on the wholesale access line provider to provide additional location-keying information to identify the specific physical access line being used. This might, for example, take the form of a DSLAM unit and port identifier.

To obtain location information about the physical line, two options are available:

1. the wholesale supplier can preemptively submit PIDF-LO location information at the start of each login session; or
2. the ISP's LIS can initiate a new downstream HELD query on demand to an LIS operated by the lower-layer supplier, using the lower-layer's location-keying information.

In either case the ISP must be able to correctly associate the Layer 3 (IP addressing information) with the appropriate lower-layer identifiers.

The former approach may be preferable to ensure timely delivery of location information to the VPC. The trade-off is that the ISP has to collect the information at the start of every single Internet access session regardless of whether the end user subsequently makes a call to the emergency services or not.

4.2.7.2 Requirement to Run an LIS

Ultimately, it will not be mandatory for any of the parties involved to implement this specific architecture. That notwithstanding, this architecture has reached consensus approval within the relevant communities, and furthermore no alternative architectures have been proposed.

The most critical component of the architecture is the ISP LIS, without which no location information may be obtained. It is highly likely that implementation of this specification will be the easiest way for an ISP to satisfy General Condition 4. However, it is as yet uncertain what the cost implications are for the ISPs, or whether there exists any opportunity for them to offset those costs by providing or facilitating location-aware services.

While the requirements of General Condition 4 only specify "technically feasible," the proportionality test from CA2003 is also relevant – a technically feasible solution that costs too much to implement would fail that test.

4.3 Implementation of VoIP 9-1-1 Services in Canada

Guy Caron, ENP
Text written in personal capacity; works for Canadian telecommunication operator

In the early 2000s, when Voice over Internet Protocol (VoIP) started to get more traction and was increasingly adopted by non-traditional telecommunications services providers (TSPs), Canada, as many other countries around the world, was faced with new challenges to the Public Safety Answering Points (PSAPs), and emergency response agencies in particular, as well as the industry as a whole.

In order to address the situation, Canada's regulatory body, the Canadian Radio-television and Telecommunications Commission (CRTC or the "Commission"), was tasked to develop an appropriate framework to ensure Canadian citizens were adequately protected when calling for help using VoIP services.

This section describes the process and solutions that Canada has investigated and adopted to achieve that goal. It is offered herein with the hope that Canada's experience will help the international community to understand the various challenges, technical, regulatory or otherwise, that the emergency services and telecom industries are facing with the advent of VoIP-based services.

4.3.1 Regulatory Framework (About the CRTC)

The CRTC is an independent public organization that regulates and supervises the Canadian broadcasting and telecommunications systems. It reports to the Parliament of Canada through the Minister of Canadian Heritage. The CRTC works to serve the needs and interests of citizens, industries, interest groups, and the government.

The CRTC's mandate is to ensure that both the broadcasting and telecommunications systems serve the Canadian public. The CRTC uses the objectives in the Broadcasting Act and the Telecommunications Act to guide its policy decisions.

In telecommunications, the CRTC ensures that Canadians receive reliable telephone and other telecommunications services, at affordable prices. The CRTC regulates only where the market does not meet the objectives of the Telecommunications Act. It regulates telecommunications carriers, including major telephone companies. This role involves:

- approving tariffs and certain agreements for the telecommunications industry;
- issuing licenses for international telecommunications services, whose networks allow telephone users to make and receive calls outside Canadian borders;
- encouraging competition in telecommunications markets; and
- responding to requests for information and concerns about broadcasting and telecommunications issues.

The CRTC also hold public hearings, round-table discussions, and informal forums to gather input from interested parties on broadcasting and telecommunications concerns.

The CRTC Interconnection Steering Committee (CISC) is an organization established by the CRTC to assist in developing information, procedures, and guidelines as may be required in various aspects of the CRTC's regulatory activities.[3]

The mandate of the CISC is to undertake tasks related to technological, administrative, and operational issues on matters assigned by the CRTC or originated by the public, that fall within the CRTC's jurisdiction.

4.3.2 Canada's Telecom Profile

Canada is a vast country spread over nine million square kilometers of land. In 2010, the total population was 33,477 million, yielding an average population density of 3.7 per square kilometer.[4] Canada's geographical situation has been a challenge in deploying telecom networks outside of the dense urban areas. Since the inception of the telephone service and with the assistance of an appropriate regulatory framework, telephone companies have been able to deploy networks allowing access to telephone service for nearly 100% of the population.

[3] Excerpt from the CISC Administrative Guidelines document version 1.1, 31 March 2001.

[4] Statistics Canada 2011 census.

Canadians now have access to advanced communications services. Approximately 94% of Canadian households can access broadband services using landline facilities. Satellite facilities extend this reach to virtually all households and are only limited by capacity constraints. Canadians can also access broadband mobile services. Approximately 91% of Canadians can access these services using handheld mobile devices [184, p. i].

In 2008, 52% of Canadian households subscribed to landline broadband Internet access services and over 99% of Canadian households subscribed to telephone service, using either landline or mobile devices [184, p. i].

The deployment of high-capacity digital networks and the emergence of Internet Protocol (IP) as the standard for data transmission and delivery have facilitated the carriage of multiple types of data on a single network; this has been a major enabler of network convergence. Today's unified data networks are capable of delivering all forms of information, be it voice, data, text or video. In 2008, over 80% of communications revenues were generated by converged companies offering both broadcasting and telecommunications services [184, p. i].

By 2011, virtually all Canadian households had access to broadband Internet services of at least 1.5 megabits per second (Mbps), delivered by landline, mobile (HSPA+ and LTE), and satellite facilities. Moreover, the availability of higher-speed broadband services (between 30 and 50 Mbps) has increased from 30% to 75% in the past two years. In 2011, 72% of Canadians had access to four broadband platforms: digital subscriber line (DSL), cable, fixed-wireless/satellite, and mobile. In 2011, 72% of households subscribed to 1.5 Mbps broadband Internet service, compared to 68% in 2010. Moreover, 54% of households subscribed to services of 5 Mbps or greater (compared to 51% in 2010) [185, p. i].

The CRTC estimated in 2009 that there are 230 broadband Internet Access Providers (IAPs)[5] in Canada, where the nine largest have over 90% of the subscribers, and the remaining 221 have the remaining 10% [186, par. 15].

Competitive Internet Access services are regulated in Canada. The Commission has adopted policies to encourage fair competition in this area. As such, the incumbents, both DSL and cable, have been directed to open their broadband access networks to third-party ISPs.

The Commission determined that the number of nomadic VoIP service subscribers declined from 161,000 in 2007 to 153,000 in 2008, and accounted for 0.8% of wireline telephone service subscribers in 2008 (20.95 million in 2008) [184, p. 199].

4.3.2.1 Canadian Emergency Services Model

For the most part, Emergency Services in Canada are enabled through tariff services provided by Incumbent Local Exchange Carriers (ILECs), which inherited the public mandate to act as Emergency Services Providers (ESPs) within their respective incumbent territories. It is up to each incumbent to build, manage, and operate an appropriate emergency services infrastructure available throughout their emergency services territory to interface with the local authorities. The emergency infrastructures allow interconnection with other Telecommunication Service Providers (TSPs) offering voice services with outbound calling capabilities in order to route their originating emergency calls to the appropriate PSAP.

[5] The IETF has defined the term "Internet Access Provider (IAP)" in RFC 5012 [7] to describe what is called "Broadband Access Service Provider (ASP)" in this section.

The 9-1-1 SPs typically recover their costs through a CRTC-approved tariff applied to wireline and wireless TSPs based on working telephone numbers. For wireline TSPs, the 9-1-1 access fee is still regulated and must explicitly appear on the customer's telephone bill. Wireless service is forborne from regulation in Canada, and as such it is up to the wireless TSPs to determine how to charge their customers for 9-1-1 access.

Competitive TSPs, facility-based and non-facility-based, must follow strict rules set forth by the Commission to interconnect with the ILEC 9-1-1 network. Those rules have been defined in the Local Competition Decision [187] and by CISC.

Canada is part of the North American Numbering Plan (NANP) and as such has adopted the dialed digits 9-1-1 for universal access to emergency services.

It is up to the local authorities to determine when to provide E9-1-1 emergency services to their citizens. In 2012, there are still some areas in Canada where 9-1-1 service is not yet available.

Local authorities are responsible for building, managing, and operating the Public Safety Answering Points (Primary PSAPs, P-PSAPs) and associated emergency response agencies (also known as Secondary PSAPs, S-PSAPs). In Canada, local authorities can be at the Provincial, county or municipal level. At the end of 2011, there was roughly 130 Primary and 400 Secondary PSAPs in Canada.

Local authorities are responsible for funding their PSAP infrastructures and ongoing operations. Funding mechanisms vary between local authorities, depending on whether they are Provincial, county or municipality based.

4.3.2.2 Canadian E9-1-1 Infrastructure

The Canadian E9-1-1 service and infrastructure, as many others around the world, were originally designed for fixed voice services. Customer Name and Address Information (CNAI) is derived from Telcos' subscriber records, which includes service (i.e., not billing) address information.

The salient features of E9-1-1 service in North America that are common between Canada and the United States are the ability (i) to selectively route a 9-1-1 call to the appropriate or designated PSAP based on the caller's location, and (ii) to provide Automatic Number Identification (ANI) and Automatic Location Identification (ALI) at call time. E9-1-1 was and will continue to be a location-based service.

The Canadian E9-1-1 infrastructure was designed slightly differently than its US cousin at the outset. First, most Canadian ALI systems function in "push mode," that is, the ALI is call-aware and thus can provide the CNAI information concurrently with the voice call being routed to the PSAP call-taker, avoiding the need for an ALI Controller at the PSAP CPE. Second, a number of call control features have been made available to the PSAP call-takers such as (i) the ability to keep a voice connection up, even if the caller attempts to hang up (known as Called Party Hold), (ii) the ability to re-ring the caller's telephone set or, if the phone was left off-hook, trigger a howler tone (known as on-hook and off-hook ringback), (iii) the ability for the call-taker to be signaled audibly of the caller's attempts to go on- or off-hook (known as Switch-hook Status) and, finally, (iv) the ability to force-disconnect a caller. In addition, some part of the country implemented a feature that allows a 9-1-1 call that was dropped before being answered at the PSAP to be forced through until answered (known as Enhanced Called Party Hold). These features are almost ubiquitously deployed and used over the PSTN in Canada and widely appreciated by the PSAP community. It should be noted however that these call control features are not supported on wireless networks and as of today, not standardized for VoIP, a significant gap for the Canadian PSAP community.

Cellular analog wireless services were introduced to Canadians in the mid-1980s. It took some time for the service to become widespread and 9-1-1 calling was unevenly supported until the time that Canadian Wireless E9-1-1 Phase I[6] was introduced in 2003 [188]. Then Canadian Wireless E9-1-1 Phase II[7] was mandated in 2009 [189] and deployed nationally within 12 months afterward.

People with disabilities have access to emergency help either indirectly though Message Relay Services (accessible through 7-1-1 dialing) or directly by dialing 9-1-1 using a TTY or TDD device. Because TTY/TDD uses voice band tones, it is supported on the E9-1-1 infrastructure without modifications. While most PSAPs are equipped with TTY/TTD devices, there is no nationwide legislation or regulation mandating their use at the PSAP, and therefore accessibility is uneven across the country.

In 2008, the CRTC initiated a proceeding [190] to explore how accessibility for people with disabilities could be improved given the advances in technologies and the growing utilization of smart phones and text messaging. This initiative resulted in the development of a unique solution allowing deaf, hard-of-hearing and speech-impaired people to converse with a PSAP agent using Short Message Service (SMS). The solution was trailed in 2011–2012. The findings from the trial and the recommendations were submitted to the CRTC in October 2012 and a decision to deploy is expected in the first quarter 2013.

4.3.3 Interim Solution for Nomadic and Fixed/Non-Native VoIP

The regulatory proceedings in regard to investigating how to provide E9-1-1 to nomadic and fixed/non-native VoIP services started in the 2004–2005 time-frame, at a time when this new emerging technology was perceived to be a fast growing market.

In 2005, the CRTC directed nomadic and fixed/non-native VoIP service providers to implement an interim solution involving a manned third-party call center until such time as a more appropriate solution providing full E9-1-1 is implemented. This interim solution was deemed necessary based on the fact that nomadic and fixed/non-native VoIP service providers could not ensure that 9-1-1 calls would always be routed to the appropriate PSAP. Consequently, the intermediary call-taker is required to verbally confirm the location of the caller prior to forward the call toward the appropriate PSAP through the 9-1-1 SP's E9-1-1 infrastructure. The call forwarding is done using a tariff service from the 9-1-1 SP that ensures adequate and up-to-date routing information. It is referred to as Basic 9-1-1 (B9-1-1) as it only provides proper routing and voice connectivity between the caller and the PSAP without additional features such as ANI/ALI, selective transfers or PSAP call control as found with the Enhanced 9-1-1 service (E9-1-1).

---

[6] The ability to route the wireless 9-1-1 call to the designated PSAP and provide cell site-based location and callback number of the handset.

[7] The ability to automatically provide the handset's latitude, longitude and region of uncertainty at a confidence level of 90%, within 50 seconds of the reception of the Wireless E9-1-1 Phase I location and callback number of the handset.

The Commission duly noted that this type of interim solution is not without shortcomings, such as [191, par. 63] the following:

- The call center agent may be inadequately trained to deal with the types of emergencies that are commonly the subject of a 9-1-1 call.
- In certain situations, the caller must be able to verbally communicate his/her location to the call center agent.
- Adding a third party increases delays related to handling of 9-1-1 calls and thus increases the risk to the individual in an emergency calling situation.
- Not all PSAPs provide a 10-digit administration number for use by third-party call centers, requiring these call centers to use alternative arrangements to transfer calls to emergency services personnel. Calls handled this way may not be given the same priority as normal 9-1-1 calls.

Figure 4.13 depicts an example architecture for the interim solution.

Here is a typical call flow:

1. VoIP caller dials 9-1-1.
2. VSP call server detects dialed digits, recognizes the call as an emergency call, and routes it (through its PSTN gateway if the third-party call center is PSTN only) to the third-party call center.

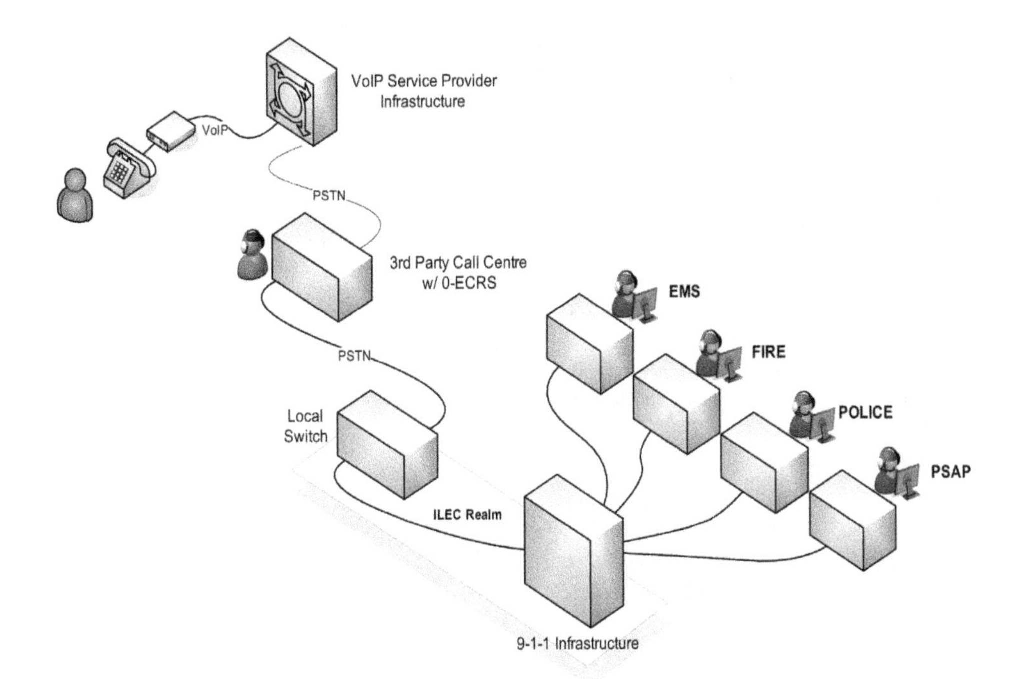

Figure 4.13 Example VoIP Basic 9-1-1 interim solution. (Reproduced and adapted with the permission of the Canadian Radio-television and Telecommunications Commission on behalf of Her Majesty in Right of Canada, 2013. Courtesy of Guy Caron, ENP.)

3. The call is answered by a live operator at the third-party call center, which requests verbally the location of the caller.
4. Upon confirmation, the operator selects the PSTN routing code pointing to the appropriate PSAP for that location, speed dials it, and bridges the PSAP call-taker with the caller.
5. The caller and the PSAP call-taker can converse.

This interim solution was subsequently amended so the third-party call center have access to the caller's telephone number and last known location as provided by the customer.

At the time of writing, seven years after being introduced as a temporary measure, this solution was still in effect and mandated for all Canadian nomadic and fixed/non-native VoIP services.

4.3.4 The (Defunct) Canadian i2 Proposal

With an interim solution in place that was less than optimal for the PSAPs, there was a sense of urgency to find more appropriate solutions. The mindset of the PSAPs at the time was "let's fix this at the outset, before it gets too big of an issue." During the course of the proceedings, various proposals were submitted to CISC in order to resolve the issues pertaining to nomadic and fixed/non-native VoIP E9-1-1 (also known as "over-the-top" VoIP). One proposal stood out, setting out what a Canadian version of the National Emergency Numbering Association (NENA) [58] i2 issue 1 [183] standard may look like. This solution, named Canadian i2, was predicated on the PSAPs' firm position that user-input location was not acceptable for 9-1-1, a position that the CRTC agreed with and subsequently adopted throughout the proceeding. As you will find later on, this proposal was eventually dismissed by the CRTC. Nevertheless, it is included in this chapter to provide the reader with some background information on what technical solutions were discussed and debated. Figure 4.14 depicts the Canadian i2 functional model.

While the NENA i2 model has been selected as the basis for the proposed Canadian implementation, four major differences between the Canadian situation and that present in the United States have been identified:

- the regulatory environment;
- the difference in size and complexity of the Public Switched Telephone Network (PSTN);
- the 9-1-1 systems framework including Public Safety Answering Point (PSAP) equipment and their interfaces; and
- the funding model.

The proposed Canadian i2 model acknowledges those differences. It was further designed with architectural solutions and operational principles appropriate to the existing Canadian 9-1-1 environment. The model, therefore:

- leveraged the existing E9-1-1 model to the extent technically and administratively possible;
- built on existing 9-1-1 infrastructures, including the routing of all 9-1-1 calls to the designated PSAPs through the existing provincial 9-1-1 infrastructures;
- minimized major modifications to existing provincial 9-1-1 infrastructures;

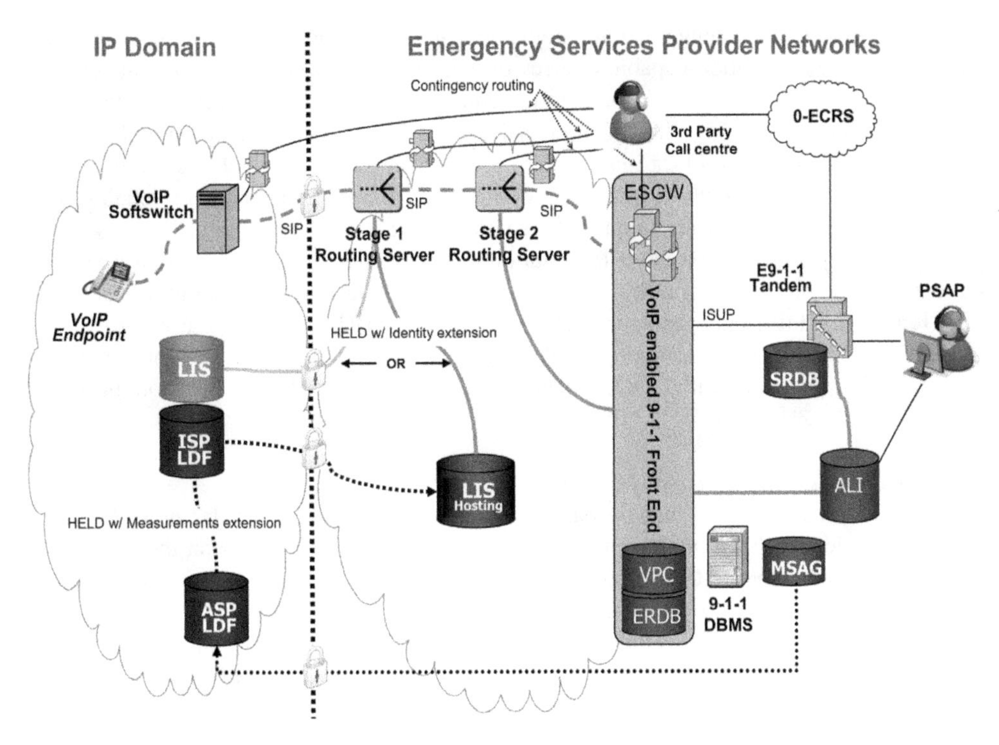

Figure 4.14 The Canadian i2 functional model. (Courtesy of Bell Canada.)

- achieved consistency with the NENA i2 model's spirit to minimize operational impacts on PSAPs; and
- introduced appropriate deviations to the NENA i2 model to achieve the mandate.

Both NENA i2 and its Canadian version faced some common challenges. The most obvious one, which was also highlighted as the most important impediment for the establishment of a solution for nomadic and fixed/non-native VoIP, was the absence of appropriate location determination and delivery functions to compute and provide trustworthy, accurate, and dependable location information in real time, without user involvement. In Canadian i2, the location determination function was proposed to be performed within the broadband access network while the location delivery function was proposed to be hosted on an LIS which could be located in the broadband access network or hosted at the Emergency Services Network level. More specifically, the Ci2 proposal exposed in some details various solutions for location-related functions in xDSL[8] and DOCSIS[9] networks, including Third Party Internet Access (TPIA) where collaboration would be required between the ASP and the third-party ISP.

[8] Digital Subscriber Line. The "x" stands for the many variants of this technology, for example, "H" for High bit rate, "A" for Asynchronous, "V" for Very high bit rate.

[9] Date Over Cable Interface Specification (CableLabs).

Another important challenge that any nomadic VoIP 9-1-1 solution was facing at the time was the absence of location-capable devices on the market, despite the vision put forward by the IETF where devices would be able to acquire and deliver their location at call time. To address this shortcoming, the Ci2 proposal included a location acquisition and delivery solution that did not rely on location-capable devices. The solution, which was based on the IP address of the caller's device, while not perfect, was believed to be generally reliable. The PSAP community unanimously supported the Ci2 solution.

Despite its apparent completeness, the solution was not well received by some parties, which led to a lengthy and time-consuming regulatory process that lasted more than four years.

In June 2010, the Commission rendered its Decision [192] based on the record of the proceedings. The following are highlights of that Decision, with some post-Decision comments from the author (indicated with superscript capital roman numerals after the decision) provided herein below.

- *Technical Issues and Technology Trends and Evolution.*

  The Commission noted that no parties indicated that Ci2 or a similar nomadic VoIP E9-1-1 service has been implemented in other countries. The Commission therefore considered that the proposed Ci2 solution was untested and that actual experience with respect to the feasibility and costs of implementing such a solution did not exist.(I)

  The Commission also noted that all parties agreed that the long-term goal in the evolution of the 9-1-1 network was to migrate customers to NG9-1-1. Further, the Commission noted that the Ci2 implementation was expected to take at least two years and that, similarly, the deployment of NG9-1-1 was expected in a few years. The Commission considered that future solutions based on NG9-1-1 would support nomadic VoIP service by using location-aware devices to determine a subscriber's location, which would make it unnecessary for ASPs to implement a location-determining platform in their networks.(II)

  The Commission considered that NG9-1-1 offers several benefits that will likely result in a technologically advanced solution: it is based on international standards; it introduces improved technologies for location detection;(III) and it provides access for multiple communications devices and technologies, on both wireline and wireless networks.

  The Commission noted that access to nomadic VoIP services is migrating from wireline to wireless high-speed Internet access through the use of wireless devices such as cellphones. However, for technical reasons, Ci2 cannot be used to determine the location of a 9-1-1 caller when a nomadic VoIP service subscriber is using a wireless connection.(IV)

  In light of the above, the Commission considered that, by the time Ci2 would be implemented, it was likely that it will have become outdated as a result of technological changes and market developments such as the introduction of NG9-1-1.(V)
- *Consumer Demand and Usage Patterns.*

  The Commission noted that, according to its 2009 Communications Monitoring Report, the number of nomadic VoIP service subscribers declined from 161,000 in 2007 to 153,000 in 2008, and accounted for 0.8% of wireline telephone service subscribers in 2008. Based on these statistics and assuming no further declines, the Commission noted that the nomadic VoIP service subscribers who would potentially use Ci2 comprise 0.4% of all 9-1-1 subscribers and 2% of high-speed Internet subscribers.(VI)

  The Commission also noted that a study on Nomadic VoIP Usage in Canada submitted as part of the proceeding indicated that two-thirds of Canadian households with nomadic VoIP service do not use it as their primary phone service. In addition, when traveling, 73% of nomadic VoIP service subscribers use their cellphone instead of the nomadic VoIP service, and only 4% use their nomadic VoIP phone adapter. The Commission noted that these statistics indicate that most nomadic VoIP service subscribers rely on other wireline and wireless services as their primary telephone service, which they can use to obtain access to emergency services. Based on the study and as noted above, the Commission noted further that access to nomadic VoIP services is migrating from wireline to wireless high-speed Internet access.

Based on its 2009 Communications Monitoring Report and other evidence, the Commission concluded that the nomadic VoIP service subscriber base represents a small portion of 9-1-1 users and was in decline. It also concluded that subscribers, whether at home or traveling, were generally able to rely on other wireline or wireless services in order to access a 9-1-1 service that is comparable to Ci2. Consequently, the Commission considered that the implementation of Ci2 would provide an additional benefit to only a small and decreasing subset of nomadic VoIP service subscribers.

- *Implementation Costs.*

  Based on the record of the proceeding, the Commission estimated that the overall cost of implementing Ci2 was more than $180 million.

  Historically, the service provider whose subscribers benefit from access to 9-1-1 service has been responsible for all costs of providing that service. Providers usually pass these costs on to their subscribers. It is through this mechanism that wireless, traditional wireline, fixed VoIP, and nomadic VoIP service providers currently cover the cost of providing 9-1-1 service to their own subscribers. The Commission noted, however, that this approach was not viable for nomadic VoIP service, as providers would not be able to generate sufficient revenues to bear the costs of implementing and operating Ci2, which are estimated at $15 per month per nomadic VoIP service subscriber.[VII],[VIII]

  Finally, the Commission considered that exempting small ASPs from implementing Ci2, while lowering the overall costs of implementation, would create public safety risks in the areas they serve. This is compounded by the fact that there are hundreds of small ASPs across Canada, so subscribers would not be aware of, or would have difficulty determining, where these gaps in service were.

Through CISC deliberations over the Ci2 proposal, other alternatives were discussed and were made part of the Report to the CRTC. Those alternatives are briefly discussed below. However, the Commission determined that none were viable alternatives at the time of the Decision.

- *IP Tracker.*

  Some parties proposed an "IP tracker" alternative that was a variation of the current VoIP Basic 9-1-1 service. This proposal would require a nomadic VoIP service provider to implement the capability to detect a major change in the IP address used to access the Internet and then automatically direct the subscriber to the VoIP service provider's call center to update their location. When the subscriber dials 9-1-1, the call center would direct the call to the appropriate public safety answering point (PSAP) and verbally provide location information, as is the case with the current VoIP 9-1-1 service.

  The Commission considered that the IP tracker proposal has a number of technical deficiencies, mainly that it may not be possible for VoIP service providers to detect major changes in the IP addresses assigned to high-speed Internet users of some ASPs, in particular those using DSL technology. The Commission also considered that the IP tracker solution would not enable providers to detect more common minor changes in IP addresses, which may result from a change in subscriber location, and therefore concluded that the IP tracker proposal was not a viable alternative.

- *Variations of Ci2.*

  Some parties proposed as alternatives a few variations to a specific part or parts of Ci2, all of which left most of the Ci2 architecture and processes intact and reduced the overall cost.

  The Commission noted that the variations of Ci2 as proposed would result in reduced functionality or removal of certain important features and processes within Ci2. The Commission considered that this situation would negatively affect the overall functioning of Ci2 and the accuracy of information provided to PSAPs. Accordingly, the Commission concluded that the proposed variations would jeopardize the overall integrity of Ci2 and that implementation of these proposals would result in public safety risks.

- *Next Generation 9-1-1.*

  Some parties submitted that future solutions based on NG9-1-1 would be able to provide location information for nomadic VoIP service providers. They submitted that such solutions would use location-aware devices (e.g., GPS-enabled devices) to determine a subscriber's location and would therefore not require the implementation of a location-determining platform in ASPs' networks.(IX) They added that vendors and the industry are currently developing NG9-1-1 technologies and standards. However, the Commission considered that, while NG9-1-1 solutions are likely to become viable alternatives in the future, the associated technologies and standards have not yet matured to the point where they can be considered viable alternatives today.

  In conclusion, the Commission noted that all parties submitted that maintaining the current VoIP 9-1-1 service would be better than implementing any of the proposed alternatives, with certain parties supporting the alternative solutions only to the extent that they are less costly than Ci2. The Commission considered that the proposed alternative solutions were not technically viable, pose public safety risks, and lack the integrity and robustness required for 9-1-1 services. The Commission therefore concluded that none of the proposals was a viable alternative solution to providing VoIP E9-1-1 service.

Post-Decision author's comments:

(I) At the time of the proceedings, NENA i2 or a variation thereof was already implemented in the United States. Unfortunately, this fact was not put on the record.

(II) To the author's knowledge, there is no location determination solution defined under NENA i3 or NG9-1-1. Rather, NENA i3 and NG9-1-1, as for NENA i2, specifically make location determination out of scope, but rely on its availability. Further, location-aware devices must rely on some form of location determination function to acquire its location. For devices operating on a wireless macro network, they may rely on a Global Navigation Positioning System (GNSS). For devices operating on any other types of network, other solutions may be required.

(III) See comment (II).

(IV) NENA had plans to evolve i2 to support mobile devices. Moreover, the underlying model that inspired the NENA i2 solution is the wireless E9-1-1 Phase II model (as specified in TIA/EIA IS-J-STD-036). Evolving NENA i2 to support mobile was believed to be a fairly simple task.

(V) As of November 2012, the interim solution is still in effect in Canada, true NG9-1-1 is not yet deployed, and i2 is still widely used in the United States.

(VI) These Canadian statistics appear to be quite low in comparison to other markets. The author believes that only registered Canadian VoIP service providers could have been considered by the Commission. Other worldwide providers like Skype may not have been accounted for, hence the rather low numbers noted.

(VII) Of course, if the number of nomadic and fixed/non-native VoIP users in Canada had been estimated higher, this cost would have been lower.

(VIII) To arrive at this monthly cost, the author believes that *all* Ci2 costs should have been imputed to the ASP, not only the location determination and delivery costs, as originally proposed by the proponents of Ci2.

(IX) See comment (II).

4.3.5 VoIP Regulatory Processes, Decisions, and Milestones

As indicated earlier, the VoIP regulatory journey started in 2004 when the Commission was asked to address the regulatory framework surrounding voice communications services using Internet Protocol. Through a series of public consultations, technical reports, and decisions, the Commission made several determinations to assist the industry in improving how 9-1-1/E9-1-1 service should be provided over VoIP technology. Below are the major determinations that paved the way throughout the process:

- VoIP services utilizing telephone numbers based on the North American Numbering Plan and providing universal access to and/or from the Public Switched Telephone Network (PSTN) have functional characteristics that are the same as circuit-switched voice telecommunications services, and as such should be regulated and treated as local exchange services, which includes the obligation to the provision of 9-1-1 and enhanced 9-1-1 service, message relay service, and privacy safeguards [193].
- Fixed (i.e., non-nomadic) local VoIP service, where the end user is assigned an NPA-NXX native to any of the local exchanges within the region covered by the customer's serving Public Safety Answering Point (PSAP), must provide 9-1-1/E9-1-1 service, where it is available from the incumbent local exchange carrier (ILEC) [191].
- The Commission found it unacceptable for nomadic and non-native VoIP services providers to route calls based on provisioned and/or user-input location, as it places undue additional burden on PSAPs, which are not equipped to deal with out-of-region callers. The Commission further noted that simply routing 9-1-1 calls to what may be the wrong PSAP would lead to a slowdown in emergency response time [191].
- VoIP services provided on a nomadic basis or with a telephone number that is not native to any of the exchanges within a customer's PSAP serving area must implement an interim solution that provides a level of service functionally comparable to Basic 9-1-1 [191].

- Minimum requirements for customer notification regarding the availability, characteristics, and limitations of the 9-1-1/E9-1-1 service offered with local voice communication service over Internet Protocol (VoIP service) were set [194].
- The Commission approved the CISC recommendations to develop a functional architecture for the implementation of VoIP E9-1-1 in Canada consistent with the NENA i2 standard, adjusted as necessary for implementation in Canada [195].
- The Commission found it inappropriate for VoIP service providers to deliver 9-1-1 calls from their fixed/non-native and nomadic VoIP customers to public safety answering points (PSAPs) using low-priority telephone lines or restricted numbers. The Commission considers that zero-dialed emergency call routing service (0-ECRS) is the only available 9-1-1 call routing method on the record that is functionally comparable to the Basic 9-1-1 service [196].
- The Commission found that, in principle, the proposed Canadian i2 functional architecture for the implementation of VoIP E9-1-1 service in Canada meets the requirements set out in Telecom Decision CRTC 2006-60 [197].
- Based on the results of the discussions surrounding the Canadian i2 functional architecture, the Commission directed parties to file economic evaluations of the costs to be incurred to develop, implement, and operate elements of the architecture as well as their proposal as to how such costs will be recovered [197].
- Based on the economic evaluations submitted, the Commission found it necessary to conduct a further proceeding to elicit better location determination platform (LDP) cost estimates [186].
- The Commission finally determined that there were no viable alternatives (including the proposed Canadian i2 solution) to the Voice over Internet Protocol (VoIP) 9-1-1 service currently provided, with nomadic and fixed/non-native VoIP service and directed nomadic and fixed/non-native VoIP carriers to make certain improvements to their current VoIP 9-1-1 service until Next Generation 9-1-1 service is implemented [192].
- To improve the interim solution, the Commission directed all Canadian nomadic and fixed/non-native VoIP service providers to provide their customers' telephone numbers to their 9-1-1 operators. The operators are to use the numbers provided as a last resort to re-establish contact with a 9-1-1 caller when the call is disconnected before the caller's location has been determined [198].
- The Commission then clarified that the requirements for service providers to notify customers and obtain their express consent, set out in previous Decisions, do not apply to fixed/native VoIP services [199].
- Finally, the Commission has planned to review the regulatory framework for Next Generation 9-1-1 services for Canada in the 2014–2015 time-frame.[10]

In summary, the current status for VoIP 9-1-1 in Canada is that fixed/native VoIP services must provide full E9-1-1 service equivalent to PSTN where available, while nomadic and fixed/non-native VoIP services must provide Basic 9-1-1 (B9-1-1) service through a manned third-party call center until such time technically and economically viable solutions emerge, including the advent of Next Generation 9-1-1 (NG9-1-1) services.

[10] At the time of writing, November 2012.

4.3.6 Lessons Learned

In many aspects, the Canadian experience is no different from others in the world. Countries that will tackle the issue of nomadic and non-native VoIP 9-1-1 services will eventually face the same challenges. While existing emergency network infrastructures and features certainly vary from country to country, there are commonalities that will likely emerge. This book is being put together to share the experiences of those countries that have lived through the process so that others who will face the issue do not start from scratch again and again. Here are a few lessons learned through the Canadian experience.

1. In 2004–2005, the over-the-top market was in its infancy and it was too early to confirm whether it would be widely adopted or would remain a niche market. This situation impeded progress, since the parties invoked the relatively low number of subscribers even when PSAPs systematically provided examples of issues they were facing on a daily basis.

   Recommendation: Ensure that you have reliable statistics on over-the-top VoIP services penetration that includes foreign providers with subscribers residing in you country. Do not hesitate to contact those providers directly.
2. Some industry players were not prepared to facilitate market entry for over-the-top service providers, sometime referring to them as "hostile providers." This position blurred the issue and took the 9-1-1 service hostage, making public safety a secondary consideration.

   Recommendation: Ensure that you have a strong regulatory framework that promotes fair competition across your communications industry, including over-the-top VoIP, before entering this journey.
3. To combine location determination solutions with the call routing solution was a bad idea, since one killed the other, thus leaving the country with a less than perfect interim solution for a long period of time.

   Recommendation: Keep the location determination, acquisition, and delivery topic and the call routing topic separate within your regulatory process. Both have distinct timelines that do not necessarily match.
4. The overall Ci2 solution was technically sound and proven but premature for its time.

   Recommendation: Whatever solution is being proposed and discussed in your country, ensure that vendors are supportive and ready to make the solution commercially available within a reasonable amount of time.
5. The complexity and challenges of breaking away from the traditional funding model were underestimated.

   Recommendation: Challenging the *status quo* is tough and risky. Your solution should adhere as much as possible to the funding model that currently exists in your country. If not possible (as was the case in Canada), strive for a proposal that includes a funding model acceptable to your regulatory body and its constituents.
6. Regulation is required, but alone is not enough. A minimum of industry willingness is necessary to move ahead.

   Recommendation: Ensure a maximum level of support from your communications industry participants, especially, but not limited to, the major players, prior to submitting your proposal to your regulatory body.

4.3.7 Conclusion

While it has been straightforward for the fixed/native VoIP services to support E9-1-1, it has been a journey through a long and strenuous regulatory process for the nomadic and fixed/non-native VoIP E9-1-1 services in Canada. Consensus among the various parties was not always achieved, sometimes leaving the CRTC with no choice other than to make its own judgment on the positions and issues raised. The implementation of automatic location determination functions in the broadband access networks, a fundamental prerequisite for any form of location-based services in the IP domain, raised the most significant issues alongside the cost recovery mechanism. The PSAP community's position has been constant and clear, namely, that a more suitable solution than the current Basic 9-1-1 service must be adopted so that emergency calls are presented in a uniform manner irrespective of the technology used. Among other things, the Commission needed to consider the perceived somewhat small base of nomadic and fixed/non-native VoIP subscribers from which some of the costs could be recovered, and the heavy customer trend in adopting mobile telecommunications technologies.

NG9-1-1 promises to solve many issues and limitations with the current E9-1-1 infrastructure. However, it relies on the availability of accurate and trustworthy location information to ensure that citizens are well protected in case of emergency. This was, and still is, a major stumbling block when it comes to communications technologies, be it mobile or fixed. Fundamentally, location technologies are associated with the type of access network they reside in. Standards development organizations (SDOs) must consider that any media can potentially support an emergency call and, as such, ensure that a suitable location platform (network- or device-based) is available. The 9-1-1 service is and will remain a location-based service. Without an accurate and trustworthy location, it is a major challenge for emergency personnel to act promptly and diligently. It has been proven may times that, when responding to an emergency call, seconds matter.

At least in North America, the wireless industry is one step ahead since it has already deployed location determination technologies and platforms for macro circuit-switched networks. Further, GPS-enabled smart phones are fast growing in the marketplace. Hopefully, these will be leveraged with VoLTE and other packet-switched services that will soon hit the market. We also see a trend in consumers substituting their traditional PSTN lines with wireless services. Whether the world will be turning fully mobile remains to be seen, but it would be imprudent to predict the end of fixed broadband. xDSL and more recently FTTx technologies are major deployments that carriers are undertaking to bring higher speeds to consumers. Conversely, there is another trend from the wireless carriers to offload some of the broadband traffic away from their macro networks toward WiFi or other broadband accesses in the home. A 9-1-1 solution that does not cover those would simply be incomplete.

In the meantime, Canadian subscribers to nomadic and fixed non-native VoIP services can only rely on the interim solution, and PSAPs will continue to do their best to serve them as promptly as possible considering the shortcomings of that solution.

4.4 US/Indiana Wireless Direct Network Project

Byron Smith
L.R. Kimball

The Indiana Wireless Direct Network (IWDN) project provided the rare opportunity to build a new emergency communications (9-1-1) infrastructure on an essentially "green field."

4.4.1 Background and History of the IWDN

The Indiana Legislature created the Indiana Wireless Advisory Board (the Board) in 1998 [200]. By State statute the Board is chaired by the State Treasurer (an elected official) and consists of six additional members appointed by the State Governor. Three of the board members are representatives of the Public Safety Answering Point (PSAP) community, and three are representatives of the Commercial Mobile Radio Service (CMRS, or wireless carrier) community.

Major responsibilities of the Board are the collection and distribution of 9-1-1 fees assessed on Indiana cellular telephone users, and the development of wireless 9-1-1 services in the State. A significant goal of the Board has been the implementation of FCC Phase II wireless 9-1-1 services throughout the State.

In 2003, the quality, reliability, and cost of Indiana wireless 9-1-1 services were concerns of the Board. The network was not redundant. Service problems occurred with unacceptable frequency, and, with multiple companies involved, were sometimes difficult to resolve promptly. Call setup times were unacceptably lengthy in certain localities, sometimes exceeding 20 seconds. The Board lacked basic statistical information, such as statewide call volume and geographical distribution information. There were problems transferring 9-1-1 calls between PSAPs served by different Local Exchange Carriers (LECs). And FCC Phase II service implementation was proceeding very slowly.

At the time, eleven wireless carriers served Indiana. These wireless carriers operated 31 mobile switching centers (MSCs) that connected to cellular towers in the State. Indiana Local Exchange Carriers (LECs) operated seventeen 9-1-1 tandems/selective routers that served the State's PSAPs. Only three of these seventeen 9-1-1 tandems had SS7 capabilities. The remaining fourteen selective routers used CAMA/MF trunks exclusively.

Each MSC had to trunk to all of the LEC-operated selective routers that served PSAPs that were within the coverage area of that mobile switch. Over 900 individual circuits were required on about 200 separate transport facilities. The total monthly recurring charges for all of these circuits and facilities consumed at least half of the 9-1-1 fees (then $0.65 per month per phone) paid by each cellular telephone customer in Indiana. Most of the balance of the 9-1-1 fees were distributed to PSAPs to assist in equipment upgrades and other costs associated with wireless 9-1-1 Phase II services.

In 2003 the Board, aware of several "wireless direct" projects elsewhere in the United States, hired L. Robert Kimball & Associates, Inc., to conduct a study to determine the feasibility of building a "wireless direct" network for Indiana. As a result of the feasibility study, the Board issued an RFI in February 2004, inviting qualified vendors to respond with proposed IWDN solutions in April 2004.

After an extensive process of review of the RFI responses, clarifications, demonstrations, and vetting, an evaluation committee recommended that the Board commission the construction of the Indiana Wireless Direct Network as proposed by INdigital Telecom, a Fort Wayne, Indiana, based Competitive Local Exchange Carrier (CLEC). A extensive contract negotiation followed, and the Board and INdigital signed a contract to build the IWDN in April 2005.

4.4.2 The IWDN Crossroads Project

The IWDN solution ultimately agreed to by the Board and by INdigital consisted of two major efforts.

The first part of the project, which ultimately became know as the "Crossroads" network, was effectively a trunk consolidation effort that used a relatively traditional SS7-based telephone network. Crossroads established two new wireless 9-1-1 tandems operating as a redundant "mated pair" selective router. These two switches are diversely located, one in the Fort Wayne area and the other in the Indianapolis area. In the Crossroads network, all the MSCs send their Indiana wireless 9-1-1 traffic to either of the Crossroads 9-1-1 tandems using diversely provisioned facilities. All MSC connections to the Crossroads switches are SS7 ISUP trunks, eliminating CAMA/MF signaling for the wireless carriers. The Crossroads switches, in turn, connected to the ILEC 9-1-1 selective routers using either ISUP or CAMA/MF trunks, as required.

The Crossroads network quickly addressed a number of the Board's concerns. It provided a redundant and diverse network all the way from all the MSCs to the ILEC tandems serving the PSAPs. It provided statewide 9-1-1 statistics. It cut the number of facilities and circuits by more than half, providing an immediate and substantial cost saving for the operation of the Indiana wireless 9-1-1 network, even as it provided full redundancy from the MSC to the ILEC selective router.

However, Crossroads did not implement the "direct" part of IWDN as requested by the Board.

Construction of Crossroads began in April 2005 with meetings with wireless carriers and ILECs concerning details of the project, the establishments of interconnect agreements with the LECs, circuit ordering, and such activities. First live traffic on the Crossroads network occurred near the end of January 2006. The Crossroads network was completed December 2006, although two of the wireless carriers did not complete total migration of their traffic to the Crossroads network until some months later.

As a result of the cost savings provided by the Crossroads network, the Board dropped the Indiana 9-1-1 fee to $0.50 per month per cell phone, even as the Board increased the PSAP distributions and provided funding for the second stage of the IWDN project, which is discussed in the next section.

4.4.3 The IN911 IP Network

The second major component of the IWDN project has become known as the "IN911 network." As of this writing IN911 remains a work in progress. This IP network provides all 9-1-1 voice, data, and signaling connections between the Crossroads 9-1-1 tandems and all PSAPs directly connected to the IN911 network in Indiana.

The IN911 network is a statewide, redundant, diverse, private, secure, and monitored IP network. It is expected that IN911 will eventually interconnect all, or nearly all, of the primary wireless PSAPs in Indiana. IN911 is intended to be a "next-generation" 9-1-1 infrastructure, capable of supporting a variety of public safety applications.

Construction of the IN911 network began in parallel with the construction of the Crossroads network. As of this writing, the IN911 network is in the "back room" of more than three-quarters of Indiana PSAPs, and construction of the IN911 network continues. The IN911 network carries voice, ANI, and ALI for wireless calls from the Crossroads selective routers directly to nearly half of the State's more rural, lower-call-volume, PSAPs. Only four LEC selective routers remain connected to the Crossroads 9-1-1 tandems.

As expected, the IN911 IP network also provides other public safety services. Examples are inter-PSAP voice calling, and data connections to IDACS for some PSAPs. (IDACS is the Indiana Data and Communications System operated by the State Police. IDACS is used for motor vehicle registration and operator license checks, criminal information, and other law enforcement activities. IDACS is a highly secure IP system that uses fully encrypted TCP connections.) Quality-of-Service (QoS) features of the IN911 ensure high-quality voice communications. The data applications, such as IDACS, operate at a lower priority than the voice channels. However, this sharing of resources provides a significant cost reduction and higher quality of service for the expenditure compared to separate networks for these services.

The IN911 network consists of IP "backbone" running on a SONET fiber ring that spans the State of Indiana. The backbone ring is provided by Indiana Fiber Network. Sixteen sites (at least one site within each LATA in the State) connect to a daisy chain of DS1 (T1) circuits that run between a small number of nearby PSAP sites. Both ends of the daisy chain connect to the IP backbone. All circuits terminate on redundant equipment. The resulting IP network has no single point of failure with respect to circuits or with respect to equipment.

Figure 4.15 shows a schematic representation of the structure of the IN911 IP network. The IP network uses the OSPF routing protocol on the backbone, and Cisco's EIGRP on the daisy chains. EIGRP provides quick IP routing re-convergence in the event of a "break" in the daisy chain. OSPF is fully compatible with open source-based servers (e.g., Linux boxes) that provide network services that are directly connected to the backbone IP network.

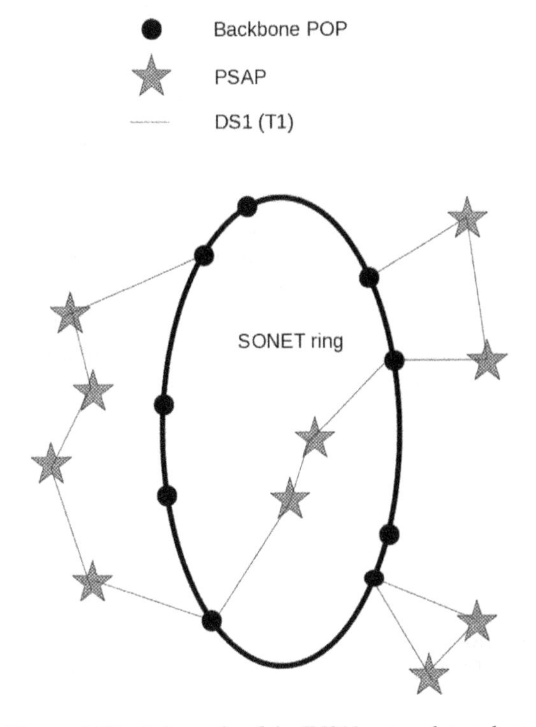

Figure 4.15 Schematic of the IN911 network topology.

All voice traffic on the IN911 network uses SIP signaling and G.711 codecs. There have been no voice quality issues with respect to the IN911 network. The network uses Cisco gateways, and a few audiocodes gateways, at most ingress and egress points. These gateways provide QoS features and echo cancelation. The only echo problems encountered in the operation of the IN911 network have been external to the network itself.

The IN911 network SIP proxies are based on an INdigital-modified open source SIP stack. A feature of the SIP User Agent in the gateways is the ability to register with at least two SIP proxies. This feature allowed the establishment of two totally independent and diversely located SIP processors in the network. If an ingress gateway does not receive a response from an SIP proxy within a short timeout period, it will send the SIP message to the redundant proxy. The egress gateways will accept SIP messages from either IN911 SIP system.

IN911 call overflow (busy) and ring-no-answer processing is controlled by using CPL (Call Processing Language) scripts that are interpreted by an SIP proxy. Many other 9-1-1 specific features of the network have been implemented in this way.

A significant challenge in implementing the IN911 network has been protocol conversion between the "next-generation" IP network and the "traditional" 9-1-1 equipment found at most PSAP sites. The two biggest challenges were meeting CAMA signaling format requirements, and call transfer capabilities in the IN911 network. Careful gateway selection and programming resolved the CAMA signaling issue.

Call transfer was trickier. Initially, INdigital spent a lot of time trying to deal with the "hook flash" signaling that is commonly used on CAMA trunks in the US 9-1-1 industry. However, it was discovered that using hook flash raised more problems than it solved, because PSAP ANI/ALI controllers, and/or LEC 911 tandems, would intercept the hook flash and perform unintended or unexpected processing on the operator's subsequent DTMF commands. This resulted in a signaling and provisioning nightmare.

Eventually, INdigital went in a different direction, by not requiring a hook flash, but by simply monitoring for mid-call DTMF tones. When a certain (programmable) DTMF sequence is detected (typically * 8 #), the call is "moved," via SIP signaling, to a specially prepared VoIP conference bridge, mid-call. This bridge then provides a low-volume dial tone to the 9-1-1 operator, who can dial strings to add on other PSAPs or destination operators to the call. The operator can continue to hear the caller throughout this process. The bridge initiates SIP connections to the additional party, and includes the ANI associated with the original call, so that the added-on party can receive 9-1-1 ALI data, if so equipped. This technique has proven to be very successful, and has been backwardly compatible with a variety of existing PSAP equipment. Some alternative signaling schemes are in place in the IN911 network in special circumstances, but the system just described is in use in 95% of the Indiana PSAPs that are currently connected to the IN911 network.

The use of SIP location conveyance for ALI data has been demonstrated in the IN911 network. A modification to an SIP proxy permits a PDIF-LO attachment to be posted in the IN911 ALI system. Then, this data is provided as the ALI response on a subsequent query by the PSAP equipment. In this way, SIP location conveyance data can be made available to PSAPs using traditional equipment. This scheme was demonstrated, in part, at the ECRIT workshop in Vienna, in October 2008, using Karl Heinz Wolf's modified "ZAP" client [201] to initiate an SIP call with a PIDF-LO attachment. The call was connected from the meeting site to the IN911 network via a secure tunnel through the Internet, and was answered by an actual

9-1-1 dispatcher in Logansport, Indiana. This system has subsequently received additional development effort at INdigital, and its viability has been demonstrated.

Several other development efforts are under way at INdigital. An SS7 ISUP to SIP signaling gateway is in place which will support the termination of ISUP trunks to any IN911 backbone POP. This, together with a modification to an SIP proxy to support traditional ANI-based 9-1-1 call routing will support the bypass and elimination of the Crossroads tandem switches, as well as offer more flexibility in the provisioning of 9-1-1 facilities. At the very least, this scheme would eliminate the need to back-haul circuits to centralized switching facilities, reducing cost and supporting a more diverse, and presumably a more survivable, 9-1-1 network. And, of course, it should be possible to terminate SIP "trunks" directly from a wireless carrier's MSC directly into the IN911 network. Several wireless carriers serving Indiana have recently expressed interest in exploring such an option.

4.4.4 Conclusion

The Indiana Wireless Direct Network has carried more than 7,000,000 wireless 9-1-1 calls from its inception. A significant component of the IWDN is the IN911 IP network, which has handled nearly 2,000,000 of the 7,000,000 calls to date. With careful design and implementation, this effort has demonstrated the viability of IP-based public safety solutions, and points a path to the next-generation networks of the future.

5

Security for IP-Based Emergency Services

5.1 Introduction

Throughout this book, we illustrate how IP networks and the SIP-based communication infrastructure create the foundation on top of which emergency services support is added via a number of building blocks. The building blocks provide the basis for (i) determining location information, (ii) recognizing an emergency call, (iii) evaluating the route toward the PSAP, (iv) establishing the SIP session setup, and (v) exchanging multimedia data traffic (such as voice, video, text messages, and real-time text). As explained in Chapter 3 there are architectural variants that impact which of these building blocks are used and how. While there are unique aspects of each of these architectural variants, for this discussion about security we will stay at a higher abstraction level to keep the description to a reasonable length.

When considering the security properties of the overall emergency services system, two observations can be made:

1. Emergency services support builds on top of the regular communication infrastructure, and therefore the security properties of the underlying infrastructure are inherited.
2. Security has to be taken into consideration for each and every protocol, since the resulting system is as secure as the weakest link. This also implies that the security analysis done for each protocol has to be taken into account by the respective users of those protocols. Considering point 1, it is easy to observe that there are many protocols involved that need to be analyzed.

Note that it is not sufficient to write interoperable protocol specifications with security consideration sections that take both deployment reality into account and offer an adequate level of protection. Implementations of these protocols also need to be secure against various attacks, such as buffer overflows, and the deployed system must be configured in the appropriate way (e.g., to withstand denial-of-service attacks or to be secured against unauthorized access). Building secure communication systems that include emergency services components

Internet Protocol-Based Emergency Services, First Edition. Hannes Tschofenig and Henning Schulzrinne.

is complex and presumes education and awareness (e.g., via employee training), good software management practices (e.g., secure coding, extensive testing), established and working processes (e.g., incident management, clearly defined responsibilities, and accountability), and so on. A detailed discussion of these tasks is, however, outside the scope of this book. Hence, we focus our description on the protocol-based mechanisms, which build the basis for any system design. Building communication systems based on global standards, as those described in this book, also ensures that those standards have been reviewed by many experts in order to avoid unexpected side-effects.

In this chapter we will discuss some of the unique security characteristics found in the emergency services environment. First, we start with the description of the communication model and the involved entities of the system, then we investigate the adversary model and security threats. Finally, we conclude with a list of countermeasures.

5.2 Communication Model

When we look at all the protocol interactions of the entire communication system that aims to interwork with legacy infrastructure, it becomes difficult to judge the threats and countermeasures. For this reason, we look at the high-level communication models, as shown in Figure 5.1.[1]

There are four main communication interactions to consider:

(I) *SIP Signaling Communication.* The SIP signaling exchange is at the heart of the emergency services communication. SIP [124] is an application-layer control (signaling) protocol for creating, modifying, and terminating sessions with one or more participants. Emergency services functionality has been added to SIP with specifications that define the ability to express emergency services functionality (by using the service URN specification in RFC 5031 [11]) and the ability to carry location objects (see [32]). This functionality for emergency call signaling that was added to the already feature-rich SIP protocol was fairly small. Most of the effort was spent on the responsibilities of the different stakeholders in the overall architecture. Chapter 3 describes these efforts, and current deployment initiatives are highlighted in Chapter 4. After the emergency services communication is terminated, it is possible for the PSAP to initiate a callback to the emergency call, for example, when details about the situation need to be clarified or the caller prematurely terminates the conversation. The call setup procedure then starts with the PSAP and the SIP session setup messages travel toward the end device that initiated the call.

(II) *Exchange of Multimedia Data.* The main purpose of the communication setup is to exchange data. For emergency calls this typically takes the form of voice packets. In the

[1] In this chapter, we use the term "Location Server" as well as "Location Information Server" (LIS). RFC 6280 [29] describes the LIS as follows: "Some entities performing the LG role are designed only to provide Targets with their own locations, as opposed to distributing a Target's location to others. The process of providing a Target with its own location is known within Geopriv as Location Configuration. The term 'Location Information Server' (LIS) is often used to describe the entity that performs this function. However, an LIS may also perform other functions, such as providing a Target's location to other entities." In this chapter we use the term LIS instead of Location Server whenever it is more appropriate.

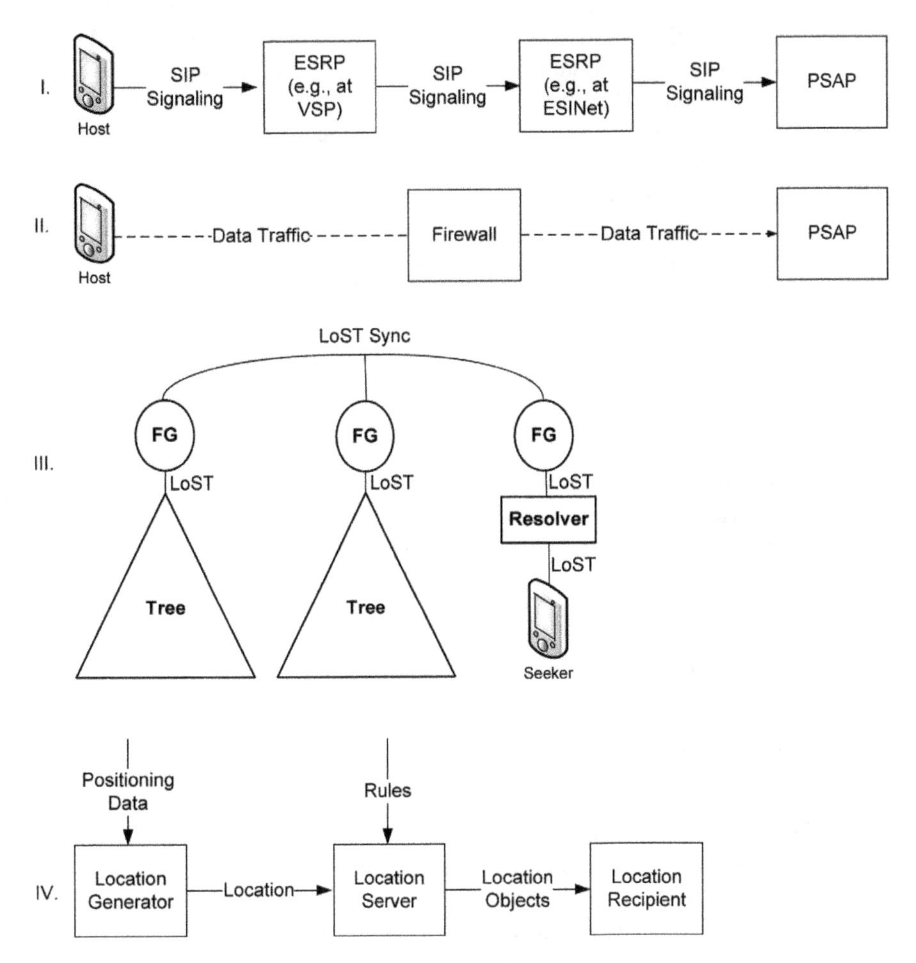

Figure 5.1 Communication model overview.

future, this will also be video, instant messages, real-time text, and pictures, supported by the capabilities of IP. SIP is able to detect and negotiate the supported functionality and to enable the exchange of these additional media streams. The used media will largely depend on the capabilities of the devices, the installed software applications, and the preferences of the emergency caller. For persons with disabilities, real-time text and video communication is likely to make a big difference, but might require a third-party relay service to be invoked, for example, for sign language interpretation.

(III) *Mapping Database.* When emergency call setup is initiated, the call routing processes must decide the next hop to route the setup messages to. A major factor in call routing is location information, since the geographically closest PSAP will often want to receive the call first. In some cases, other factors, such as the load situation at a PSAP or the ability to accept certain media types, also influence the decision. For location-based routing, the distributed mapping database that uses the protocol LoST is used, as illustrated in [132] and described in section 3.3. The design is similar to the DNS and allows end hosts and ESRPs to ask for the appropriate PSAP based on a given location input.

(IV) *Location Infrastructure.* The location retrieval interaction starts with an entity issuing a request for a location object. The requesting party is the Location Recipient, which may be the end host, an ESRP, or the PSAP. Each of these entities have a legitimate interest in receiving location information for the determination of the appropriate emergency dial strings, for emergency call routing, or for the dispatch of first responders. When a location server, which is commonly assumed to be operated by an ISP, receives a location request, it may not always have the current information cached. The location server then starts a location determination process whereby location generators collect positioning data. The location server may provide location information as a one-shot transaction, or it may provide multiple updates as location determination techniques compute better location over time or when the location of the target device changes. For security and privacy purposes, location may not be shared without a prior authorization check, which is invoked based on a rule set available to the location server. In Chapter 2, several location configuration protocols are described in detail, and RFC 6280 [29] illustrates a location architecture and introduces terminology.

5.3 Adversary Models and Security Threats

In a discussion about security threats, it is important to keep the anticipated capabilities of an adversary in mind. We distinguish between three types of adversaries:

- *External Adversary Model.* This type of adversary is the most commonly considered adversary in communication networks. The adversary is external to the analyzed system and interferes from the outside. In such a threat model, it is assumed that none of the emergency service infrastructure elements has been compromised. Consider the case where a person makes an emergency call. An adversary could be located along the communication path between the end host and the location server, or between the end host and the PSAP. Without proper communication security, this adversary could eavesdrop on the communication or prevent the emergency caller from successfully calling for help.
- *Malicious Infrastructure Adversary Model.* With this type of adversary model, we assume that some entities in the emergency services infrastructure either are misconfigured or compromised and thereby disrupt the normal emergency system operation. Such misconfigured or compromised elements can include the LIS, the Location to Service Translation (LoST) infrastructure [68, 132], and call routing elements, which may be manipulated to respond with false information.
- *Malicious End Host Adversary Model.* With this adversary model, we assume that the end system is compromised. Although this type of adversary model, is a subcategory of the malicious infrastructure adversary model, it deserves special attention since end systems, like tablets and desktop PCs, are often more vulnerable due to their poor software update policy and lack of proper administration. Furthermore, a human interacting with the emergency services authorities may have intentions that are not aligned with the call-takers or the first responders. We will discuss these problems in the context of hoax calls.

In our descriptions of individual security threats, we will not explicitly call out the different adversary models, but we will explain the assumptions that need to be met for a successful attack.

5.4 Security Threats

As a starting point, we have to think about the motivations of an attacker and to speculate about their available resources to launch such an attack. While some attacks will "carry over" from the existing telephony system, others will be new due to the additional capabilities offered by IP-based emergency services systems.

Attackers may direct their efforts either against a portion of the emergency response system or against an individual. Attacks against the emergency response system have two main objectives:

- To deny system services to all users in a given area. The motivation may range from thoughtless vandalism, through wide-scale criminality, to terrorism.
- To convey prank calls to PSAP call-takers and first responders. A fairly small number of prank calls may consume an enormous amount of resources from the emergency services infrastructure. While there are many variations of these prank and hoax calls, a severe version has been called "swatting," where an adversary simulates the commission of a punishable offense (such as a hijacking situation) and at the same time fools the emergency services system in using wrong location information for dispatching police units.

One can certainly imagine other, more exotic types of attacks. For example, an attacker could do the following:

- Divert emergency calls to non-emergency sites. This is a form of denial-of-service attack that can be very confusing for the emergency caller since he or she expects to talk to a PSAP operator but instead gets connected to someone else.
- Gain fraudulent use of services, by using an emergency identifier to bypass normal authentication, authorization, and accounting procedures. The goal of the adversary with such an attack is to gain a financial advantage.
- Get elevated priority treatment in the hope to gain faster service by blocking others' competing calls for help.

Attacks against an individual may have the following motivation:

- to prevent an individual from receiving aid;
- to gain information about an emergency that can be applied either against an individual involved in that emergency or to the profit of the attacker;
- to deliver unwanted messages to a non-emergency caller using technologies developed for emergency services purposes, for example, the PSAP callback mechanism, to bypass authorization policies;
- to use technology designed for emergency services to turn a user's phone into a recording device;

- to use sensitive personal data (e.g., health records) for non-emergency services purposes, or to retain personal data beyond an acceptable retention period; or
- to disclose location information without prior consent to third parties by bypassing the access control mechanism of a location server.

In the subsections that follow, we describe a few attacks in more detail. The list is not exhaustive, but illustrates concerns frequently raised.

5.4.1 Denial-of-Service Attacks

Emergency services have at least three finite resources subject to denial-of-service (DoS) attacks: the network and server infrastructure; call-takers; and first responders, such as fire fighters and police officers. The task of protecting network and server infrastructure shares similarities with high-value e-commerce sites, and lessons can be learned from these environments particularly regarding capacity planning, load balancing, DoS prevention techniques, and abuse reporting. Call-takers are a far more limited resource; even large cities might only have PSAPs with only a handful of PSAP call-takers on duty. Even if the call-takers attempt to question the caller to weed out prank calls, they would be quickly overwhelmed by even a small-scale attack. Finally, first responder resources are limited as well; and since every assignment takes an extended period of time, their resources are quickly dried up.

5.4.2 Attacks Involving the Emergency Identifier

The overall process of establishing an emergency call begins with the person in need of help dialing the emergency dial string. The exact sequence of digits depends on the infrastructure to which the device is connected. Although 1-1-2 has become the emergency services number for Europe, and 9-1-1 for the United States, many countries still provide emergency numbers in addition to the 112 and/or 911. Furthermore, many large enterprises, universities, and hotels prefix the emergency numbers with additional digits, such as 0-112. Therefore, it is important for devices that can be used in different environments to automatically detect which dial string triggers an emergency call. Pre-programming the list of numbers in use worldwide is not possible due to the overlap of emergency and non-emergency numbers.

With dial strings there are two challenges to solve:

1. A device needs the ability to learn the emergency services numbers available for a specific attachment point. LoST provides a mechanism for obtaining the emergency dial string for a given location. 3GPP radio networks also allow a device to be provisioned with emergency numbers.
2. For unambiguous processing of protocol messages, it is useful to replace the actual dial string with a symbolic name. This is accomplished with service URNs; see RFC 5031 [11]. Calls marked with this emergency identifier are then treated as emergency calls by the call routing entities by giving them special treatment. An example of a service URN is "urn:service:sos.police." In combination with LoST's ability to learn the dial string for a given location, this allows a device to dynamically translate emergency dial strings to service URNs for usage in emergency calls.

However, users do not "dial" an emergency URN itself. Instead, the entered emergency dial strings are translated to corresponding service URNs and then carried in the Request-URI of the INVITE request. This translation should ideally be done at the endpoint so that location can be conveyed in the signaling or call features (e.g., noise cancelation) can be disabled. Once a call is marked with a service URN, call routing entities give it preferential treatment.

The main possibility of attack involves use of the emergency identifier to bypass the normal procedures in order to achieve fraudulent use of services. An attack of this sort is possible only if the following conditions are true:

- The attacker is faking an emergency call.
- The call enters the domain of a service provider, which accepts it without applying normal procedures for authentication and authorization because the signaling carries the emergency identifier.
- The service provider routes the call according to the called address (e.g., SIP Request-URI), without verifying that this is the address of a PSAP (noting that a URI by itself does not indicate the nature of the entity to which it is pointing).

If these conditions are satisfied, the attacker can bypass normal service provider authorization procedures for arbitrary destinations, simply by reprogramming the emergency caller's device to add the emergency identifier to non-emergency call signaling. If the Resource Priority headers for emergency communications [202] are used, a call routing device may also be tricked into using this information as the sole basis for bypassing authentication, or authorization procedures if it uses these headers as the sole basis for its decision. This would also allow an adversary to use this indication to gain preferential treatment of marked traffic over other network traffic.

5.4.3 Attacks Against the Mapping System

This section considers attacks intended to reduce the effectiveness of the emergency response system for all callers in a given area. If the mapping operation is disabled, then the correct functioning of the emergency call routing infrastructure cannot be guaranteed. As a consequence, the probability that emergency calls will be routed to the wrong PSAP increases. Routing calls to the wrong PSAP may have two consequences: emergency response to the affected calls is delayed; and PSAP call-taker resources outside the immediate area of the emergency are consumed due to the extra effort required to redirect the calls. Alternatively, attacks that cause the client to receive a URI that does not lead to a PSAP have the immediate effect of causing emergency calls to fail.

Three basic attacks on the mapping process can be identified: denial of service, impersonation of the mapping server, or corruption of the mapping database. Denial of service can be achieved in several ways:

- by a flooding attack on the mapping server;
- by taking control of the mapping server and either preventing it from responding or causing it to send incorrect responses; and

- by taking control of any intermediary node (e.g., a router) through which the mapping queries and responses pass, and then using that control to block them – an adversary may also attempt to modify the mapping protocol signaling messages, or, additionally, be able to replay past communication exchanges to fool an emergency caller by returning incorrect results.

In an impersonation attack, the attacker induces the mapping client to direct its queries to a host under the attacker's control rather than the real mapping server, or the attacker suppresses the response from the real mapping server and sends a spoofed response.

The former type of impersonation attack itself is an issue of mapping server discovery rather than the mapping protocol directly. However, the mapping protocol may allow impersonation to be detected, thereby preventing acceptance of responses from an impersonating entity and possibly triggering a more secure discovery procedure.

Injecting fake mapping entities into the distributed mapping database may lead to an inconsistent state of the distributed database, and also lead to denial-of-service attacks. If the mapping data contains a URL that does not exist, then emergency services for the indicated area are not reachable. If all mapping data contains URLs that point to a single PSAP (rather than a large number), then this PSAP is likely to experience overload conditions. If the mapping data contains a URL that points to a server controlled by the adversary itself, then it might impersonate PSAPs.

5.4.4 Attacks Against the Location Information Server

A LIS provides information to end hosts and other entities in the system, which is then used for routing of emergency calls and for dispatching first responders. The LIS itself often has to obtain information from other sources to compute the location of an end system in an iterative fashion.

An adversary who wants to provide false location to a PSAP has a number of choices. Tampering with location information is one possibility, and interfering with the location determination procedure executed by an LIS is another possible approach.

The process of determining location information heavily depends on the specific network deployment, but, at an abstract level, the process is fairly simple. A Location Recipient, like the end host or an ESRP, transmits a request for location information to an LIS. Some information about the device to be located has to be provided in that request to allow the LIS to do its job. Unfortunately, there is no unique device identifier available that ideally fulfills that purpose. The Location Recipients have a few identifiers to choose from; the IP address is commonly used. In other cases, more identifiers are available; for example, those listed in [35] include an MAC address, a Network Access Identifier, and the DHCP Unique Identifier. When an LIS receives such a request for location information, it associates the obtained identifier with information available in its databases. The data may have been manually provisioned, but will typically be collected automatically from normal network operation, such as network management, network attachment procedures, and mobility protocols. For example, an LIS located in a DSL network receives a request asking for location related to a specific IP address. The IP address may be allocated from a pool of addresses maintained by the Authentication,

Authorization, and Accounting (AAA) server, and therefore the AAA server has to be queried. The AAA server may know the current attachment point of the end host or may use available information about the DSL Access Module (DSLAM) and Access Node (AN) and the related identifiers (e.g., Ethernet VLAN tags, Layer 2 tunnel identifiers, virtual port ID (VPI) and virtual circuit ID (VCI)) to determine the position of the end host. Further examples of such measurement identifiers are provided in [81].

Consequently, there are various methods by which an adversary can interfere with the process of resolving a chain of identifiers to obtain location information. If the adversary succeeds in feeding incorrect information in the lookup step, it may be able to fool an LIS into providing wrong location information. Examples of interfering with identifier mapping include the sending of a false MAC address or an IP address to obtain different location information. It should also be noted that wiremap maintenance is prone to errors, thereby resulting in wrong information being provided out even in the absence of malice.

5.4.5 Swatting

Prank calls have been a problem for the emergency services dating back to the time of street-corner call boxes. Individual prank calls waste scarce emergency service resources and possibly endanger bystanders or emergency service personnel as they rush to the reported scene of a fire or accident. Risks to human life are not a typical security threat within communication protocols. Emergency services are, however, different in this regard. Recent 9-1-1 "swatting" incidents with life-threatening consequences have captured media attention [203]. Some of these incidents have involved spoofing of the originating phone number when calling 9-1-1, leading to a false location being obtained via the phone number-to-location lookup used in fixed line environments. This could result, for example, in an armed SWAT team being dispatched to the location of a completely innocent citizen if the swatter is able to convince the call-taker that a serious crime is under way. The FBI has warned about the increasing prevalence of swatting incidents [204].

Legacy emergency services rely on the ability to identify callers as well as on the difficulty of location spoofing for normal users to limit prank calls. The ability to ascertain identity is important, since the threat of severe punishments reduces prank calls. Mechanically placing a large number of emergency calls that appear to come from different locations is difficult. Calls from pay phones are subject to greater scrutiny by the call-taker. In the current system, it would be very difficult for an attacker from country "Foo" to attack the emergency services infrastructure located in country "Bar."

In countries that do not allow SIM-less emergency calls, that is, emergency calls that are made without any authentication, the identity of most callers can be ascertained, so that the threat of severe punishments reduces prank calls. As a comparison, in countries where SIM-less emergency calls are allowed, prank calls may be as high as 50% [205].

There is a fear that VoIP systems further simplify these types of prank calls when identities and location information can easily be crafted. It may even be possible for attackers located in one country to attack the emergency services infrastructure located in a different country or to mechanically (with the help of botnets) initiate a large number of emergency calls that appear to come from different locations.

5.4.6 *Attacks to Prevent a Specific Individual From Receiving Aid*

If an attacker wishes to deny emergency service to a specific individual, the mass attacks described earlier will obviously work provided that the target individual is within the affected population. Except for the flooding attack on the mapping infrastructure, the attacker may also want to focus on a specific individual. To guarantee effectiveness, an adversary may attack the end device directly rather than the emergency services infrastructure.

The choices available to the attacker are as follows:

- to take control of any intermediary node (e.g., a WLAN router at the user's home) – if further security mechanisms are absent, this allows the adversary to modify requests and responses or even to block the entire communication;
- to interfere with the communication of the end device and other emergency service entities, for example, over the WLAN home network; and
- to infect the user's device with malware and consequently to have full control over the device.

In general, these type of attacks are difficult to prevent by an emergency services system itself and require improvements for Internet security in general.

5.4.7 *Attacks to Gain Information About an Emergency*

This section discusses attacks used to gain information about an emergency. The attacker may be seeking the location of the caller (e.g., to effect a criminal attack) or to use information to link an individual (the caller or someone else involved in the emergency) with embarrassing information related to the emergency (e.g., "Who did the police take away just now?"). Finally, the attacker could take profit from the emergency, perhaps by offering his or her services (e.g., a news reporter, or a lawyer aggressively seeking new business). The primary information that interceptions of mapping requests and responses will reveal are a location, a URI identifying a PSAP, the emergency service identifier, and the addresses of the mapping client and server. The location information can be directly useful to an attacker if the attacker has high assurance that the observed query is related to an emergency involving the target. The type of emergency (fire, police, or ambulance) might also be revealed by the emergency service identifier in the mapping query. The other pieces of information may provide the basis for further attacks on emergency call routing. The attacker may gain information that allows for interference with the call after it has been set up or for interception of the media stream between the caller and the PSAP.

Finally, the attacker may gain access to the conversation between the emergency caller and the call-taker rather than just the metadata about the incident.

5.4.8 *Interfering With the LIS and LoST Server Discovery Procedure*

Many entities in the emergency services architecture are configurable to offer some amount of flexibility. Dynamic discovery procedures have been developed to avoid manual configuration. The LIS and the LoST server discovery are examples.

The primary attack against the discovery step is impersonation. When there is no natural *a priori* relationship between the two devices, then the attack surface is increased. End devices, for example, are not supposed to be pre-configured manually with an LIS located in their ISP's network. Particularly, for mobile devices that frequently change their point of attachment, the LIS needs to be dynamically discovered. In the case of LoST, however, an end device may be statically provisioned to use a single LoST server all the time. Similarly to DNS, however, it is possible to discover and use a local LoST sever, which would provide improved resilience.

An attacker could attempt to compromise LIS and LoST discovery at any of three stages:

1. providing a falsified domain name to be used as input to U-NAPTR;
2. altering the DNS records used in U-NAPTR resolution; and
3. impersonating the LIS.

U-NAPTR is entirely dependent on its inputs. In falsifying a domain name, an attacker avoids any later protections, bypassing them entirely. To ensure the reliability of the access network domain name DHCP option, it is necessary to prevent DHCP messages from being modified or spoofed by attackers.

Once a client has been tricked into talking with the wrong LIS or LoST server, subsequent steps for emergency service protocol execution may fail or be manipulated in favor of the adversary.

5.4.9 Call Identity Spoofing

If an adversary can place emergency calls without disclosing its identity, then determining the source of prank calls may be more difficult. Neither the PSAP call-takers nor the emergency services authorities authenticate emergency callers themselves directly. However, there are at least two separate layers of authentication and authorization:

- authentication at the link layer or at the network layer (e.g., using the Extensible Authentication Protocol (EAP) [141]);
- authentication at the application layer, for example at the VoIP application.

Note that this split is the result of the separation between the ISP and the VSP. Since two different stakeholders manage the identity space, the identities used during authentication are different as well. While not all architectures assume such a separation of roles, they are nevertheless common on the Internet today.

Even with proper authentication at the VSP or the ISP, there is the question of how strong the prior identity proofing step was, that is, what identity information did the customer provide in order to create an account with an ISP or VSP to use their communication servers. The quality of identity proofing must not be ignored since it links the digital identity to the identity of a person in the real world.

In case of misuse, the emergency services authorities must contact the VSP and/or the ISP to identify a particular individual. In certain cases, there is no authentication procedure executed and hence this re-identification can be challenging. This might, for example, be the case with an open IEEE 802.11 WLAN hotspot. While the owner of the WLAN hotspot can be determined, this may be insufficient for determining the adversary utilizing this WLAN network.

Given the importance of authentication for ensuring accountability in case of misuse, mistakes made in the legacy telephony network can be fixed. In particular, the ability to make emergency calls without any form of authentication or by utilizing caller-id spoofing can be prevented by various technical means, by education about the negative side-effects, and by regulatory frameworks. Mandating authentication for emergency calls, and even the introduction of special credentials, for example, an emergency certificate, is imaginable. Needless to say, dedicated security mechanisms increase costs, introduce an administrative overhead, and are only, from a security point of view, useful when widely used.

5.5 Countermeasures

In the previous section we illustrated a few attacks on the emergency services system. In this section, we highlight a number of techniques to mitigate these threats.

5.5.1 Discovery

Physical or link layer security are commonplace methods for securing the initial communication link between an end device and the network infrastructure to reduce the possibility of attack, particularly at the early phases of the communication establishment. These link layer security mechanisms are particularly useful in those cases where location information is directly exchanged via DHCP, as described in section 2.2. While DHCP offers its own security mechanism (see RFC 3118 [206]), it is impracticable for deployments. However, for the interaction with HELD or LoST, additional security capabilities are available at higher protocol layers since these protocols run over HTTP and TLS.

LoST and HELD intentionally share a very similar discovery mechanism that can be used by end devices as well as by intermediate entities like emergency services routing proxies. Details of the discovery procedure for HELD are described in section 2.4 and for LoST in section 3.3.

In the case of a dynamic discovery of an LIS and a LoST server in the access network, an end device performs the following steps:

1. Acquire the access network domain name. DHCP can provide the end host with a domain name.
2. Use this domain name as input to the DNS-based resolution mechanism. For LoST, this resolution mechanism is described in RFC 5222 [68]. For HELD, it is defined in RFC 5986 [71]. In both cases, the URI-enabled NAPTR specification [72] is used.

An ESRP may also need to discover the network domain name for an LIS (as in the case in transition scenarios where no Location by Reference is available), but it cannot rely on DHCP for that purpose. Sections 4.3 and 4.2 illustrate alternatives for discovering the access network domain of an LIS. For a LoST server discovery, the ESRP is pre-configured with a domain name, which simplifies discovery since every LoST is able to receive a correct answer from the distributed mapping database, as explained in section 3.3.

To avoid an attacker modifying the query or its result of any interaction with an LIS or a LoST server, Transport Layer Security (TLS) is strongly recommended. For LoST there is the additional concern that operating without TLS allows cache poisoning since LoST has a built-in caching mechanism.

The entity interacting with a LoST server or an LIS has to check the TLS server's identity, as described in Section 3.1 of RFC 2818 [207]. The server identity check should only be omitted when getting any answer, even from a potentially malicious LIS or LoST server, is preferred over closing the connection (and thus not getting any answer at all), since it allows an attacker to masquerade as an LIS or LoST server.

The domain name that is used to authenticate the LIS or LoST server is the domain name in the URI, that is, the result of the U-NAPTR resolution, and it is compared against the server's certificate. To phrase it differently, when the access network domain name is "example.com" and the U-NAPTR lookup leads to a https://lostserver.example.com URL, then the certificate provided by that server has to contain the lostserver.example.com name. Therefore, if an attacker modifies any of the DNS records used in the resolution process, this URI could be replaced by an invalid URI.

DNS Security (DNSSEC) [208] can be used to protect against these threats. While DNSSEC is not yet completely deployed, users should be aware of the risk, particularly when they are requesting NAPTR records in environments where the local recursive name server, or the network between the client and the local recursive name server, is not considered trustworthy. Security considerations specific to U-NAPTR are described in more detail in [72].

LoST deployments that are unable to use DNSSEC and unwilling to trust DNS resolution without DNSSEC cannot use the NATPR-based discovery of LoST servers as-is. When suitable configuration mechanisms are available, one possibility is to configure the LoST server URIs (instead of the domain name to be used for NAPTR resolution) directly.

Note that this procedure is different from the security consideration recommendations in RFC 3958 [209]. RFC 3958 suggests that the certificate be compared to the input of NAPTR resolution to the certificate, not the output (host name in the URI). This approach was not chosen because, in emergency service use cases, it is likely that deployments will see a large number of inputs to the U-NAPTR algorithm resolve to a single server, typically run by a local emergency services authority. Checking the input to the NAPTR resolution against the certificates provided by the LoST server would be impracticable, as the list of organizations using it would be large, subject to rapid change, and unknown to the LoST server operator. The use of server identity leaves open the possibility of DNS-based attacks, as the NAPTR records may be altered by an attacker. The attacks include, for example, interception of DNS packets between the client and the recursive name server, DNS cache poisoning, and intentional modifications by the recursive name server; see RFC 3833 [210] for a more comprehensive discussion.

Using the domain name in the URI is more compatible with existing HTTP client software, which authenticate servers based on the domain name in the URI.

A LIS or LoST server that is identified by an "http:" URI cannot be authenticated. Use of unsecured HTTP also does not meet requirements in HELD for confidentiality and integrity. If an "http:" URI is the product of discovery, this leaves devices vulnerable to several attacks. Lower-layer protections, such as layer 2 traffic separation, might be used to provide some security guarantees against adversaries on the wireless interface, but should not be seen as a replace application layer security.

Generally, LoST servers will not need to authenticate or authorize clients presenting mapping queries. If they do, an authentication of the underlying transport mechanism, such as HTTP basic and digest authentication, may be used. Basic authentication should only be used in combination with TLS. The usage of TLS with mutual certificate-based authentication is another option for server-to-server communication.

5.5.2 Secure Session Setup and Caller Identity

The Session Initiation Protocol (SIP) [124] and the Session Description Protocol (SDP) [211] are used to set up multimedia sessions or calls. SIP messages travel from the emergency callers device via the intermediate SIP infrastructure toward the call-taker's device at the PSAP.

The SIP protocol suite offers solutions for securing SIP signaling as well as conveying caller identity information to the called party. The mechanisms can be clustered into three categories:

1. The process of verifying the user's identity. The VSP's infrastructure elements authenticate the user. This can, for example, happen via the basic SIP authentication mechanisms (such as digest authentication).
2. The process of asserting the previously verified identity to a third party. The authenticated identity is important not only for the VSP but also for end-to-end communication to the remote party. The VSP can assert this identity toward other parties using mechanisms such as SIP Identity, described in RFC 4474 [212], or P-Asserted-Identity, specified in RFC 3325 [130].
3. SIP signaling security. This ensures that an adversary cannot inject fake signaling messages, eavesdrop on the communication, replay messages, and so on. Transport Layer Security (TLS) is used for providing authentication, integrity, and confidentiality protection between neighboring SIP nodes. Since SIP signaling supports multiple transport protocols, not just TCP, Datagram TLS [213] was introduced to allow TLS functionality to datagram transport protocols.

There are two main technologies for communicating identity information in SIP:

- *P-Asserted-Identity (RFC 3325).*
  After authenticating the user, the VSP's SIP proxy adds the P-Asserted-Identity (PAI) header to the SIP message. This header carries the authenticated identity (SIP URI) of the user. The P-Asserted-Identity header is protected only in a hop-by-hop fashion between the SIP proxies along the path. The mechanism can only be used within a trust domain in which the SIP proxies and UAs communicate securely and the proxies are mutually trusted. The design of PAI is therefore based on a chain of trust rather than on a cryptographic end-to-end security solution.
- *SIP Identity (RFC 4474).*
  SIP Identity extends the PAI concept with a cryptographic identity assurance. SIP messages are sent to an Authentication Service, which is responsible for verifying that the user agent software knows a shared secret, for example, using the HTTP Digest authentication protocol. Based on successful user authentication, identity information is written

into the From header of the SIP request. This part is identical to the PAI scheme. Then, the Authentication Service adds a digital signature to a new SIP Identity header before forwarding it to the final recipient. Within the forwarded SIP request, the Authentication Service also provides a reference (using an HTTP URI in the Identity-Info header) to its own domain certificate. The recipient of the SIP message, for example, the call-taker's SIP user agent software, performs the following actions to verify the authenticated identity. First, it fetches and validates the certificate of the Authentication Service. Then, it verifies the signature of the SIP message and the identity of the user. Finally, it checks the value of signed Date header to protect against replay attacks.

5.5.3 Media Exchange

The main goal of the communication establishment explained in the previous section is in the exchange of multimedia data between two (or more) SIP endpoints. The offered security services must not only protect the communication setup but also ensure the protection of the media exchange. SDP, which is responsible for negotiating the media, is a versatile protocol. SDP not only allows the setup of Real-Time Transport Protocol (RTP) [174], which is used to transmit real-time media on top of UDP and TCP [214], but also allows the setup of TCP [215] and additionally TCP/TLS connections for use with media sessions [216].

The Secure RTP (SRTP) [217] is the established standard for securing RTP. In order to allow SRTP to offer its service, cryptographic keys need to be established between the involved communication parties via a key exchange protocol. Various key exchange protocols have been proposed and analyzed [218], and among all the options, DTLS-SRTP, described in RFC 5763 [219, 220], was chosen as the preferred IETF mechanism. However, the precursor to DTLS-SRTP, the Security Description (SDES) protocol [221], is widely used today despite its inferior security characteristics.

In order to protect against a number of attacks, it is therefore necessary for the SIP communication endpoints to implement and use a key exchange protocol (and DTLS-SRTP and for today's deployment environments also SDES) and SRTP itself.

5.5.4 Mapping Database Security

In section 3.3 we described the IETF architecture, including the use of the distributed mapping database that uses the LoST protocol and LoST Sync for distributing mappings between server nodes.

With the protocol exchange of mapping information, a minimum requirement is to authenticate neighboring server nodes using available HTTP security mechanisms, such as HTTP Digest [222], HTTP Basic [222] over TLS, or plain TLS with client and server certificates [223].

The setup of the LoST server relationships requires some manual configuration and hence the choice of the security mechanisms used between the two entities is a deployment-specific decision. Nevertheless, the usage of certificates is an attractive option since it allows the use of a Public Key Infrastructure (PKI) with separate trust anchors. Whenever a new server infrastructure element is introduced, a new certificate is obtained and signed by the corresponding Certificate Authority. This certificate is then ready for use by the other infrastructure elements without additional administrative configuration burden. In any case, the two communicating endpoints must authenticate each other and utilize the established secure communication channel (i.e., an integrity protected exchange of data with the help of the TLS Record Layer) to avoid the possibility of injecting bogus mappings.

A malicious entity could, however, intentionally modify mappings or inject bogus mappings. To avoid one entity claiming a service boundary belonging to some other, any node introducing a new service boundary must digitally sign the mapping and thereby protect the data with an XML digital signature. This ensures that a new mapping is associated to a particular owner with non-repudiation properties. In the absence of any automatic procedures, a system administrator must approve the received mapping prior to its inclusion in the database. Determining who can speak for a particular region is inherently difficult unless there is a small set of authorizing entities that all other participants can trust. Receiving systems should be particularly suspicious if an existing coverage region is replaced with a new one containing different contact points. With this end-to-end security mechanism, it is nevertheless guaranteed that mappings are modified by servers forwarding them as part of the synchronization procedure.

6

Emergency Services for Persons With Disabilities

Gunnar Hellström
Omnitor

This chapter is focused on the emergency call situation for people with communications-related disabilities. That means people who have little or no use of voice communication, such as people with deafness, deaf-blindness, hearing impairments, speech impairments, aphasia, and so on. A globally agreed policy is that people with disabilities shall be provided with equivalent access to emergency services as provided to users of telephony.

6.1 What Is Specific With Communication for People With Disabilities?

In order to discuss communication for people with disabilities and the requirement for equivalent services, we need to start by reviewing the characteristics of regular voice telephony, as used for mainstream access to emergency services.

6.1.1 Important Characteristics of Regular Voice Telephony

- *Global Interoperability in Addressing.* Anyone with access to voice telephony can call anybody else globally by just dialing the other person's number.
- *Global Interoperability in Conversation Transmission.* When the call is established, a voice conversation in real time can take place, regardless of the make and technology of the two parties' equipment and services. The medium is audio.
- *Interoperability with Mobile Telephony.* The users of telephony at a fixed location can exchange calls also with all users of mobile telephony.
- *Emergency Service Access.* The device and service used for everyday calls can also be used for emergency calls. Emergency calls are made to a regional short number.

Internet Protocol-Based Emergency Services, First Edition. Hannes Tschofenig and Henning Schulzrinne.

6.1.2 Important Characteristics of Accessible Conversational Services Suitable for People With Disabilities

People with communications-related disabilities require communication in other modes and media in addition to voice:

- *Voice Replacements and Complements.* People with communications-related disabilities need to replace or complement voice communication with communication in modes and media that suit their capabilities.
- *Real Time Text.* Some may find it most suitable to have access to text conversation in real time, where characters are transmitted as soon as they are typed so that the communication parties can be in synchronism with their thoughts, just as in a voice call.
- *Video.* Some may find it most suitable to have access to video communication for use of sign language communication, for lip reading, recognition or for showing things, features or feelings.
- *Wide-Band Audio.* Traditional telephony has far from ideal audio characteristics. High audio frequencies are cut off, so that even fully hearing users can barely hear the difference between an "s" and an "f" and other high-pitch language sounds. By use of wide-band audio, this situation can be approved, to great benefit for many users with hearing impairments.
- *Specific Combination of Media.* Each call may require a specific combination of most beneficial media and modes of communication. Access to all three conversational media of real-time text, video, and audio in the same call provides this opportunity in an uncomplicated fashion, and allows the users to mix and match media as most suitable for the moment.
- *Total Conversation.* A service providing the three real-time media communication means together has been defined and named Total Conversation. It was first defined by the International Telecommunications Union, ITU and has since been recognized by other organizations and included in service deployments. Total Conversation is best implemented in broadband networks, where the video medium can be assigned sufficient bandwidth for suitable quality for sign language and lip-reading conversation.
- *Relay Service.* Relay services provide translation between different modes and media, in order to fulfill the need by people with disabilities to have calls with other modes and media. These relay services form an important backbone for communication with all users of voice telephony. Relay services are defined for sign language users, for text users, for users of weak speech or speech that is hard to understand, for speech users with support of real-time captions, and for deaf-blind users mixing real-time text and voice or sign language communication.
- *Emergency Access.* Emergency service access for people with disabilities can be provided either directly between user and PSAP when they can use common modes and media, or with relay services included when the PSAP needs support for mode or media translation. An often used solution is to let text calls be handled by the PSAP operator directly, while sign language calls are handled with relay service support for translation between sign language and voice. For all cases it is valuable for the efficiency of the communication to let media be shared with the PSAP, even if a relay service is involved for translation of the main conversation. Emergency services need to be provided through the same number as for everybody else in the region.

6.2 Reality Today

Most persons with disabilities do not have equivalent services today. In some countries, there has been an ambition since the 1970s or 1980s to at least provide a limited real-time text service in the PSTN, called text telephony (TTY in North America).

Text telephony does not meet all requirements for equivalent services. Its importance is decreasing even in countries where it has been widespread. Instead, the users now use a patchwork of services providing at least some of the media communications they require, but not the global interoperability. Various kinds of video telephony and text communication services are replacing the text telephone service.

Users are accepting the use of a range of "free" services without the opportunity to call emergency services the day they need it. In a few countries Total Conversation is emerging, but still in these not all users have Total Conversation, so the users need to use a whole range of services to reach all their friends and colleagues.

Short Message Service (SMS) has become an important service for people with communications disabilities because of its global deployment. It is slow to compose messages by SMS, but it provides a way to communicate to nearly everywhere. Therefore, it has become common also to require access to emergency services through SMS, and the number of countries implementing SMS access to emergency services is increasing. It has been reported that emergencies have been solved after initiating the emergency communication by SMS. It seems reasonable to include SMS access in emergency services while still informing users about its inferior functionality for this purpose.

6.3 Interpretation of the Term "Equivalent Service"

Considering the characteristics of voice telephone services, and accessible conversational services, it can be deduced that the term "equivalent service" shall be interpreted as follows:

- *Real-Time Text.* People who need to type or read for their communication must be provided with real-time text in the service. Only by this means can the service be equivalent in that it carries the conversational medium of choice for text users.
- *Video.* People who need to use sign language, lip reading or other visual means for their communication must be provided with video communication in their service. Only by this means can the service be equivalent in that it carries the conversational media of choice for these users.
- *Voice.* People who need other media may very well also need voice communication. Equivalent communication sometimes mean complementing voice communication, sometimes replacing voice communication.
- *Relay Services.* People who have no, or limited, use of voice communication directly must be provided with relay services suitable for conversion between voice and their way of communicating. Only by this means can the service be equivalent in the sense that it can be used for calls with the enormous population of voice telephone users.
- *Total Conversation Access.* People have varying needs and capabilities which may vary from call to call. Therefore the needs must be met by Total Conversation services offering both video, real-time text, and voice simultaneously. Only by this means will the service provide the global interoperability needed between different users.

- *Applying Standards for Total Conversation.* Implementation of Total Conversation needs to follow established standards so that it can be used for calls to other total conversation users as well as relay services and emergency services. Only by this means can the services be equivalent in that they can provide global interoperability in media.
- *Addressing Through the International Number System.* The Total Conversation implementations need to be able to use phone numbers according to the international phone number system. Only by that means can they set up calls with all voice phones to fulfill the need for global interoperability in addressing.
- *Emergency Service Access.* People need to be offered the ability to have emergency service calls to the regional emergency number, using the media and modes and equipment they use for everyday communication, as well as invocation of a possible relay service they usually might have. Only in that way can the emergency service comply with the emergency access requirements to provide equivalent services to what other users have.

6.4 Sad History

Conversational services accessible by people with disabilities have a sad history of incompatibility and low functionality. Text-phone systems were introduced in the 1970s and 1980s, using different technologies so that interoperability problems appeared. The result was that, while voice telephone users could call each other globally, people with disabilities had very limited interoperability regions. Only a few countries have introduced relay services and only a few countries have arranged for emergency services accessible for people with disabilities. This clearly shows that market forces themselves are not sufficient to ensure provision of the services making accessible communication equivalent to other communication services. Policy actions and regulation support to create equivalent services are clearly needed.

6.5 Policy and Regulation Support

The inferior situation regarding access to communication services and emergency services by people with disabilities has been realized by governments and organizations and now many strong and well-formulated policy expressions and regulations exist. They express the rights of people with disabilities to be provided with communication services including emergency service access equivalent to what is provided to other people.

6.5.1 UN Convention on the Rights of Persons With Disabilities

The highest level of policy support is found in the United Nation's Convention on the Rights of Persons With Disabilities. A large number of countries have signed this convention, and its intentions are on the way to being implemented in national laws and regional regulations. Many articles in the Convention relate to rights to access to emergency services in suitable modes. As an example, Article 9, Accessibility says:

> States Parties shall take appropriate measures to ensure to persons with disabilities access, on an equal basis with others, to the physical environment, to transportation, to information and communications, including information and communications technologies and systems, . . . These

> measures, which shall include the identification and elimination of obstacles and barriers to accessibility, shall apply to, inter alia:
>
> . . .
>
> b) Information, communications and other services, including electronic services and emergency services.

With these conventions as a background, many countries and regions have their own laws and regulations supporting accessible conversational services in general, and emergency services.

6.5.2 The European Union Universal Service Directive

The European Union has a Universal Service Directive [224] adopted in 2009, that all member states are required to implement. Article 26 "Emergency services and the single European emergency call number" (paragraph 4) requires access for people with disabilities both in their main place of residence and when traveling in Europe:

> Member States shall ensure that access for disabled end-users to emergency services is equivalent to that enjoyed by other end-users. Measures taken to ensure that disabled end-users are able to access emergency services whilst travelling in other Member States shall be based to the greatest extent possible on European standards or specifications published in accordance with the provisions of Article 17 of Directive 2002/21/EC (Framework Directive), and they shall not prevent Member States from adopting additional requirements in order to pursue the objectives set out in this Article.

6.5.3 The Telecom Act and Public Procurement Act in the United States

In the United States, two law texts dominate the communication accessibility area, "Section 255" and "Section 508." Section 255 belongs to the Telecom Act and regulates accessibility of telecommunications products to be marketed in the United States. Section 508 belongs to the Public Procurement Act and regulates accessibility of telecommunication products procured by US government.

The main requirement that currently influences emergency services is that the US type of textphone for PSTN, called TTY, shall be supported. Even if a very large number of TTY devices are in operation in the United States, their number is now decreasing, and users move over to a patchwork of other kinds of communication that provide either higher functionality or better mobility. The two law sections are under revision to accommodate this move to modern communications environments and a new proposal was made in December 2011. In that revision, support for TTY is maintained for PSTN, while for IP networks, use of, or interoperability with, SIP with real-time text is required. Accessibility of video communication and wide-band audio are also included in the revision proposal.

6.5.4 Americans With Disability Act

The base for legal requirements to access emergency services for people with disabilities in the United States is found in the Americans With Disability Act (ADA). A brief explanation can be read in [225].

6.5.5 *Relay Service Regulation in the United States*

The relay services in the United States operate under regulation called "Ten-Digit Numbering and Emergency Call Handling Procedures for Internet-Based Telecommunications Relay Service (TRS)" [226]. One major reason for this regulation was to enable location provision from relay service users to the PSAP. By this regulation, users who use IP-based equipment and relay services use a 10-digit number from the North American Number Plan, so that they can be called through the same kind of relay service they use themselves for calls. The users also use the emergency number 9-1-1 directly, and not any relay service address, with the result that calling emergency services is fairly equivalent to other users calling 9-1-1.

6.6 Good Opportunities in IP-Based Services

In the shift toward IP-based communication, there are excellent opportunities to fulfill user needs by good multimedia communication in broadband networks. The broadband networks provide the base where the functionality of conversational services can be extended from plain voice to full Total Conversation, still maintaining interoperability and affordability. It is the environment where equivalent services can be provided.

The main issue in standardization of emergency services for mainstream users is the provision and use of location information when moving to IP-based user terminals and services. At the same time, the main issue for people with disabilities is to agree on a harmonized way to handle the extra media and mode requirements from disabled users, and the needs of some of the users with disabilities to have relay services included in the calls.

The current state of standardization in the Internet Engineering Task Force is as follows.

RFC 5012 defines requirements for IP-based emergency services in the IETF. It contains many requirements on call routing, but also some related to media handling, such as "Emergency calling must support a variety of media. Such media should include voice, conversational text (RFC 4103 [227]), instant messaging, and video."

The main document related to IP emergency services in the IETF is [123] with the associated architectural specification in [40]. These IETF documents describe how IP-based emergency calls shall be placed. Accessibility aspects are covered throughout these specifications through the protocols for call control, routing, location conveyance, and inclusion of relay services as well as though inclusion of the codecs for real-time text, audio, and video, as well as message protocols. In particular, the following requirements, taken from [123] and [40], are essential.

- *Media Requirements for Endpoints*

  Endpoints MUST send and receive media streams on RTP [174].

  Normal SIP offer/answer [172] negotiations MUST be used to agree on the media streams to be used.

  Endpoints supporting voice MUST support G.711 A-law (and mu-law if they are intended be used in North America) encoded voice as described in [178]. It is desirable to include wide-band audio codecs, such as AMR-WB, in the offer.

Silence suppression (Voice Activity Detection methods) MUST NOT be used on emergency calls. PSAP call-takers sometimes get information on what is happening in the background to determine how to process the call.

Endpoints supporting Instant Messaging (IM) MUST support both RFC 3428 [228] and RFC 4975 [229].

Endpoints supporting real-time text MUST use RFC 4103 [227]. The expectations for emergency service support for the real-time text medium, described in Real Time Text over IP Using the Session Initiation Protocol (SIP) [230], Section 7.1, SHOULD be fulfilled.

Endpoints supporting video MUST support H.264 as per RFC 3984 [231].

- *Media Requirements for PSAPs*

  PSAPs should always accept RTP media streams [174]. Traditionally, voice has been the only media stream accepted by PSAPs. In some countries, text, in the form of Baudot codes or similar tone-encoded signaling within a voice band, is accepted ("TTY") for persons who have hearing disabilities. Using SIP signaling includes the capability to negotiate media. Normal SIP offer/answer [172] negotiations should be used to agree on the media streams to be used. PSAPs SHOULD accept real-time text [227]. All PSAPs should accept G.711 A-law (and mu-law in North America) encoded voice as described in [178]. Newer text forms are rapidly appearing, with instant messaging now very common. PSAPs should accept IM with at least "pager-mode" MESSAGE request [228] as well as Message Session Relay Protocol (MSRP) [229]. Video may be important to support Video Relay Service (sign language interpretation) as well as modern video phones.

- *Mid-Call Behavior*

  During the course of an emergency call, devices and proxies MUST support REFER transactions with method=INVITE and the Referred-by: header [232] in that transaction. Some PSAPs often include dispatchers, responders or specialists on a call. Some responder's dispatchers are not located in the primary PSAP; the call may have to be transferred to another PSAP. Most often this will be an attended transfer, or a bridged transfer. Therefore, a PSAP may need to REFER request [232] a call to a bridge for conferencing. Devices that normally involve the user in transfer operations should consider the effect of such interactions when a stressed user places an emergency call. Requiring UI manipulation during such events may not be desirable. Relay services for communication with people with disabilities may be included in the call with the bridge. The UA should be prepared to have the call transferred (usually attended, but possibly blind) as per [233].

- *Caller Refer Preferences*

  SIP Caller Preferences [234] MAY be used to signal how the PSAP should handle the call. For example, a language preference expressed in an Accept-Language header may be used as a hint to cause the PSAP to route the call to a call-taker who speaks the requested language. SIP Caller Preferences may also be used to indicate a need to invoke a relay service for communication with people with disabilities in the call.

- *PSAP Callback*

  Devices device SHOULD have a globally routable URI in a Contact: header which remains valid for 30 minutes past the time the original call containing the URI completes unless the device registration expires and is not renewed. Callbacks to the Contact: header URI received within 30 minutes of an emergency call must reach the device regardless of call features or services that would normally cause the call to be routed to some other entity.

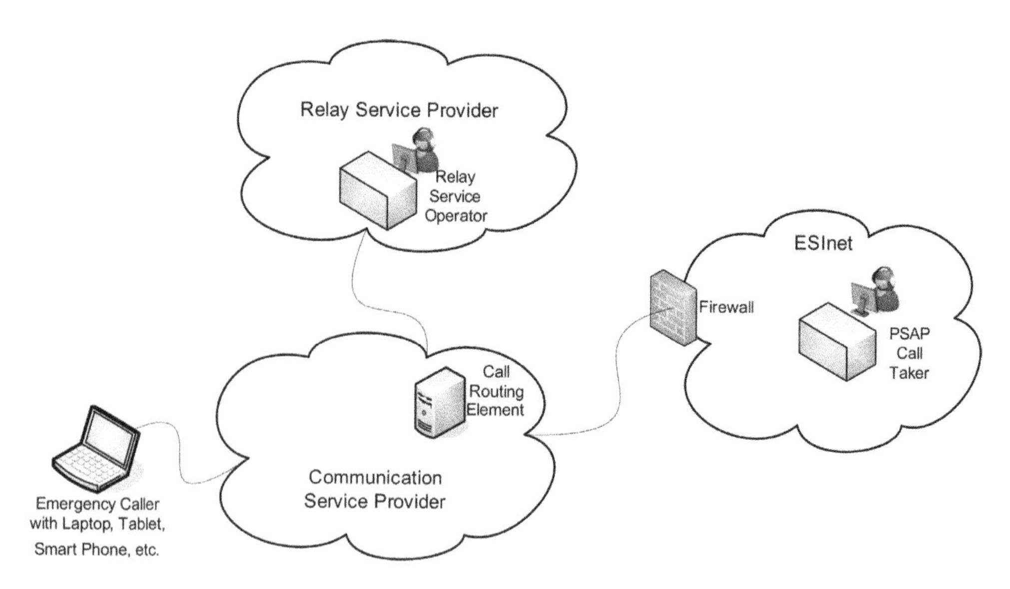

Figure 6.1 Total Conversation relay support.

6.7 Implementation Experience

One way to include a sign language relay service suitable for emergency service application is through a three-party Total Conversation conference bridge. Figure 6.1 shows an architecture where a relay service is invoked by a Communication Service Provider.

A relay brings persons with disabilities a big step closer to being equivalent to regular voice emergency callers. This in line with both policy requirements and mainstream standardization.

7

Regulatory Situation

Regulation has to fulfill a number of roles in telecommunication. The universal service ensures that the necessary communication facilities are available to everyone, even in remote areas. Price regulation also guarantees widespread availability. Disability access allows and improves participation in society for those who are handicapped. For emergency services, regulators often intervene to deal with concerns about market failure and anti-competitive behavior.

Persons with a deep technical background often see regulation as a component they do not need to care about. We encourage you to see it as a way to distribute responsibilities among different stakeholders in a distributed system.

In this chapter the US and European emergency services regulations are explained. We selected these two regions because these write-ups provide a good supplement to the technical deployments described in other chapters. Additionally, our familiarity with the stakeholders in Europe and in the United States made it easier to find knowledgeable persons to share their insight with us.

In section 7.1 Christopher Libertelli provides a history of emergency services support in the United States, and explains the challenges that arise with the introduced split between the access and the application service provider. The description may also help those who are unfamiliar with some of the basic technical concepts to better understand the NENA i2 architecture, for example.

In section 7.2 Margit Brandl and Samuel Laurinkari offer insight into the European emergency services regulation, taking the relationship between the different functions of the European institutions into account.

The regulatory situation has a significant impact on the deployment status of emergency services support for Voice over Internet Protocol in many countries, and these two sections provide the necessary background to understand where the obligations currently reside. While for some it is seen as desirable that regulation is "technology-independent," the differences between legacy fixed, legacy cellular, and Internet communication makes it clear that technology also introduces new stakeholders and additional challenges that lead to the need to revisit previously made assumptions.

Internet Protocol-Based Emergency Services, First Edition. Hannes Tschofenig and Henning Schulzrinne.

7.1 Regulatory Aspects of Emergency Services in the United States

Christopher Libertelli
Skype

7.1.1 Introduction

Telephony services have served as the link between distressed Americans and those authorities that provide emergency services. Over the years the nature of this connection has evolved a great deal. Today, anyone claiming to offer a "phone" or landline replacement service to consumers will be under pressure not only to provide a connection to the 9-1-1 emergency service network, but also to provide this 9-1-1 network with a callback number and location information in an appropriate format.

We will look at the evolution of 9-1-1 service requirements, identify the challenges that new technologies face when attempting to meet these requirements, and detail how these services have (and have not) responded to recent legislative and regulatory requirements relating to 9-1-1 services.

7.1.2 Background

7.1.2.1 Nationwide 9-1-1

In the early days of the telephone in the United States, there was very little national coordination between emergency services providers. Traditionally, emergency services were provided by the local authorities. As such, it did not seem unreasonable that each local authority should have its own number for citizens to dial in case of an emergency. It was not until the late 1960s that the federal government, facing a growing and increasingly mobile population, decided to establish a single emergency services number. The vision was for every American to be able to connect to the appropriate Public Service Answering Point (PSAP) by dialing the same three digits nationwide. The PSAP would work in tandem with local authorities and could dispatch the appropriate personnel depending on the type of emergency. In 1967, Congress, after consultation with the American Telephone Company (AT&T), designated 9-1-1 as the nationwide emergency service number. There are more than 6000 PSAPs answering 9-1-1 calls today.

Connecting callers to the appropriate PSAP presented a technical challenge for the network call switching process, however. The phone network was designed to route the call to the end receiver based on the information provided by the number that the caller entered. The amount of information that the caller enters when he or she presses 9-1-1 is not sufficient for the network to know where to route the call. To deal with this problem, network operators put intelligence into the network in the form of "selective routers," routers programmed through databases to know which lines ought to connect to which PSAP. The selective router receives the 9-1-1 call and routes it to the appropriate PSAP over a dedicated 9-1-1 trunk group. The gateway to the 9-1-1 network was effectively owned by the telephone company that owned the local exchange office.

In the early days of the deployment of 9-1-1 services, AT&T not only had a monopoly over long-distance telephone services, but also controlled the vast majority of local exchange services as well. Because of their control over local exchange networks, AT&T passed the costs of the central office modifications on to subscribers.[1] But even though access to the 9-1-1 gateway was paid for by the American consumer, ownership belonged to AT&T. And when AT&T was forced to divest its interests in these local exchange services as the result of an antitrust case, ownership of the emergency services gateway passed on to the newly formed "Baby Bells" or incumbent local exchange carriers (ILECs).

Although a competitive local exchange carrier (CLEC) that wishes to compete with incumbent carriers has the right to interconnect with the incumbent's 9-1-1 gateway trunk, the CLEC must partner with the ILEC that controls the trunk that connects the PSTN to the PSAP. The details of this partnership vary. Sometimes the ILEC carries the call from the CLEC's central office, sometimes the CLEC delivers their calls to the ILEC's central office, and sometimes a third-party carrier is used. However, whether the CLEC chooses to purchase a dedicated trunk port, interconnection transport or switch ports at the ILEC's end office, the CLEC must pay the ILEC an "infrastructure charge" to cover the costs of hardware modifications needed in order to support the CLEC's 9-1-1 services, on top of any 9-1-1 surcharges that are levied by local authorities. For more information, the interested reader should see, for example, Verizon's E9-1-1 wholesale offering [236].

7.1.2.2 Enhanced 9-1-1

The early 9-1-1 service only provided a connection to the appropriate PSAP; it did not provide the PSAP with a callback number or the location of the caller. If the call was disconnected or the caller was unable to provide his or her location to the emergency service operator, emergency services were unable to respond to the caller location.

In time, local telephone companies developed the ability to send Automatic Number Identification (ANI) information along with the call. They programmed the selective routers to use the ANI to query a selective router database to determine the appropriate PSAP. The call was sent to the PSAP, along with the ANI. The PSAP then uses a separate data link to query an Automatic Location Identification (ALI) database in order to obtain the location information of the caller.

These databases are generally maintained by the ILEC that controls the emergency service gateway, and are populated with data from subscriber records. When a CLEC chooses to partner with an ILEC to provide E9-1-1 services, the CLEC must either provide the ILEC with the information, or must enter the information into the database itself, depending on the details of the arrangement. Both the CLEC's and the ILEC's ability to provide this information rests on the fact that they provide a service to a fixed address. The ability to automatically provide the location of the caller to the PSAP is challenged when a telephony service does not always terminate at the same address.

7.1.3 E9-1-1 Requirements

By the 1990s, pressing 9-1-1 on most home phones would connect you to the most appropriate PSAP via the 9-1-1 trunks and send the PSAP your address as well as a callback number should you get disconnected. This improved the effectiveness of emergency services in the

[1] "A Bell System policy was established to absorb the cost of central office modifications and any additions necessary to accommodate the 9-1-1 code as part of the general rate base" [235].

United States. People could "ask" for help even when they could not speak. The connection now had a way of working around call drop complications. Emergency services were only as far away as the nearest telephone.

But this new E9-1-1 paradigm faced challenges as new types of telephony services began to compete with traditional wired services. The growth of the cellular phone industry in the 1990s, and of the VoIP industry in the 2000s, has meant that many Americans no longer feel the need to be in close proximity to a wired telephone. In their nascent stages, neither of these services offered E9-1-1 services. PSAP operators increasingly found themselves requiring verbal directions from callers before they could dispatch emergency services as these new types of telephony services grew. Congress and the Federal Communications Commission (FCC), the US regulator in charge of communication services in the United States, took action to ensure the continuation of effective E9-1-1 services.

7.1.3.1 Implementation of Wireless E9-1-1

The FCC began pushing for wireless adoption of E9-1-1 services in 1996. Initially, wireless customers had to verbally provide a PSAP with his or her location and callback number. The FCC adopted rules that forced wireless providers to move toward E9-1-1 in two separate phases. By 1st April 1998, wireless providers were asked to provide the PSAP with a callback number and the location of the base station through which the call was connected. After 1st October 2001, PSAPs could request a wireless carrier to provide the PSAP with accurate (within 50–100 m) coordinates of the handset that made the call. The wireless provider then had six months to comply with the request [237].

Wireless operators found it challenging to provide emergency operators with the precise location of their customers because their customers are nomadic. It was up to the wireless providers to come up with ways of locating the device. They had a choice of network-based solutions, handset-based solutions, or some hybrid of the two. Most chose to use a method of triangulating the location of a handset based on where it stood in relation to a number of cell towers [238].

Although many wireless carriers are able to provide E9-1-1 services today, the implementation of wireless E9-1-1 did not go as smoothly as the FCC hoped. All of the major carriers filed waivers asking for extensions on the 1st October deadline [239]. Even when the carriers did have technological solutions at hand, the PSAPs often did not have the resources to upgrade their systems to receive this new type of data. In a 2002 report for the FCC, noted wireless expert Dale Hatfield shed light on the sorry state of wireless E9-1-1 implementation, citing "the total number of stakeholders involved, the complexity of the inter-relationships among the stakeholders, and the incentives and constraints on those stakeholders" as possible explanations for the slow progress [240]. Hatfield was commissioned to do a follow-up report on wireless E9-1-1 implementation difficulties, but was forced to terminate the unpublished study in 2006 after presenting his findings to officials at the FCC [241].

7.1.3.2 Implementation of VoIP With E9-1-1

As broadband penetration grew, entrepreneurs realized that they could provide telephony service by routing calls over customers' existing broadband lines. Vonage, founded in 2001, began marketing itself as the "Broadband Phone Company." Customers began switching from traditional wired voice services. This caused quite a stir for emergency service providers. Not only could Vonage not provide E9-1-1 solutions, but they could not even provide basic 9-1-1 services reliably because of the way the legacy 9-1-1 infrastructure is designed, and the way Vonage's VoIP implementation routed their calls.

Both wired and wireless E9-1-1 solutions used the physical infrastructure that they owned, the copper wire and the cell tower, in order to determine the geographic location of their customer. When a VoIP call comes from the Internet to the VoIP provider's point of presence, the provider has no way of determining what path those packets took over the public Internet before they arrive at that point of presence. Many VoIP providers offer "nomadic VoIP," allowing customers to use their service from any broadband connection.

If a VoIP service is forced to provide location information through the legacy system of selective routers and ALI databases, then, in the absence of Internet Service Providers disclosing location to the endpoint or to the VoIP provider, the only way for a VoIP provider to offer user locations to PSAPs is by asking their customers to constantly update their own location information. With accurate self-reported data, a VoIP company could partner with others to develop more inclusive ALI databases for the PSAPs to interact with. However, convincing customers to manually update their location information on a regular basis would likely prove extraordinarily difficult, especially as VoIP moves to mobile broadband networks.

Despite these challenges, VoIP providers recognized the need to figure out 9-1-1 solutions for VoIP customers. In January 2005, the Voice on the Net (VON) Coalition partnered with the National Emergency Number Association (NENA) to put out a progress report and detail a way forward for emergency services. However, in June 2005, the FCC preempted those initiatives and intervened. It adopted a rule stating that any VoIP service that could be used to make and receive calls to the PSTN, or "interconnected VoIP," must be able to connect to E9-1-1 and provide a callback number and location information for the PSAP using only the network currently in place (the PSAPs would not have to update their own hardware as they did in minor ways when implementing wireless E9-1-1) [242].

Unlike wireless operators, who only needed to arrange interconnection agreements with the ILECs in the areas they operate, VoIP providers were forced to arrange agreements with virtually every ILEC operator because their customers could use the service anywhere where he or she had a broadband connection. And, unlike wireless operators and CLECs, VoIP providers are not considered "telecommunication carriers." At the time the rule was established, only telecommunication carriers had a right to interconnect to the ILECs selective routers. VoIP companies were therefore expected to rely on third-party carriers or the ILEC themselves to provide interconnection. This forced VoIP providers to rely on the very companies that they were competing with. In November 2005, when a number of interconnected VoIP providers missed the deadline, they claimed that they were still waiting for the cooperation from those companies that controlled access to the 9-1-1 network [243].

In 2008, Congress passed the New and Emerging Technologies 9-1-1 Improvement Act, which reiterated VoIP's E9-1-1 obligation, but also gave them interconnection rights, meaning they no longer had to rely on third-party carriers to connect them to the 9-1-1 gateway if they did not wish to [244]. Most VoIP providers today offer some level of E9-1-1 coverage, but because the location they report to the ALI database is often based on self-reporting, customers still must verbally confirm their location to ensure emergency personal are dispatched to the correct location.

In addition, the New and Emerging Technologies 9-1-1 Improvement Act charted a course for the government, notably the Department of Transportation working in partnership with the Federal Communications Commission, to facilitate a next-generation emergency calling infrastructure. Most industry commentators agree that this next-generation architecture will be based in large measure on the NENANENA NG9-1-1 i3 standard. The promise of the NENA i3 standard is that it largely avoids the technical complexity and competitive concerns of incumbent LEC control over selective router infrastructure and PSAP access. Many standard-setting groups, including the IETF ECRIT, have worked to evaluate the i3 standard and to ensure that it is promulgated and deployed in a way that makes sense for IP and Internet infrastructures. In addition, the prospect of altering the architecture in a way that acknowledges the separation of access provider and VoIP/application service providers holds out the promise that manual updates of location information will be a thing of the past. In this way, software-defined services as well as network operators could collaborate to deliver multimedia streams of communications directly to NG9-1-1 systems and PSAPs, aiding their ability to reach callers in distress with first responders at the most critical times.

7.2 Regulatory Aspects of Emergency Services in the European Union

Margit Brandl and Samuel Laurinkari
Nokia Siemens Networks

7.2.1 Introduction

The European Union is a union of independent states that features the characteristics of a federal state, on the one hand, and an intergovernmental body, on the other. Depending on the policy area in question, the founding treaties confer powers on the European Union to adopt legislation that is binding upon all Member States.

Any piece of legislation adopted by the European Union needs to have its legal base in the founding treaties. In addition, the legislative act needs to adhere to the principles of subsidiarity and proportionality, which means that the European Union should only legislate in areas where Member States could not sufficiently tackle the problem in question by acting independently.

Article 95 of the EC Treaty[2] provides a general legal basis for any measures that aim to harmonize such policy areas, where differences in legislation may undermine the effectiveness of the internal market. This has allowed the EU institutions to legislate on a wide variety of subject matters, as any difference in legislation may potentially be seen as discouraging businesses or consumers to engage in activities in Member States other than their own.

The European Union-wide regulation of the telecommunications markets has also been based to a large extent on Article 95 of the EC Treaty. After the "liberalization"[3] and "Europeanization"[4] of the telecoms markets in the early 1990s, the European Union has adopted a plethora of legislation regulating, *inter alia*, the access to infrastructure, the obligations

[2] The entry into force of the Treaty of Lisbon brought about changes to the substance and numbering of the EU founding treaties. However, as this section deals with legislation adopted under the old framework, reference is made to the numbering and substance at the time of adoption.
[3] "Liberalization" of the telecoms markets refers to the process under which traditionally state-owned monopolies were struck down by EU legislation, which mandated open markets by allowing and facilitating competition by setting obligations for operators with significant market power. For example, local loop unbundling (mandated access for competitors to the incumbent's infrastructure) has successfully resulted in more competitive markets.
[4] "Europeanization" refers to the process of a field of law or policy becoming governed by the *acquis communautaire*, thus EU legislation instead of and/or in addition to national legislation.

of incumbent operators, national competition, consumers' rights to universal service, and emergency service obligations. This package is called the EU telecoms regulatory framework.

7.2.2 Regulatory Development of Emergency Services Under EU Law

Traditionally, all European countries have regulated and governed their emergency services and the access to those services on a national level. Despite the fact that the organization of the emergency services still remains within the competence of the Member States, the European Union has laid down some conditions, which have to be fulfilled. With the Europeanization of the telecom markets, the European Union-wide emergency number 112 was first introduced by a decision [245] of the Council of the European Union[5] in the early 1990s. This decision required Member States to adopt the European emergency number 112 as the preferred number in addition to or instead of any national emergency numbers. The Member States had until December 1992 to implement the 112 obligation. Certain grounds, for example, high implementation costs, were justified reasons to postpone the implementation to December 1996. In addition to the mere obligation to provide access to emergency services through 112, Member States had to ensure that emergency calls were received correctly and routed to the appropriate emergency control center according to technical feasibilities existing in the public networks.

In the Authorization Directive 97/13/EC [246], the EU legislator required operators to enable emergency calls even if other telecommunications services had been barred (e.g., because of unpaid bills). One year later, the European Union adopted Directive 98/10/EC [247], which laid down more detailed requirements for handling emergency calls through 112: emergency calls had to be possible from public telephones, and calls to 112 had to be free of charge and without using any means of payment [248].

Currently, the obligation to provide access to emergency services through 112 and a number of other European Union-based obligations are laid down in the "Universal Services Directive," which was adopted in 2002 as part of the EU telecom regulatory framework.[6]

7.2.3 Current Legal Framework

7.2.3.1 EU Law Grants for Minimum Harmonization of Emergency Service Regulation

The Universal Service Directive[7] established a regime under which both the Member States and the private industry actors are placed under a number of obligations regarding emergency

[5] The Council of the European Union, also known as the Council of Ministers or the Council, is the main legislative body of the European Union (often co-legislating with the European Parliament, in some policy areas also as the single veto body). The Council is composed of ministers of the Member States, depending on the policy area in question.

[6] The current regulatory framework is composed of the following legislative acts: Directive (2002/21/EC) [249] on a common regulatory framework, Directive (2002/19/EC) [250] on access and interconnection, Directive (2002/20/EC) [251] on the authorisation of electronic communications networks and services, Directive (2002/22/EC) [224] on universal service and users' rights relating to electronic communications networks and services, Directive (2002/58/EC) [252] on privacy and electronic communications, Directive (2002/77/EC) [253] on competition in the markets for electronic communications services, and Regulation (2000/2887/EC) [254] on unbundled access to the local loop.

[7] Directive 2002/22/EC of the European Parliament and the Council of 7th March 2002 on Universal Service and Users' Rights Relating to Electronic Communications Networks and Services [224], hereinafter also USD.

services. In particular Articles 2, 6, 7, 23 and 26 of the Universal Service Directive are of relevance with regard to emergency services.

These provisions include the obligation to provide emergency services through the European emergency number 112 free of charge, due care in handling incoming calls, providing caller location for the emergency authorities, and raising awareness of the European emergency number 112.[8] As the Directive sets out only certain conditions that need to be transposed into national legislation, emergency services are harmonized at a minimum level in the European Union. That concept is called "minimum harmonized." In other words, the EU legislator has set the minimum conditions that need to be fulfilled, but the Member States remain competent to adopt more detailed and regulated legislation and may freely choose how to implement the requirements arising from the Universal Service Directive. Hence, the precise provision of emergency services depends on the Member State and its policies.

To illustrate, if one wants to find out the precise emergency service legislation in Germany, one needs to consult the German telecom law and related regulations. In this concrete example, the German legislator has adopted a law (Telecommunications Act [255]) to transpose the Universal Service Directive and further empowered the German Regulatory Agency to adopt a bylaw (Bylaw on Emergency Call Connections [256]) regulating the details of emergency services.

7.2.3.2 Technological Changes Required Adaptations in the Legal Framework

As outlined above, the European Union started regulating emergency services in 1992. At that time, most voice traffic was circuit-switched fixed line telephony. The rapid evolution of telecommunications toward mobile solutions has also had its impacts on the regulation of emergency services. Whereas obligations such as callback functions or caller location were relatively readily implemented for fixed lines, mobile network operators had to develop suitable solutions. The situation is getting even more complex as, for example, Voice over Internet Protocol (VoIP) services are taking over key functionalities of traditional telephony. Hence, emergency service regulation needs to be capable of answering questions like which (telephony) services need to provide access to emergency services.

7.2.3.3 Circuit-Switched and Packet-Based Voice Services, Fixed and Mobile Voice Services

Emergency service provisions were originally designed for the circuit-switched voice world. Early formulations of emergency service provisioning described above just stated that telecom licenses can have conditions regarding access to emergency services and that end user calls to emergency services need to be free of charge. The concrete implementation was up to the competent authorities in the Member States.

As a fixed phone at a fixed location can easily be connected to person(s) in a building or a household, thus location information for the PSAP was relatively easy to process.

With the uptake of mobile telephony granting emergency services became more complicated. SIM-less emergency calls, calls with inserted SIM cards but (in the case of prepaid subscriptions) with no credits left were to be dealt with together with solving the problem of delivering appropriate and usable user and location information to PSAPs. Furthermore,

[8] Universal Service Directive, Article 26, Subparagraphs 1, 2, 3, 4, respectively.

the user and location information of normal mobile subscriptions may also be challenging, depending on the PSAP's access to the subscription databases. Whereas in some countries PSAPs use a shared database of all operators, in other countries PSAPs have first to identify the relevant operator before deriving user data from the individual database. What seemed to be rather straightforward in the fixed world suddenly added another layer of complexity to the mobile domain.

Now that circuit-switched voice telephony is being replaced more and more by packet-based voice services, new issues arise. Where managed packet-based voice services are used as a substitute for traditional circuit-switched telephony, the connection to a given user and address may be comparable to traditional offerings unless the service can be used nomadically. Pure best effort VoIP services like PC-to-PC VoIP or even PC-to-phone VoIP have to be looked at separately.

With more and more voice services on the market, which differ in terms of capabilities, features, and quality, one of the basic questions to be answered is who is obliged to provide access to emergency services? In the following section, two of the obligations laid down in the Universal Service Directive, access to emergency services and caller location, will be discussed in greater detail. Emphasis will be put on the problems addressed above while anticipating the effects of the changing EU telecoms framework.

7.2.3.4 The Problem With a Circular Definition

One of the problematic aspects of the current framework is the circular definition of the emergency service obligation. It is not entirely clear which providers and which technologies fall under the (EU minimum) obligation to provide those services. In Article 2(c) a "publicly available telephone service" is described as follows:

> "publicly available telephone service" means a service available to the public for originating and receiving national and international calls and access to emergency services through a number or numbers in a national or international telephone numbering plan, and in addition may, where relevant, include one or more of the following services: the provision of operator assistance, directory enquiry services, directories, provision of public pay phones, provision of service under special terms, provision of special facilities for customers with disabilities or with special social needs and/or the provision of non-geographic services.

This definition is relevant in the light of Article 26 of the Universal Services Directive, which stipulates that Member States must ensure that end users of "publicly available telephone services" are able to call the emergency services free of charge. This puts the national telecom authorities under the obligation to oblige the service providers that match the "PATS" (Publicly Available Telephone Service) definition provided for in Article 2 above to offer emergency services free of charge to their customers.

The wording of the Directive has left much room for interpretation, as PATS are described as call services including emergency services, and emergency service obligations are imposed on PATS providers. In other words, the definition seems to run in circles, or even seems unintentionally[9] to leave the choice whether or not to provide access to emergency services and thus fulfill the PATS definition to the operator.

[9] There are other obligations connected when offering publicly available telephone service, like price transparency, carrier selection and pre-selection, and network integrity requirements, to name but a few.

Going through Article 2(c) of the Universal Service Directive point by point, one can see the following:

- Providers of publicly available telephony services are obliged to grant access to emergency services. That means that those providers that offer their telephony services to closed user groups only are not obliged to offer access to emergency services. Publicly available suggests that anybody must get the chance to subscribe to the service. Closed user groups like company networks therefore do not fulfill that criterion.
- Providers of publicly available telephony services must offer their services for originating and receiving national and international calls. A pure VoIP offering that offers PC-to-PC voice services does not fulfill that requirement. Neither does a PC-to-phone service, even if the user can call national and international numbers (originate calls) unless he is given an E.164 number and can therefore also receive national and international calls.
- The previous cumulative requirement gets even more complex, as it has to be read together with the requirement of accessing emergency services through a number or numbers in a national or international telephone numbering plan.

So, in conclusion, only a service that is available to the public that grants its users the possibility to both make and receive calls to national and international (E.164) numbers has to offer access to emergency services.

While this applies to traditional PSTN[10] and mobile voice services, VoIP has to be analyzed separately. Where VoIP services are offered as a substitute for traditional PSTN services, that is, in the form of managed services that offer, among other things, the same functionalities as traditional services, using, for example, an E.164 number where calls can be originated and received (even if this service is used in a nomadic way[11]), then access to emergency services must be granted.

In the case of mere PC-to-PC or PC-to-phone applications, access to emergency services is not a mandatory requirement for the provider of these services.

7.2.3.5 Who Pays for Granting Access to Emergency Services?

Those applications that fulfill the requirements of the PATS definition must provide access to emergency services free of charge for the user (Article 26 of the Universal Service Directive). The Directive remains silent on the implementation of this requirement and whether or not the service provider has or has not got the right to reclaim costs from, for example, the State. Member States are free to transpose the Directive and clarify issues according to their national particularities.

Most often, national rules require operators to hand over emergency call details – in particular, location information details – in a specific data format. Also this requirement, and the related question whether operators can claim their costs for implementing special requirements, remains subject to national transposition (i.e., national law).

[10] Access to emergency services also needs to be granted from public pay telephones.

[11] For example, the user can log-in at any VoIP terminal he/she chooses and is then able to make and receive call under "his/her" E.164 number.

7.2.3.6 Privacy Issues in Emergency Service Regulation

When user information and location data are automatically provided, privacy issues clearly become relevant. The question was dealt with in the EU framework by adopting the Data Protection Directive,[12] which in paragraph (36) stipulates that:

> Member States may restrict the users' and subscribers' rights to privacy with regard to calling line identification where this is necessary to trace nuisance calls and with regard to calling line identification and location data where this is necessary to allow emergency services to carry out their tasks as effectively as possible. For these purposes, Member States may adopt specific provisions to entitle providers of electronic communications services to provide access to calling line identification and location data without the prior consent of the users or subscribers concerned.

This, together with paragraph (10) (laying down the legal wording), effectively exempts emergency service providers from restrictions regarding data protection and privacy.

7.2.3.7 Actual Implementation: Some Examples

The circular definition of publicly available telephony services and its implications on the obligation to provide access to emergency services discussed in section 7.2.3.4 has not resulted in many problems regarding the traditional circuit-switched telephony, but has met some debate in relation to fixed and nomadic VoIP services. Regulators in different Member States have interpreted the obligation in different manners, which have resulted in a disharmonized application of EU law. According to a recent Commission report, 24 Member States oblige VoIP services offering PATS services to offer also access to Member States. The interpretation of the definition, however, varies heavily from one Member State to the other [257].

For example, in a decision concerning the classification of a VoIP service provided by TeliaSonera, the Finnish Communication Regulatory Authority (FICORA) held that the service should be classified as PATS, as it was available to the public, users were able to originate and receive national and international calls, users could access emergency services, and the service was available under the Finnish numbering plan. TeliaSonera was thus obliged to comply with the obligations set for PATS in the Universal Service Directive, *inter alia* access to emergency services free of charge [258].

The Italian regulator, on the other hand, has announced that it is to classify VoIP services with features analogous to traditional telephony services rendered on the PSTN and "nomadic" VoIP services as PATS and therefore bring them under the regulation. Other types of VoIP will remain outside of the scope of the regulation [259].

As the Directive only sets out the obligation to a minimum set of services that are to provide emergency services, Member States remain free to broaden the number of services on which the access to emergency services is mandated. The UK regulator OFCOM adopted a new policy in 2008, according to which such VoIP services that do not offer the user the ability to receive calls must also provide free access to emergency services despite the fact that they fall outside of the PATS definition [260].

[12] Directive 2002/58/EC of the European Parliament and of the Council of 12th July 2002 concerning the processing of personal data and the protection of privacy in the electronic communications sector.

7.2.3.8 Caller Location Obligation in the EU Framework

The Universal Service Directive requires Member States to ensure that caller location information is made available to authorities handling emergencies.

As outlined above, for fixed networks, where a given terminal is registered at a given location and can be connected with an address, a (centralized) database or directory may be enough, but with the introduction of mobile telephony, caller location information for emergency services has become even more relevant in the process of steering emergency medical assistance, police or fire brigade to the appropriate location where the emergency is taking place (in case the caller is unable to provide sufficient information about his/her location).

The importance of precise caller location information in emergency services becomes apparent by looking at statistics provided by CGALIES (Coordination Group on Access to Location Information by Emergency Services), according to which every year in EU Member States 3.5 million emergency calls are handled ineffectively because callers have provided inaccurate information about their location. Furthermore, it is also estimated that emergency services are not able to dispatch a rescue team for approximately 2.5 million calls because of the absence of sufficient location information [261].

Article 26(3) of the Universal Service Directive [224] stipulates that [emphasis added]:

> Member States shall ensure that undertakings which operate public telephone networks make caller location information available to authorities handling emergencies, *to the extent technically feasible*, for all calls to the single European emergency number 112.

According to the Commission, caller location means "in a public mobile network the data processed indicating the geographic position of a user's mobile terminal and in a public fixed network the data about the physical address of the termination point" [262].

Similar to the discussion on the definition of the emergency service obligation above, the caller location obligation wording seems to leave much room for discretion for the National Regulatory Authorities, which are in charge of the implementation of the provisions of the Directive. However, the European Commission[13] has chosen a strict line on the caller location obligation, and has taken 14 Member States to the European Court of Justice for failure to implement Article 26(3) (caller location obligation) correctly.[14] Hence, the Commission interprets the clause "to the extent technically feasible" in such a way that anything that *can* technically be done, *should* be done.

The wording of the Directive is technology-neutral in the sense that it does not make a distinction between fixed and mobile. Hence, the obligation applies equally to both, having caused some implementation troubles in the Member States. The majority of the cases the Commission has brought before the European Court of Justice have concerned the non-implementation of caller location information of mobile calls. Even in 2009, three Member

[13] The European Commission is the "working arm" of the European Union. It has the sole power of initiative and is responsible for monitoring the implementation of EU legislation. If a Member State fails to fulfill its obligations under EU law, the Commission may initiate proceedings under Article 226/EC and take the Member State to the European Court of Justice, which may issue penalty payments until the obligations are fulfilled.

[14] The Commission asked the Court of Justice to fine Italy for not providing caller location information for 112 calls; more is available in ref. [263].

States were still unable to ensure full caller location information to emergency authorities for all mobile emergency calls [263].

The Commission has used Recommendations, which under the EU telecom regulatory framework are in principle non-binding but require a good reason not to be followed by the National Regulatory Authorities, to advance its preferences in the implementation process in the Member States. In a 2003 Recommendation [262], the Commission outlined its preferences regarding the implementation of the caller location obligation. Before the Recommendation was issued, operators in Member States had used two different methods to provide the emergency authorities with caller location information: the "push" and the "pull" methods.

Whereas in the "push" method the information is automatically pushed with the initial call together with information contained in the calling line identity (CLI), in the "pull" method the information is transmitted on demand, using the CLI and preferably the emergency location protocol [248]. In the recommendation, the Commission clearly stated its preference for the "push" method. This caused some debate among the EU Member States, which had so far assumed that the implementation methods would be left entirely at their discretion.

An example of the problems caused by the regulatory provisions regarding caller location information is the case of the Commission versus Lithuania,[15] where the Commission took Lithuania to the European Court of Justice for failing to implement caller location obligations properly. The Member State offered in its defense that Lithuanian operators had invested in the pull method that was not yet operational. Lithuania understood the 2003 Recommendation as an obligation to switch to the "push" method at the expenses of the "pull" method and claimed that there was legal uncertainty.

As will be seen below, the New Regulatory Framework[16] will not necessarily simplify the current provisions but will indeed introduce changed wording and hence new interpretation of necessary implementation measures.

The methods used to provide caller location information vary considerably depending on whether the call originates from a fixed line or a mobile phone. According to the Commission, the majority of Member States have set up central databases from which emergency service authorities receive address information for fixed calls. Practically, this means that the calling line identity is used to retrieve the address information of the fixed line from the central database. Often, this database functions also as the general directory operated by the fixed incumbent operator. When considering the effectiveness of the database system, recall has to be made to the comprehensiveness of the database and the update mechanisms.

Hence, it depends on who is responsible for updating the database. Even if it is centrally managed, for example, by the incumbent operator, every other operator is responsible for communicating updates, for example, on new customers or customers that have used number portability functionalities. Every party can jeopardizes the proper functioning of the caller location information system when information is not processed timely enough.

Another problem that might occur in certain directory-based databases is that, if a customer has chosen not to include personal data in the directory, the location information is not available to the emergency authorities either. However, currently only three Member States are unable

[15] Case C-274/07 Commission v. Lithuania.

[16] Since late 2007, the EU Commission, Council, and Parliament have been working on a modernization of the legal framework for telecommunications in Europe, the so-called New Regulatory Framework. More can be found in ref. [264].

to provide caller location information of fixed calls if the customer has chosen not to include details in the database [257].

In mobile emergency calls the system is rather different. Here, the caller location is usually carried out on the basis of the GSM subscriber connection number. Following this number, the subscriber's home network operator uses its location information server to provide caller location information based on the last cell identity forwarded by the mobile device. The accuracy of caller location information retrieved in this way varies from 50 meters in optimal urban conditions to 40 kilometers in maritime areas.[17] Hence, it is anticipated that the more precise satellite-based GPS system will take over as prevailing technique as well-equipped mobile phones become more common [257].

When it comes to mobile location information, there is a huge debate here whether the translation of cell identities into actual addresses has to be done at the operator's side or at the PSAP. In Germany,[18] for instance, the provision dealing with location information says that the geographic location has to be as accurate as state-of-the-art commercially used localization services. As long as this is state of the art, the cell identity needs to be provided. This is the name of the cell and the geographic location of the mast that is serving this cell. The operator has to provide the PSAP also with all information necessary to translate the name of the cell into an actual geographic location.

7.2.4 New Legal Framework

In accordance with the provisions concerning the periodic review of the telecoms legislation, the update of the existing regulatory framework was initiated in 2007. The Commission drafted a reform package, consisting of three Amending Directives, which was adopted by the Council and Parliament in November 2009.[19]

7.2.4.1 Who Needs to Provide Access to Emergency Services in the New Framework?

The package stipulates a number of changes compared to the previous legislation. First, the Directive will no longer use the circular definition referred to above, but will oblige Member States[20] to "ensure that undertakings providing end users with an electronic communications service for originating national calls to a number or numbers in a national telephone numbering plan provide access to emergency services." This will most likely ease the discussion on the classification of different technologies used in the telecommunications industry. Following the provision, any regulated service ("electronic communications service," ECS) that facilitates calls to a number (therefore excluding PC-to-PC, for example) will be required to provide access to emergency services.

---

[17] According to the website of the Finnish Emergency Service Authority, available in Finnish at http://www.112.fi/index.php?pageName=matkapuhelinpaikannus, last visited 15th September 2012.

[18] Article 4, para 7, nr 3 and 4 of ref. [256].

[19] The three documents (namely, Regulation (EC) No 1211/2009 establishing BEREC and the Office; Directive 2009/136/EC amending the Universal Service Directive 2002/22/EC, the ePrivacy Directive 2002/58/EC and the consumer protection Regulation (EC) No 2006/2004; and Directive 2009/140/EC amending the Framework Directive 2002/21/EC, the Access Directive 2002/19/EC and the Authorization Directive 2002/20/EC) can be found in refs. [265], [266], and [267], respectively.

[20] Article 26(2) in ref. [268].

7.2.4.2 Caller Location in the New Framework

Also the regulation of caller location information changes. In the future, in Article 26(5) the Universal Service Directive will state that [emphasis added]:

> Member States shall ensure that undertakings concerned make caller location information available free of charge to the authority handling emergency calls *as soon as the call reaches that authority*. This shall apply to all calls to the single European emergency call number "112". Member States may extend this obligation to cover calls to national emergency numbers. Competent regulatory authorities shall lay down criteria for the accuracy and reliability of the location information provided.

The clause "free of charge" is a major victory for PSAPs. Without this addition, PSAPs have in some countries had to pay even up to 0.15 euro per location information given. However, this does not mean that operators must bear the costs of location information – many countries have adopted certain compensation schemes to reimburse the operators providing access to emergency services.

The wording above pays tribute to the technology shift from traditional telephony to a more complex world. According to the previous formulation in the Universal Service Directive, the *public telephone network operators* are responsible for providing caller location information. However, in the new wording, the *undertakings concerned* will have to make caller location information available. This presumably broadens the number of obliged undertakings significantly. It could, for example, mean that, besides "traditional" operators, also those falling in the category of "electronic communication service," such as some VoIP providers, providers of calling solutions for corporate networks, providers for closed user groups, and other service providers, would need to find solutions for emergency services and for making available location information.

The interesting part of the changes to the wording is the clause *as soon as the call reaches that authority*, which needs to be interpreted in the light of the "push" and "pull" methods discussed above. As outlined above, the Commission has clearly stated its preference toward the "push" method in its 2003 Recommendation.

According to a Commission report [257] on the implementation of emergency service obligations, location information is provided within 15 seconds to 1 hour after the PSAP (Public Safety Answering Point) has requested the information if the "pull" method is used, whereas "push" information is available immediately the call reaches the PSAP. Considering the new wording, *as soon as the call reaches that authority*, one may conclude that all 21 Member States who are applying the "pull" method at the moment will be obliged to apply the "push" method in the future in order to fulfill the obligations arising from the revised Directive.

The wording also effectively excludes practices where PSAPs are required to acquire location information by a voice or fax query.

Another interesting recommendation appears in the Consumer Rights Directive, although it is in a so-called "Recital" (no. 40), so it is not binding on all Member States to implement it. This recommends the development of international standards to provide emergency calling accurately and reliably even for regulated network-independent providers (such as some VoIP providers):

> Network-independent undertakings may not have control over networks and may not be able to ensure that emergency calls made through their service are routed with the same reliability, as they

> may not be able to guarantee service availability, given that problems related to infrastructure are not under their control. For network-independent undertakings, caller location information may not always be technically feasible. Once internationally-recognised standards ensuring accurate and reliable routing and connection to the emergency services are in place, network-independent undertakings should also fulfil the obligations related to caller location information at a level comparable to that required of other undertakings.

7.2.5 Emergency Regulation Outside of the EU Telecom Regulatory Framework

In addition to the provisions in the EU telecoms framework, the European Union has also used self-regulatory approaches to harmonize certain issues relating to emergency service obligations. In the self-regulatory process, the European Union "encourages" private industry actors to come to a voluntary agreement on an issue the European Union would be competent to regulate through legislation.

In September 2009 the European Commission announced that the car industry, the telecom industry, and the GSM association had reached an agreement to implement the eCall project. eCall refers to a technology solution, where an automatic call to the emergency authorities is initiated by a car that has had an accident. What this means in practice is that a GSM-based mobile device integrated in the car will automatically get in contact with the emergency authorities under the European emergency number 112 and provide the accident location following the caller location obligations in the Universal Service Directive. According to the Commission, eCall will save 2500 lives each year within the European Union/EEA area.

7.2.6 Conclusion

European emergency service legislation has a long history, first on the national level and later within the EU framework. As EU legislation takes the form of minimum standard directives, national implementation plays a great role. Hence, national law needs to be consulted for the precise regulation of emergency services. This regulatory dualism allows the European Union to harmonize a minimum set of services, whereas the Member States remain competent to decide how to implement those provisions and how to transform obligations into national law in greater detail. In other words, Member States remain competent to regulate emergency services as long as they remain within the boundaries of the EU framework – also in the future.

This division of competence has shown its effectiveness in the regulatory uncertainty brought about by the emergence of IP-based voice telephony and the steep uptake of mobile telephony. As we have seen in the examples of VoIP regulation in the United Kingdom, Finland, and Italy, national regulators are often able to react more flexibly on technological developments than the EU legislators. Therefore, it is important to keep regulatory discretion in the Member States. That ensures that emergency service regulation is kept updated and matches the most recent developments.

This discretion, however, often leads to disharmonized solutions within the European Union. As any difference in national legislation may be seen as a potential threat to the proper functioning of the internal market, the European Union will also need to update its framework to keep the differences limited.

It is unclear whether the changing EU telecoms framework will provide clear solutions for this problem. As an example, one might think of the "push/pull" discussion above. It remains to be seen whether the Commission will take Member States to Court in case operators do not provide caller location information as soon as the call reaches the PSAP but only after 10 seconds, say. If the Commission takes a strict line with the "push" method, some Member States will probably need to amend their laws on privacy; for example, in Finland, the emergency authorities are at the moment allowed to request location information details only if they consider that there is an imminent danger to life or limb.

Many provisions will need to be clarified by Commission Recommendations and/or case law through infringement proceedings.

8

Research Projects and Pilots

In this chapter we would like to present ongoing as well as completed projects on emergency services, since they provide a wealth of information.

"REsponding to All Citizens needing Help" (REACH112) [269] is an ongoing pilot project funded by the European Commission. Explicitly aimed at creating a pan-European pilot deployment with a focus on persons with disabilities, it is certainly distinct from the other projects discussed in this chapter. It is interesting to see that the obstacles described in the write-up in section 8.1 were seen as a challenge for Total Conversation already before the start of the project and the situation has not changed. Particularly, the lack of Quality-of-Service (QoS) deployment for real-time communication services (especially noticeable with high-quality video) was also discussed in a workshop in July 2012 organized by the Internet Architecture Board (IAB) and the Internet Research Task Force (IRTF) [270].

The "IP-Based Emergency Applications and ServiCes for Next Generation Networks" (PEACE) [271] project was also funded by the European Commission. It differs from the REACH112 project because it focuses on research rather than deployment of emergency services in the participating Member States, which had an impact on the selection of the consortium members. PEACE also has a wider focus than REACH112 and only one part of the project's efforts are described in section 8.2, namely those focusing on regular 1-1-2 emergency calls. For those interested in the IP Multimedia Subsystem (IMS)-based emergency services solutions, PEACE definitely provides valuable input, also in the form of open-source software.

Finally, we also wanted to cover a completed project from the United States on the next generation of 9-1-1 emergency services architectures (section 8.3). The US Department of Transportation's Next Generation 9-1-1 (NG9-1-1) [272] initiative is a completed project that aimed to study the functional requirements, to develop a system architecture, and to investigate a transition plan for deploying IP-based 9-1-1 networks. The project involved a selection of PSAPs and focused on a number of multimedia emergency services technologies. NG9-1-1 is similar in style to the project REACH112, which started years later, although REACH112 is much smaller and is focused on providing services for persons with disabilities only. In comparison to PEACE, the US DOT NG9-1-1 project did not focus only on research and did not focus exclusively on the IMS architecture.

Internet Protocol-Based Emergency Services, First Edition. Hannes Tschofenig and Henning Schulzrinne.

A few remarks on the authors of these sections are in order. The section about REACH112 was written by Jim Kyle and John Martin. Jim and John are both members of the REACH112 project. Yacine Rebahi contributed the section about PEACE. Yacine is a researcher working in the PEACE project, and is responsible for many of the research results in that project. To cover the large US Department of Transportation NG9-1-1 project, we had to ask several project members, namely John Chiaramonte, Gordon Vanauken, Wonsang Song and Jong Yul Kim, to share their project experience with us.

Since all these projects publish source code, software modules, and various studies and white papers, the short write-ups are only meant to raise your interest. Please visit the project websites to learn more about their list of publications, potential collaboration possibilities, and ongoing activities.

8.1 REACH112: Responding to All Citizens Needing Help

Jim Kyle[1] and John Martin[2]
[1] *University of Bristol*
[2] *AuPix Limited*

8.1.1 Outline

Although telephony was invented in the 19th century and has had an enormous enabling effect for some groups of people, for example, deaf people, the advent of telephony became an additional burden and a cause for discrimination especially in employment. Although new technology has existed for some time to give deaf people at least some access to this facility, we are only now able to work in a pan-European pilot deployment, which offers real visual access.

The goal of the EU-funded REACH112 (REsponding to All Citizens needing Help) pilot project [269] is to make "telephones" accessible for people with disabilities. The solution is to add video and real-time text to the calls, forming Total Conversation (TC), so that interoperability of voice telephony between service providers is maintained. The provision of video means that sign language, lip reading, and face recognition are possible. In conditions of limited bandwidth, text can be used, which is also of general value to those with reduced hearing. The relevance of having voice in the calls is that many people with disabilities have some use of voice.

In REACH112, a broad partnership of telephony operators and stakeholder groups implemented and deployed pilot services in five countries in Europe. Major priorities for REACH112 were to increase person-to-person communication and to provide access to 112 emergency services. In this write-up, we summarize the project objectives and some of the obstacles we have had to overcome, and provide some indications of the successes of the project.

The concept of Total Conversation is presented in Figure 8.1. Users are able to see each other when they call, are able to use text and video relay services, and are also able to contact 112 services directly, that is, without using a relay service. Although this is an Internet-based solution, they are able to use standard telephone numbers, and to interact with textphone users and with hearing people.

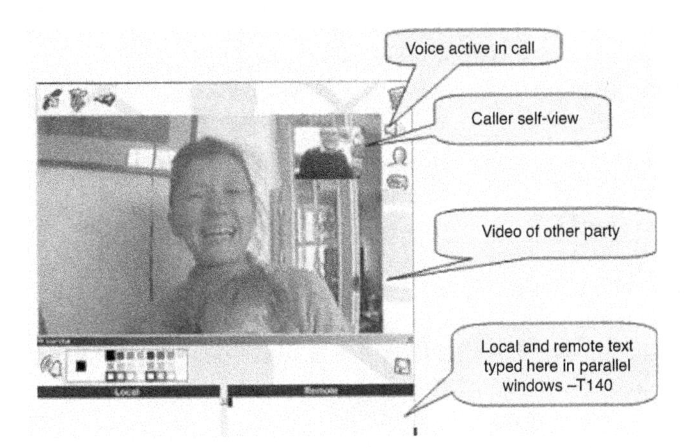

Figure 8.1 Total Conversation concept: user interface example.

Total Conversation (TC) is a specification for telecommunication that mandates video, text, and speech simultaneously and in real time, between parties in the interaction. It is defined by ITU-T in the standard F.703 Multimedia Conversational Services Description. There is a difference between the T.140 text medium (which Total Conversation uses) and most other services including text, in that the text flows as it is typed. The effect is that the users are in continuous contact and do not experience the delays between messages that make other text systems frustrating and less immediate.

The target users are those for whom communication through visual means is a necessity. REACH112 focuses on the needs of the most challenging users, who fall in the following groups:

- deaf sign language users;
- those who are deafened (use speech for communicating to others, but need visual means to understand others' speech);
- persons with a hearing loss, but who use speech and text;
- those who are deaf-blind and need enhanced video, voice, and text;
- hard-of-hearing people (including those who are elderly and isolated), who may use speech, but need amplification and other visual cues;
- those with speech disabilities, who may need a relay service; and
- persons with learning disabilities.

However, we believe that, if we solve the issues of these groups, we will significantly aid others and mainstream groups, such as elderly people. The achieved target user group was 7800 people in the five countries in which the pilot program took place.

The endpoint devices are available in different forms:

- fixed dedicated videophones with text capabilities;
- desktop PCs and laptops running Windows and Linux;

- tablets with Android and iPad/iPod Touch;
- TC clients on smart phones, with Android and iPhone; and
- TC functionality as an application running in a Web browser.

It should be noted that most of the videophones that are commercially available are not Total Conversation devices as they do not have the real-time text functionality.

REACH112 has stipulated that inter-service communication should use industry standards for call control (based on the Session Initiation Protocol, SIP), video (H.263 or H.264), voice (G.711), and text (T.140). ENUM [273] is used for telephone number discovery between service operators. Service providers are free to use whatever protocols they see fit within their services, but must adhere to REACH112 protocols to consider themselves interoperable with other REACH112-compliant providers.

The REACH112 system is typically controlled by a call distribution server in each service provider's network, which manages and tracks the TC call setup and terminations. Interpreter or video relay agents are typically distributed in this system and are issued with devices that allow simultaneous video conversations and VoIP conversations supported by text. This software will connect with existing technology for text communication and ensures that the installed base of text users can be integrated into the system.

8.1.2 Emergency Service Access

The approach chosen by REACH112 is to create a connection between callers and the emergency service operator at the public safety answering point (PSAP), that is, the PSAP agent for the police, fire or ambulance service. Their role is not only to raise an alarm and to send service personnel to the site, but also to ascertain the extent of the problem, to give continuing advice until the service personnel arrive, and to ensure as far as possible the health and safety of those on site. In this role, the agent engages in conversation with the user.

At present, users of a telephone are able to use a single number and be routed to a call handling agency where a call-taker then routes them to the specific emergency service in the location of the user.

REACH112 utilizes existing fixed and mobile access networks. The technology to achieve the goals of REACH112 has been proven and has been implemented in the pilots. However, there remain significant challenges to service development, which we will discuss below.

In REACH112, it is intended that TC capability is available to all persons, from user, through relay, call handling, and PSAP operators (see Figure 8.2), and this applies especially to emergency service operators, where specially trained communication staff, including sign language users, are able to manage the incoming call.

Agents at PSAPs and call-takers (i.e., those who receive the first contact when the user "dials" 112) are able to control all elements of the call – by being able to see and converse with the user in speech, use text (T140 standard), and video (for sign language and speech reading), as well as simultaneously being able to see relay service interpreters. The TC display at the PSAP or in a centralized call handling center can display several participants with live video, at the same time (the number of participants displayed will depend on the capacity of

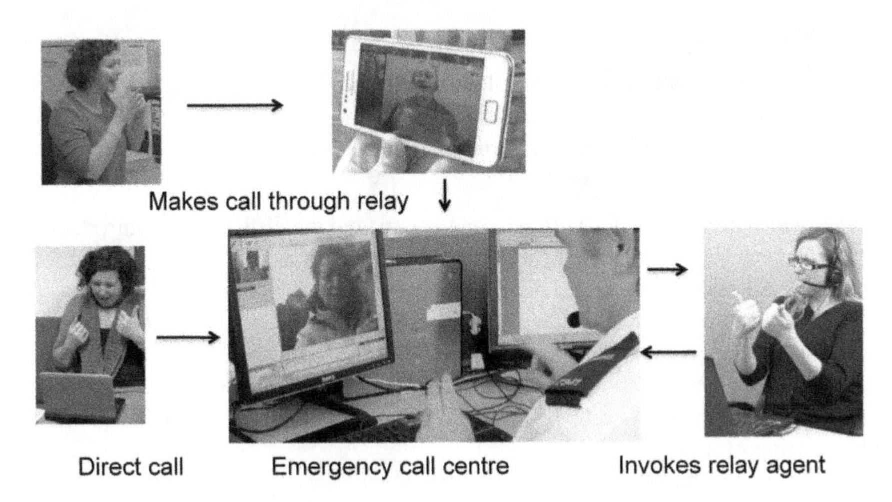

Figure 8.2 All users in the chain have access to Total Conversation facilities.

the network connection). This allows the agent an overall view of the incident and a better means to coordinate the emergency service response.

The agent's Total Conversation facility is IP-based and connects seamlessly with VoIP services and breaks out to traditional PSTN and mobile phone systems, through in-built gateways, although with some loss of information when such legacy interworking takes place.

Within the REACH112 project we had a balanced consortium of:

- PSAPs – police (United Kingdom, The Netherlands), fire (United Kingdom), ambulance (France), and multi-agency (Spain, Sweden);
- call handling (Sweden, The Netherlands, Spain);
- relay services (United Kingdom, Sweden, France, Spain, The Netherlands);
- disability organizations (United Kingdom, France, Europe-wide);
- international telecoms operators and suppliers (Siemens Enterprise, Nokia, France Telecom);
- researchers (University of Bristol); and
- Total Conversation providers (Sweden, France, United Kingdom).

The project work began in July 2009 to create the national infrastructures and to provide interoperability across countries. User installations and targets for P2P interaction were followed by relay service interaction and then direct PSAP contact. The funding for the project ran until mid-2012.

During the pilot implementation period when all services were in operation and open to the users, over 970,000 Total Conversation calls were made. There were over 120,000 relay calls and over 70 real emergency calls. Call patterns varied in each pilot program, with active users making significantly more calls in Sweden (a mature market) than in the United Kingdom (a completely new market set up for this project).

8.1.3 The Obstacles

Despite the advanced state of the technology itself and the overall success, there remain considerable obstacles to the further development, with which we are actively engaging. These obstacles are discussed mainly in terms of the UK situation, which we know best.

1. The nature of the communities with whom we work and the difficulties of introducing new technologies.
2. Training of service personnel, and the development of the support service.
3. The ethical issues concerning acquisition of user information and the personal security and privacy issues relating to video communication.
4. The difficulties to obtain network-provided location information on callers to a 112 service. Lacking the deployment of network-provided IP location requires the project to investigate alternative approaches.
5. Quality of service over IP networks, which remains elusive.

8.1.3.1 Communities (or Lack of Business Model)

It is fairly obvious that people who do not hear require a different means of distance communication from voice telephony. Since deaf people in the deaf community in the United Kingdom have a developed visual language, which has been described since at least the 16th century, then it is obvious that the means of distance communication has to be by visual telephony. Despite the ubiquitous provision of voice telephony, video telephony is still in its infancy. Typically, of course, because video telephony is a latecomer to the party, users are expected to meet the costs of the innovation. While, for hearing users, video is an option, for the users in REACH112 video is a necessity.

Most attempts to establish video telephony (apart from in Sweden) have been in sporadic and incomplete projects since 1997 mostly involving installations in public service locations. These projects have often entailed disparate video-conferencing systems and hardware, which inter-works rather poorly. Very few of these projects have involved ongoing support and training after the initial installation period. At the present moment, if one was to call all the numbers on the list of supposed video installations in deaf locations in the United Kingdom (around 1000), the likelihood of reaching another person who can sign, or of receiving any sort of reply at all, will be very low – we estimate less than 10%. Once installed, most public location videophones have been unused and unloved.

As broadband Internet connections have become more widely available, the cost of video calling is greatly reduced, and this brings more users into the video telephony community. Just as the increase in use of telephones has altered the perception of the hearing community about interaction and information, it can be predicted that there will be a huge social change among deaf people. This development is well advanced in the United States, where a more effective funding model has provided free installations and free support. In parts of Europe, Web clients of various formats are used for bilateral communication and for small group interaction, but these are not supported effectively.

The issue in introducing a new form of video telecommunications is the perception that, because it uses the Internet, then it should be free. Since there is typically no government funding, other business models have to be developed to support the service. As a result, there are significant problems in reaching the user groups and in registering them.

In order to do this satisfactorily, REACH112 involved the stakeholders and user associations and continues to work directly with the local users, through workshops and public demonstrations.

8.1.3.2 Training

One aspect recognized by the European Commission in supporting REACH112 has been the need for training of all links in the service chain, from home user to emergency service agents. In the United Kingdom, this chain is complicated by the need to engage with around 250 PSAPs, using different protocols for call handling and dispatch. In the case of disabled users, the implementation of training is hampered by the negative experiences that these groups have had in regard to training and the real problems of finding a suitable medium for such training.

REACH112 approached this by involving stakeholders throughout and at all levels, and then by ensuring that all outward-facing materials and information were presented in sign language and in simplified and visually appropriate text. The project websites offer service and support for end users.

At the same time, models for implementation of the systems for businesses, public services as well as emergency service and relay agents had to be developed.

8.1.3.3 Data Protection and Privacy Requirements

REACH112 developed a framework for ethical matters relating to the registration of end users and was mindful of the data protection and privacy requirements in each country. This was a significant consideration of the project in the early stages of the development, and, after a thorough examination of data protection legislation across Europe, checklists for project contact with all users and services were formulated and put into place.

There are also issues regarding the nature of video communication. Once you accept a call, you are faced with a visual image that has a greater potential to offend than does a voice-only call. User policies and practices had to be explained carefully and appropriately in the language of the users in order to reduce such problems, and strict procedures for dealing with complaints were instigated. This aspect was tackled first in the terms and conditions of use and then in the training.

8.1.3.4 Non-Availability of Location Information

While the project aspiration was to ensure that emergency services were able to locate IP-based devices using network-provided location services, it is clear that there is not a system in place to achieve this to the extent required. While the project did not have the scope to deploy such a system, alternative approaches were utilized.

It is common practice in any case for all emergency service agents to ask the caller about their location and to confirm this. In REACH112 this was backed up by stored data on the user supplied at the time of registration. This is a solution that works reasonably well for users who are not nomadic. For nomadic and mobile usage, the location services offered by modern mobile phones (such as GPS, and third-party-provided location databases) were utilized.

8.1.3.5 Lack of Quality of Service

This is perhaps the most difficult aspect to manage, and we will spend the remainder of this section on considering its impact on video telephony.

"Best Effort" Not Reasonable for Supporting Interactive Real-Time Services

The service that "best endeavors" Internet provides may be unique in the history of technological development in that it has never provided an explicit assurance in performing the functions it claims. This level of service may be acceptable where the value of information from the Internet outweighs the frustration, slow performance, and breaks in service. To some extent, this is reflected in the charging model, that is, connectivity only. These aspects may be a peculiarity of the UK broadband services, but the need to move from a Universal Broadband Commitment to an obligation would require the network operators to recognize the need to have metrics of outcome (and not speed) and to make transparent the network parameters and performance. This aspect was already identified by Cisco in 2005[1] (albeit from a point of view of preserving the profitability of the network operator):

> Regardless of how conservatively a service provider designs its network infrastructure, it cannot economically avoid over-subscription of network bandwidth.... The reality is that the cost to deliver a broadband Internet connection greatly exceeds the price a consumer would be willing to pay. For this reason, providers assume network over-subscription to reach an acceptable price.... Without a service control solution, providers have few alternatives for managing this situation:
>
> - Add more bandwidth in reaction to customer complaints, a solution that only increases network expenditure without a corresponding increase in ARPU and return on investment.
> - Ignore the problem and let subscribers manage with whatever bandwidth is available, a solution that results in significant customer turnover.

Both approaches have been taken, but these are not solutions. These are also approaches that seriously affect time-critical applications, such as videophone conversations.

Galbraith *et al.* [274] proposed to tackle the QoS issue from the upstream side as well as the downstream, which in their view would bring benefits to applications, such as video telephony. Many other QoS solutions have been proposed by researchers and some were even standardized. Unfortunately, they have not been deployed.

Reliable Service

Using sign language in video telephony demands the highest levels of Internet connection reliability. Hearing users can fall back to voice-only VoIP when video quality becomes unusable

[1] "Cisco Service Control: A Guide to Sustained Broadband Profitability," Cisco Systems White Paper, pp. 3–4. This white paper was accessed by the authors at http://www.democraticmedia.org/PDFs/CiscoBroadbandProfit.pdf at the time of writing, but it has since been removed.

or even to PSTN for a voice connection. In regard to the video for hearing users, temporary visual artifacts are merely irritating. Where the visual field is the only medium that conveys all of the information, then predictable and reliable service that assures delivery is essential.

The provision of video telephony to deaf or hard-of-hearing users becomes part of the important interactions of their lives, from their business and personal interaction to contact with emergency services. They should expect a quality of experience that is sufficient for their needs and which is consistent and predictable.

Requirements to Achieve an Acceptable Quality of Experience

Even though Web browsing may seem satisfactory when the user has a poor Internet connection, the user's experience with video telephony can be problematic. We installed Internet videophones in the Bristol area in the middle of 2005 and, initially, people were happy. However, by the beginning of 2006, people started to complain: "The videophone was freezing, jumping, was blurring, there were strange boxes on the screen." And so on.

The same results were coming from other users – on different equipment and also from people using webcams. It was just not reliable – at some times it was OK, but not at other times.

We carried out tests and recorded what happened at each end of the video call. We looked at the picture quality and we examined the data packets that were sent across the Internet. We examined the picture quality and analyzed which data was being passed through the network and which was not.

We found that there was considerable delay in sending some packets of data across the network – up to 20 seconds at times – and also that some packets of data were being lost. When certain videophones receive only part of the picture, they freeze and wait for the next packet. After a certain period of time (sometimes as great as one-tenth of a second) the videophone has to assume the packet is lost and must continue to process the next packets in the stream. However, if the waiting delay is great, then the users may see a frozen image and then the attempted frame rate sent to them will drop. We found that with even a small loss of packets (less than 2%, which is not unusual in today's Internet in the United Kingdom) and delays in other packets of more than a tenth of a second, the videophones would often assume that the packet loss was the "fault" of the videophone itself and would reduce the amount of data it tried to send; hence everything slows down and interaction between the users worsens. The effect was more obvious at certain times of day (and applies to all devices).

Although less than 2% of packets were being lost in some calls, the cumulative damage (including the reduction in throughput and delay) could be estimated to be considerably more because of the consequent reduction in data transmission and then the psychological effects on the user's signing. Not surprisingly, the deaf users we worked with said it was impossible to have a conversation under these conditions.

While this data was taken from older videophones (and there are now improvements), the underlying problem of the network remains almost the same. While we can deliver better-quality images, there is still no guarantee that user experience with the necessary frame rate and picture quality for sign language can be guaranteed.

However, REACH112 had to work within this framework. The network traffic issues vary from country to country, and from operator to operator. We believe that the performance for interactive real-time communication will gradually improve as the attention of the Internet community increases.

8.1.4 Conclusion

REACH112 has been an ambitious project for Total Conversation, active in five countries in Europe. There is considerable pressure from the European Commission, and the end users, to exploit and to continue to offer the service. In the REACH112 project we showed that 112 calls with Total Conversation can be made successfully. The feedback from end users as well as the involved PSAPs was very positive. There is, however, room for improvement. REACH112 worked on these challenges and up-to-date results can be found on our project website [269].

8.2 PEACE: IP-Based Emergency Applications and Services for Next-Generation Networks

Yacine Rebahi
Fraunhofer FOKUS

8.2.1 Introduction

The transition to next-generation networks is often coupled with the vision of innovative services providing personalized and customizable services over an all-IP infrastructure. To enable a smooth transition, next-generation all-IP networks need to support not only more services but also current vital services, namely emergency services. PEACE (IP-Based Emergency Applications and ServiCes for Next Generation Networks) is a research project funded by the European Commission that aims to provide a general emergency management framework addressing extreme emergency situations, such as terrorist attacks and natural catastrophes, as well as day-to-day emergency cases based on the IP Multimedia Subsystem (IMS).

To achieve its goals, the PEACE project addresses two major technological challenges:

1. A general solution for secure multimedia communication in extreme emergency situations will be provided. This will often involve the establishment of an *ad hoc* networking environment for communication among first responders. In this context, the PEACE project will be devising mechanisms for fast and lightweight establishment of keying material between various first responders in order to ensure the security of their communication. Furthermore, to enable multimedia communication in such environments, current centralized services, such as VoIP call routing and name translation, need to be supported in an *ad hoc* networking environment.
2. The PEACE project will investigate the provision of day-to-day emergency communication in next-generation all-IP networks. Owing to the different structure of IP and PSTN networks, it is not possible simply to reuse PSTN emergency services standards for emergency services communication in IP networks. This involves location information management, emergency call identification, and routing.

The PEACE project started in September 2008 and lasted for 27 months. The Consortium, shown in Table 8.1, is a healthy mixture of SMEs, manufacturers, operators, and reputable research organizations. For more information about PEACE, we refer to the project website [271].

Table 8.1 PEACE Consortium

Partner name	Country
PDM&FC Portugal	Portugal
Instituto de Telecomunicações IT Portugal	Portugal
Kingston University	United Kingdom
Fraunhofer FOKUS	Germany
University of Patras	Greece
Thales THC France	France
Telefonica I+D TID Spain	Spain
PaleBlue PB Sweden	Sweden

8.2.2 Project Scope

To be compliant with the scope of this book, we will only focus on the daily emergency services.

For the underlying communication infrastructure, the members of the PEACE project have chosen the IP Multimedia Subsystem (IMS) architecture specified by the 3GPP. IMS is based on SIP, and the choice was made because IMS is the preferred choice for telecommunication operators.

Today's emergency services are heavily regulated. This has led to some requirements that the different standardization bodies working in the emergency area took into account when specifying the related solutions. This also required that the PEACE project have a strong link to the standardization activities (and to contribute to them, if possible) that have been undertaken particularly in 3GPP [275], IETF [57], and OMA [59]. The roadmap for that has been the evaluation of the mentioned standards and accordingly their implementation if they are appropriate, or suggestion of new solutions if the standards are not appropriate or if the issue to be investigated was not addressed. This will also ensure the usability of the project output by academia and industry, as this is one of the main objectives of the PEACE project.

The sections below explain the project goals in more detail.

8.2.2.1 IMS Emergency Services Prototype

The PEACE daily emergency framework is based on the 3GPP specification release 7. However, to the best of our knowledge, no open-source implementation of such emergency support framework exists so far. Having a working model that can be used and tested will show not only that the standards are a theoretical description of the concept but also that they can be transformed into software in the form of a proof of concept. A working demonstration will be essential to the acceptance of the mentioned standards.

8.2.2.2 Caller Location Retrieval

Caller location information retrieval is one of the main pillars of the emergency systems because first responders cannot be dispatched if the call-taker does not know (at least approximately) the caller's location. The work of 3GPP in this area is mainly reusing the standards defined by the IETF groups ECRIT [121], Geopriv [46], and SIPCORE [276], and the SIP location conveyance [32] and the PIDF-LO [17] specification in particular.

In the context of location information retrieval, and in addition to the use of some of the mentioned standards, the PEACE framework focuses on the following approach.

In a mobile environment, a caller's current location retrieval is based on the cell ID or on the Global Positioning System (GPS). Passing the cell ID information to the PSAP through TDM is not possible. This has pushed a number of operators (at least in Spain) to develop solutions that allow a PSAP to retrieve location via proprietary protocols. To get rid of the complexity and the rigidity of these components, we suggest in the PEACE project the specification and the development of the Global Location Enable (GLE), which estimates the caller location based on his or her access network details (3G cell coverage shape, WiFi access point coordinates, etc.) and passes this information to the IMS platform in the PIDF-LO format over the LOCSIP protocol [277]. In this case, no gatewaying technology is needed.

As conventional standalone GPS has difficulty providing reliable positions in poor signal conditions, Assisted GPS (A-GPS) can address the corresponding problems by using network elements in the form of assistance servers. This assistance generally falls into two categories: (i) information used to more quickly acquire satellites, or (ii) calculations done remotely. In the PEACE project, we will develop a solution that allows the user device (in particular, the IMS client) to contact the A-GPS server over the OMA Secure User Plane Protocol (SUPL). This does not require any modification in the operator's network (since SUPL runs on top of IP) unlike earlier control plane location solutions. Nevertheless, it requires the SUPL network infrastructure to exist. Location in roaming also needs interaction with the roaming user's home network, but this scenario is not dealt with in the PEACE project. As a summary, A-GPS improves Time To First Fix (TTFF) and positioning errors compared to conventional GPS.

8.2.2.3 Security Framework for Emergency Services

Like any other network service, emergency services are prone to misuses and attacks. In the literature, the work on security for emergency calls is very scarce. So far, only RFC 5069 [278] has discussed the security threats as well as the setting of requirements to deal with them. RFC 5069 is without doubt a good starting point for securing emergency services provision. However, we would like to mention the following related limitations:

- RFC 5069 was targeting "open" environments (SIP-based VoIP networks) and not "private" environments, such as IMS networks. In addition to that, this RFC remains a high-level work and some of the scenarios described there might fail to occur in special environments like IMS.
- RFC 5069 describes the security concerns in the emergency environment as well as some high-level requirements.

In the PEACE project, the IMS emergency services architecture has been analyzed with respect to their security threats and to suggest solutions.

8.2.2.4 Support for Persons with Disabilities

Persons with disabilities should also be provided with access to emergency services. Within this context, the PEACE project will specify and develop the following solutions:

- Provision of a friendly user interface supporting audio, video and live text in any combination. Live text is a real-time text communication, where text flows character by character as typed. Live text is an improvement of the traditional Instant Messaging, which is not adequate for reporting an emergency, as described later on.
- Secure conveyance of the caller's medical information. Persons with severe illnesses, like diabetes and heart problems, are included in the category of persons with disabilities.

8.2.2.5 Prioritization of Emergency Calls

There is a common belief that emergency calls must be prioritized with respect to normal calls, especially in the presence of network congestion. For efficient call prioritization, signaling traffic as well as data traffic has to be considered. In the literature, prioritization is usually achieved through some queues in which packets or messages are classified according to their importance. Unfortunately, adding such queues to SIP servers will increase their load, as more processing is needed. This will certainly lead to a degradation in the performance of the SIP servers. As a consequence, the PEACE project members suggest the concept of virtual queuing, where a virtual queue is built based on sending some "retry after" messages to the sender. Various factors are taken into account, namely the arrival time of the messages, the time needed for the current message processing, and the length of the virtual queue. In addition to that, an interface to the policy enforcement point will ensure the assignment of the appropriate QoS class. A more detailed discussion can be found in section 8.2.3.5.

8.2.3 Development Status

8.2.3.1 IMS Core Enhancement

So far, a preliminary version (ready for public testing) of the emergency enhancement related to the IMS core is available as open-source software under the GNU GPL licence; (see [279]). It is implemented in conformance with the 3GPP specifications TS 23.167 in C language and tested in a Linux environment.

Our IMS emergency support software is expected to be tested in public by allowing the NGN community to download the "enhanced" IMS core from the portal http://www.berlios.de, and submit it to functional, stress, and performance tests. The "enhanced" IMS core can be downloaded and installed as described in [280]. The P-CSCF, I-CSCF, and S-CSCF have been enhanced to be compliant with the emergency services specification, regarding emergency registration and emergency call recognition and routing. The code for the E-CSCF and the LRF entities, and testing scenarios, have also been included.

The E-CSCF is an SIP stateful Back-to-Back User Agent (B2BUA) that protects the PSAP. All the messages sent by the caller will contain a universal emergency identifier, based on the type of service recognized by the UE or the P-CSCF. When sending messages to the PSAP, the E-CSCF will replace the emergency identifiers with the real PSAP SIP URI in the From or

To headers. Moreover, we intend to implement the same mechanism for the Contact address as well, by statically or dynamically allocating a Contact address by the E-CSCF to PSAPs. In this way, the caller will never be able to know the SIP URI (or the associated SIP URI if a CS PSAP) or the Contact address of the PSAP, which might contain the IP address, thus being unable to directly call and attack it.

8.2.3.2 IMS Client Enhancement

The project MONSTER [281] provides an extensible plug-and-play framework developed by Fraunhofer FOKUS (Figure 8.3). This toolkit enables the creation of rich terminal applications compliant to NGN, IPTV, and Web standards. It has also been enhanced with a location-aware emergency services add-on. The extension includes several modules and has the following roles:

- *Caller Side.* To recognize the emergency calls and integrate the current location of the user (if available) in the message flow so that the network can map the service request to the nearest Public Safety Answering Point (PSAP).
- *PSAP Side.* Displaying location information of the caller in a user-friendly way is necessary, in order to minimize the delay between the time when the PSAP receives the emergency call and the time when the first responders are dispatched.

The Emergency Add-on registers a listener to outgoing calls and thus it will be alerted about any outgoing session before the initial request of the new call is sent to the network. It can then verify if the call is destined to an emergency service by comparing the remote URI to both universal emergency identifiers, or emergency URNs, for example, "urn:service:sos.ambulance" [11], and emergency telephone numbers, for example, 112. In this way, all the generated calls, coming from the Emergency Add-on or other modules (Contact list, Call log, Dialing interface) can be intercepted and recognized as emergency calls.

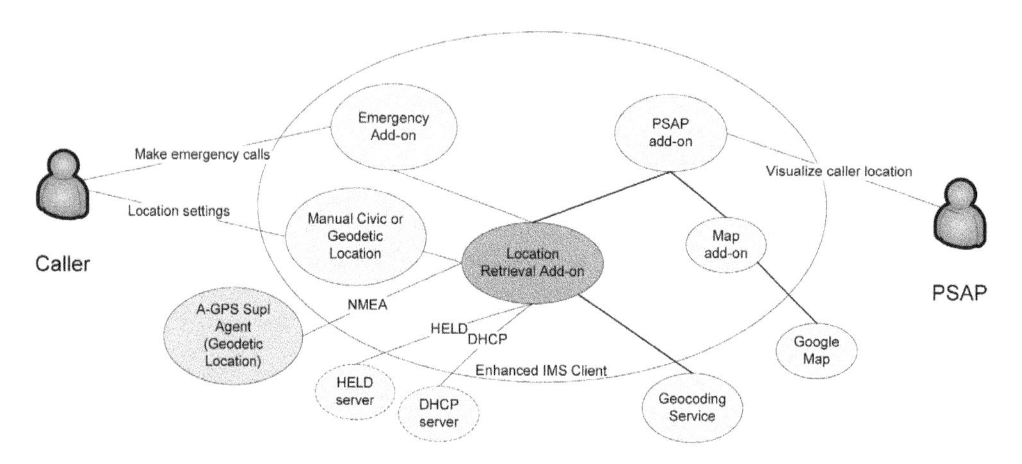

Figure 8.3 Functional blocks of the Monster IMS Client.

When recognizing an emergency call:

- The Location Retrieval Add-on will be queried for the most accurate location information. This information, if available, will be added to the initial request of the emergency call in a PIDF-LO format [17]. It can later be used by the network to route the call to the most appropriate PSAP.
- If the users entered an emergency service number, for example, 112, it will be replaced with a matching emergency service URN, for example, "urn:service:sos.ambulance." In this way, even if the user is dialing 112 and the ambulance is not registered at this number, the network will be able to properly recognize the call as directed to the ambulance.

The Location Retrieval Add-on is a module that can gather the user's civic or geodetic location by supporting a variety of retrieval methods including civic or geodetic information statically assigned by the user, using a NMEA connection [282]. Other methods, like using a HELD server [31] or a DHCP server, can be easily integrated [16, 27]. The module also prioritizes the location information depending on the freshness of the information and the accuracy of the method. For example, the A-GPS Agent is considered more accurate than manual configuration.

The PSAP module intercepts the emergency incoming calls. If the initial request contains location information, this will be decoded by the Location Retrieval Add-on. Decoding is the process where the useful information is obtained. For example, from the PIDF-LO civic location information XML fragment `<A3>Berlin</A3>`, the string "Berlin" represents the important piece of information. The result is a string formed by adding all the acquired information. Then, if the location information had a geodetic type, representing the coordinates of a point, the Map Add-on, based on a Google Maps API [283], is instructed to display a new target on the map at those specific coordinates with a label containing the caller SIP URI.

A different approach was taken for the case when the location was originally in a civic format. To translate the civic address into geographical coordinates, an interface with the Google Geocoding Service [284] has been developed. Thus the PSAP Map Add-on can query the Geocoding Service as to which coordinates correspond to the decoded location information and then instruct the Map Add-on to display the resulting point in the graphical user interface.

Another important feature at the PSAP is to make a callback if the call is suddenly interrupted or for transmitting further information to the caller. By using the menu of the history call of the Monster client, it is possible to generate a call using the SIP-URI of the caller. If the user is registered, his or her home network will forward the call to the visited network and the client will receive the call. No further enhancements were needed on the IMS core side to support callback calls for registered emergency callers.

The NMEA [282] interface with the GPS-enabled mobile equipment is implemented as a NMEA client that connects through TCP to the IP and port of the NMEA server and waits for location geodetic coordinates. The IP and port of the connection are set through the same user preferences as the civic location information. The user preferences allow the user to enable or disable the interface to the GPS-enabled mobile equipment before or after signing in.

8.2.3.3 Global Location Enabler (GLE)

The goal of this enabler is to estimate caller locations over a range of access networks. Implementation has been carried out on a Linux environment and tested at the Telefonica I+D facilities in Madrid (Spain).

The IMS core requests the caller's location from the GLE via a LOCSIP interface, which is an SIP-based SUBSCRIBE/NOTIFY protocol standardized by the Open Mobile Alliance (OMA) [277] and is built on top of the IETF presence event package RFC 3856 [285] and location filters that limit asynchronous location notifications [34]. Positions are formatted in geodetic coordinates (latitude and longitude) or civic addresses. The kind of access network the caller is using is determined examining the request, and a different algorithm is performed depending on the network.

For WiFi access networks, caller locations are estimated from their IP address. The GLE knows the positions of the WiFi hotspot antennas and receives information from the DHCP server about IP assignment. For mobile 2G/3G access networks, the GLE reads caller numbers and queries a Gateway Mobile Location Center (GMLC) for fetching caller positions using the Mobile Location Protocol (MLP), standardized by OMA. Support for Assisted GPS was implemented using OMA's Secure User Plane Location (SUPL) 2.0 protocol. GLE's role will be that of an SUPL server, providing assistance data to GPS-enabled handsets.

In short, the GLE uses the operator's network details in order to estimate emergency caller locations. Different access networks need different location algorithms, but this is hidden from the IMS core, which sees the GLE as a single node.

8.2.3.4 Assisted GPS

The purpose of the developments of Assisted GPS (A-GPS) is to reduce the required time to obtain the location of a particular call (in this case, emergency call). Therefore, it is necessary to develop some mechanism into the mobile equipment to allow the communication exchange with the SUPL 2.0 server available in the GLE, as described in the previous section, in order to obtain the assistance data for GPS-enabled handsets. The server-side implementation of SUPL 2.0 supports both GPS and Galileo satellites, and is backwards compatible with off-the-shelf A-GPS handsets, which are based on SUPL 1.0 (supporting only GPS). As soon as Galileo is fully deployed, the A-GPS mechanism developed in PEACE will be ready for it. The client side is implemented in Java language and ported and tested in a Nokia Symbian platform.

The mobile equipment implements a GPS interface to request the Assistance Data of Location. This interface implements the Userplane Location Protocol (ULP), which is used between the GLE (SUPL Location Platform known as SLP) and an SET (SUPL Enabled Terminal).

Depending on which SUPL Agent initiates the dialogue, an SUPL INIT message is sent to the SET (network initiated, also known as Mobile Terminated, MT), or an SUPL START message is sent to the SLP (SET initiated, also known as Mobile Originated, MO). The implementation of ULP is over a secure IP connection (SSL) on top of TCP, with the following exception: the SUPL INIT message is transported over WAP Push.

8.2.3.5 Emergency Call Prioritization

Owing to the intrinsic nature of emergency voice services, common sense implies that calls to emergency services need to be prioritized over non-emergency calls. This assumption takes even more importance in the case of a general emergency situation, such as a natural disaster or in a situation where system overload and congestion may occur.

Emergency calls are established using the SIP protocol and sent from the access network to the IP Multimedia Subsystem (IMS). The SIP protocol provides retransmission of calls after a certain timeout. In case we are in an overload situation where not all the calls can be processed instantaneously, call retransmission will then lead to a higher system load, which will degrade the quality of the service exponentially.

Summing up, our main objectives are to:

1. prioritize emergency calls over non-emergency calls; and
2. control the load in the system, avoiding unnecessary retransmissions.

The way to address these issues is to use a virtual queue implemented in the first entry point of the IMS infrastructure: the P-CSCF. SIP calls first arrive at the Proxy-CSCF, which will parse the call's SIP header to find out whether the call is an emergency call or not. Once this is determined, the next step is to schedule the incoming call in the virtual queue to determine when this call will be served.

The algorithm to place the calls in the queue is depicted in Figure 8.4. This algorithm ensures that emergency calls will be served always before any other non-emergency calls (objective 1). Now, we still have to deal with avoiding unnecessary retransmissions to reduce the SIP traffic (objective 2). This corresponds in the algorithm to the "if necessary notify caller" box.

According to the normal flow in an SIP call, the proxy receiving the INVITE request will respond with a provisional 1xx response, usually a 100 (Trying), to prevent the caller resending the SIP INVITE over the time. Upon reception of the provisional response, the caller will set up a timer that will trigger an error when it expires. In the RFC 3261, this triggering is called Timer C, and it must be set to more than three minutes. After this time, the caller takes the session as terminated. On the other hand, if there is no provisional response, the caller will resend the call, issuing unnecessary traffic.

According to this, the value of Timer C represents the maximum rescheduling time inside the virtual queue. In order not to have the caller dismiss the current session, there are two main alternatives:

- Use of the Retry-After header together with responses such as 480 (Temporarily Unavailable), 486 (Busy Here), 503 (Server Busy) or 600 (Busy Everywhere) providing the estimated number of seconds where it is foreseen that the call will be able to be processed by the queue.
- Issuing a 182 (Queued) response. This response may include the estimated number of seconds that will be spent in the queue. After this time, the call will be processed. The server may issue more than one 182 responses to update the caller about the status in the queue.

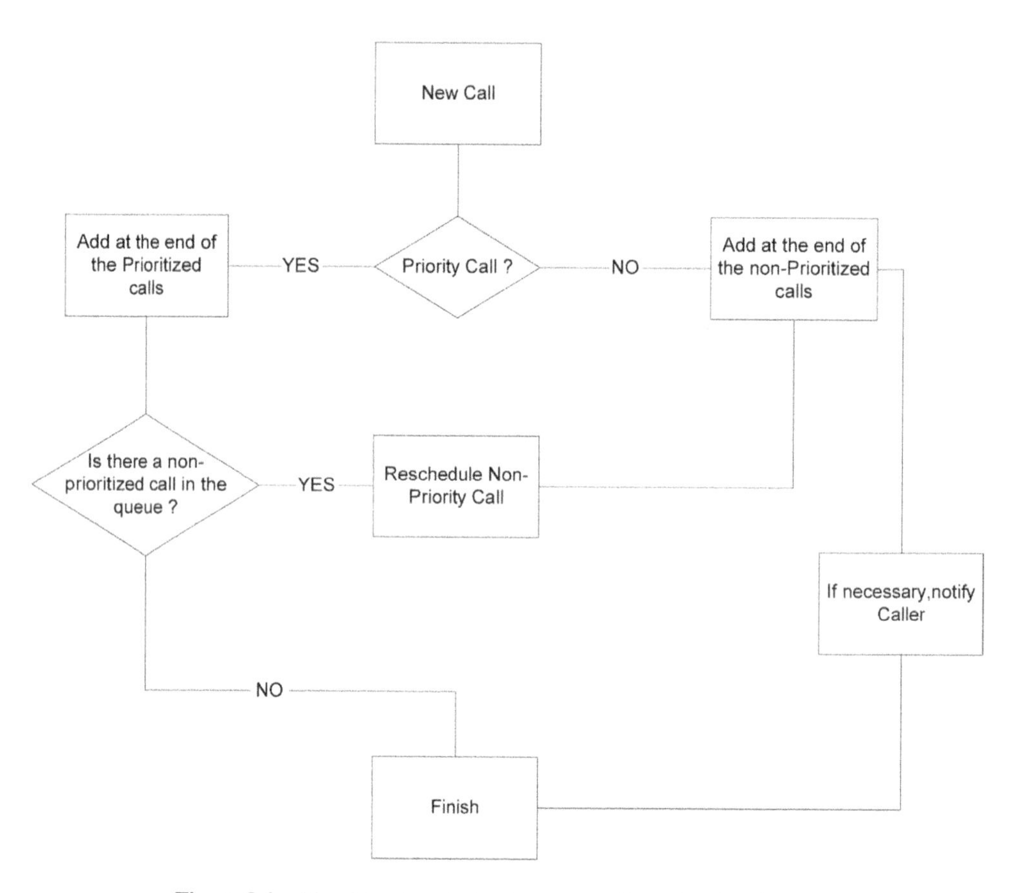

Figure 8.4 Algorithm for placing incoming calls in the virtual queue.

The differences between the two approaches is that 182 queuing is stateful (the call information will remain in memory) and will issue less traffic from caller to proxy, since the caller will be updated with the current status of the call periodically, whereas the Retry-After queuing is stateless (the call is dismissed) and will issue more signaling between the caller and the proxy: at least one resend and a provisional response.

If, after dismissing one call, a caller tries to resend it later in a different session, then the call will have lost the grace priority of having already been waiting in the queue once: the call will have a new SIP Call-ID and will be treated as new. For this reason, an internal call identifier needs to be defined to be able to map between past calls that have been queued and new calls from the same user.

The proposed queue ID is based on a hash function as shown below:

```
queueID = md5sum { RESPONSE/REQUEST : SIP method : From : To }
```

This will ensure that the re-issue of the call after a Retry-After will be treated with priority compared to other calls that have never been in the queue.

Another issue is how to estimate the waiting time inside the queue. This is done by keeping some statistics of the real duration of the calls inside the queue. Each call is time-stamped when it arrives and when it leaves, and the duration is used to estimate the duration of the next call. This involves having a long-term estimation (more than one day) with lots of past calls that is used as a reference each time the system is started, and a short-term estimation (last 100 calls or last minute – depending on the server load, this might be configurable), which is used to weight the current conditions of the system provided that, in a congestion or overload situation, call processing will take more time than in a normal situation and the system needs to adapt the estimation to the current conditions.

Prioritizing the emergency calls media data traffic is also taken into account by the Policy and Charging Control (PCC) architecture specified by 3GPP [286]. The P-CSCF can use the Diameter [143]-based Rx interface to interconnect with a Policy Control and Charging Rules Function (PCRF) to manage IP-CAN sessions with respect to service authorization and resource reservation. The P-CSCF informs the PCRF of the triggered requests of resources for IMS registrations and IMS sessions and can subscribe to notifications of signaling path status and/or change of IP-CAN type. The PCRF creates the policy rules from the information provisioned by the P-CSCF and enforces them. At the same time the PCRF will trigger notifications to the above-mentioned events toward the P-CSCF in case there is a loss of bearer, for example, based on information from the lower layers.

The Open IMS Core testbed provides support for the Rx interface. The P-CSCF differentiates between non-emergency calls and emergency calls. Thus, including a Service-URN AVP as the type of emergency service [11] along with the session information in the case of an emergency one, the P-CSCF handles the notifications received on the Rx interface, releasing the allocated resources when necessary.

The Rx interface can be used to plug in the Open IMS Core in an Evolved Packet Core (EPC) [287] environment. One of the first prototypes of EPC is the Open EPC testbed [288], and the interaction has already been tested.

8.2.3.6 The Live Text Solution

The Live Text solution is implemented as an extension to the IMS client developed by Fraunhofer FOKUS. It integrates new functionality, providing a means for communication where oral dialog is not possible. This includes, for example, the communication between a mute person and the PSAP. But other than just being a form of text communication, this extension enables the typing to be monitored by the receiver. This is an essential requirement for an emergency communications application, given the fact that, under a critical scenario, every character of information received from the person in distress is likely to be important. Therefore, it is not acceptable that a sentence is not received because the person for some reason is unable to press the "Enter" key. In the same way it is important for the person behind the PSAP to capture the typing pattern of the caller, such as hesitations and typing mistakes followed by corrections, and so on. The level of writing difficulties can suggest the level of stress or emotional condition of the person typing. A conventional chat functionality that only sends entire sentences is unable to provide this type of information. Only a live-text functionality enables such data to be made available for the person reading the chat.

Our solution is based on the Message Session Relay Protocol (MSRP) described in RFC 4975 [229]. MSRP is a protocol for transmitting a series of related instant messages in the context of a session. Message sessions are treated like any other media stream when set up via a rendezvous or session creation protocol such as SIP. The IMS client already supportsMSRP and it was modified in the context of the PEACE project to meet the live-text requirements discussed earlier. The corresponding functionality has been created without compromising network efficiency in spite of the more frequent usage of the communication link when compared to the conventional chat. While the quality of the network can put a constraint in the timely reception of the characters, this is not usually an issue, since a delay of 1 or 2 seconds is still acceptable for a text communication.

Other than physical impairment as a motivation to use this functionality, it is also useful in situations where the network availability is limited and voice communications are either difficult or not possible. An example would be a catastrophic scenario of total landline and terrestrial radio communications outage, where the only backup solution available would be emergency low-bitrate satellite links. Given the small bandwidth requirements of live text, it would work properly under these conditions, while still allowing clear information to be exchanged between first responders and coordinators or from civilians in distress to rescue centers.

8.3 US Department of Transportation's NG 9-1-1 Pilot Project

John Chiaramonte,[1] Gordon Vanauken,[2] Wonsang Song,[3] and Jong Yul Kim[3]
[1]*Booz Allen Hamilton*
[2]*L. R. Kimball*
[3]*Columbia University*

8.3.1 Overview

The US Department of Transportation's (DOT) Next Generation 9-1-1 (NG9-1-1) [272] Initiative sought to define and document a vision for the future of 9-1-1. The goal of the US DOT NG9-1-1 Initiative, a research and development project established in late 2006, was to complete a concept of operations, functional requirements, and system architecture. Additionally, it served to develop a transition plan that considers responsibilities, costs, schedule, and benefits for deploying Internet Protocol (IP)-based 9-1-1 networks.

The NG9-1-1 project is seen as a transition enabler, to assist the public in making a 9-1-1 call from any wired, wireless, or IP-based device, and to allow the emergency responders to take advantage of enhanced call delivery, multimedia data, and advanced call transfer capabilities. To accomplish these goals, the ideas and needs of both public and private 9-1-1 stakeholders were incorporated.

The NG9-1-1 Proof of Concept (POC) developed and deployed software and network components demonstrating the desired capabilities of the NG9-1-1 system. The focus of the POC was on the 9-1-1 call, from origination to delivery and handling by a public safety call-taker. This included the following.

- Call origination using:
  - IP User Agents, such as laptop computers, IP telephones, and IP wireless devices (for transmitting audio, data, text, and streaming video).

  - Cellular devices (for transmitting both audio and Short Message Service, SMS).
  - Third-party call center (for transmitting audio and vehicle telematics data).
  - Devices demonstrating the needs of the deaf and hard-of-hearing community (e.g., real-time text and video).
- Call support and processing using:
  - Standard IP access networks.
  - NG9-1-1 network components, such as Emergency Services Routing Proxy (ESRP) and transition gateways.
  - NG9-1-1 databases, such those related to business rules, the Location to Service Translation (LoST) protocol, Location Information Servers (LIS), and logging of call records.
- Call termination at the Public Safety Answering Point (PSAP) using:
  - IP-based Automatic Call Distribution (ACD) systems.
  - IP telephones and workstations.
  - Human–machine interface.

Three test laboratories and five PSAPs hosted the basic infrastructure for the NG9-1-1 POC demonstration. The laboratory tests focused on the call setup and routing of calls, while the PSAP-based testing focused more on the call termination and handling. The following laboratories housed the equipment and functionally operated as a single system:

1. Booz Allen Hamilton Center for Network & Systems Innovation (CNSI), Herndon, Virginia;
2. Texas A&M Internet2 Laboratory (TAMU), located in College Station, Texas;
3. Columbia University Next Gen Laboratory, located in New York, New York.

After completion of configuration and testing at the laboratories, equipment and software was installed and tested at five PSAP locations. At the PSAPs, professional call-taker, dispatch, and supervisory personnel were trained to assist with the POC testing. POC participating PSAPs included:

1. City of Rochester Emergency Communications Department, Rochester, New York;
2. King County E-911 System, Seattle, Washington;
3. Metropolitan Emergency Services Board – Ramsey County Emergency Communications Center, St. Paul, Minnesota;
4. State of Montana – Public Safety Services Bureau, Helena, Montana;
5. State of Indiana – Office of State Treasurer, Wireless 911 Board, Kosciusko County, Indiana.

These PSAPs were selected from over 50 applicants, using objective criteria developed by the NG9-1-1 team. While the field included many impressive applicants, resulting in a very close competition, the NG9-1-1 Initiative was limited by funding and a tight schedule, requiring a decision to limit participation. The selected PSAPs were geographically diverse and serve populations from the relatively small community to mid-sized metropolitan areas and even an entire state. They were seen as a representative sample of the 9-1-1 PSAP community at large.

Figure 8.5 shows the POC setup graphically. As can be seen, the different networks are interconnected over the Internet using the Generic Routing Encapsulation (GRE) [289]

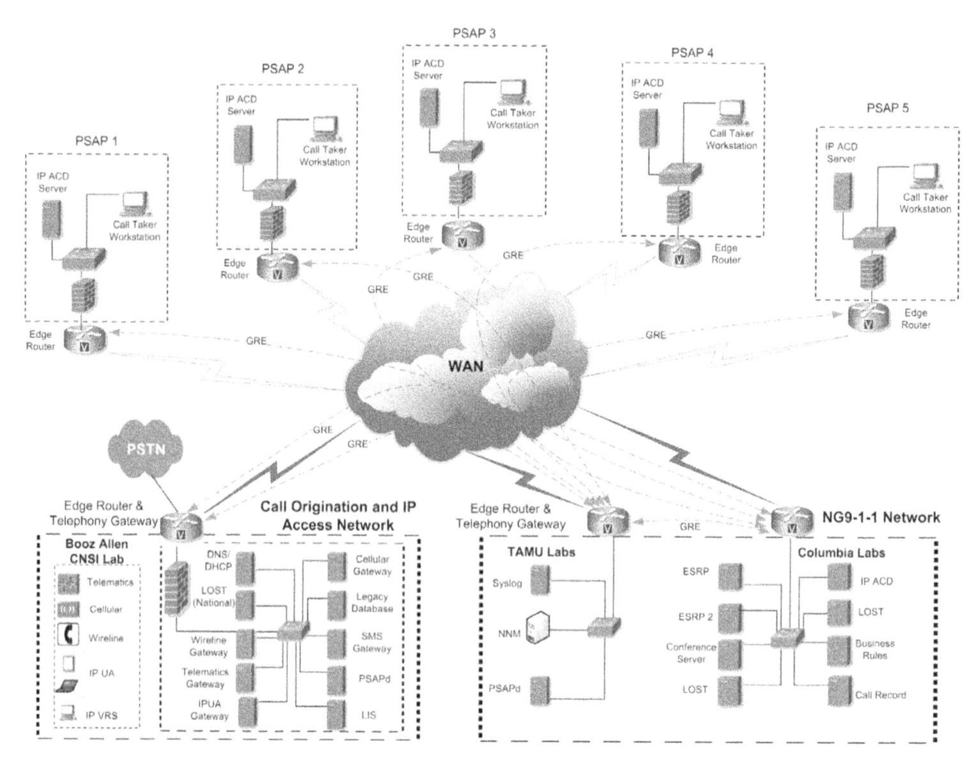

Figure 8.5 US DOT NG9-1-1 POC system architecture.

mechanism. Similarly to a Virtual Private Network (VPN), this ensures separation of the traffic used by the POC for emergency services purposes from the rest of the Internet traffic.

8.3.2 Proof-of-Concept Description

The NG9-1-1 Proof-of-Concept system (NG9-1-1 POC system) is divided into two parts, namely, the call origination network, and the NG9-1-1 network. The call origination network is the network in which emergency calls are initiated, such as cellular network, Public Switched Telephony Network (PSTN), or Voice over IP (VoIP) network. The NG9-1-1 network is an IP-based network shared by a group of Public Safety Answering Points (PSAPs). Each PSAP serves users within its jurisdictional boundary. Therefore, each NG9-1-1 network covers a region that is the union of the boundaries of the PSAPs participating in that NG9-1-1 network.

The NG9-1-1 network contains Emergency Services Routing Proxies (ESRPs), Location to Service Translation (LoST) servers [68], and IP-based PSAPs. ESRP is an SIP routing entity that forwards the call to the most appropriate PSAP. LoST server is a location resolution entity that maintains the PSAP boundaries and PSAP URLs. An ESRP needs a LoST server in order to determine the PSAP URL based on the caller's location. Together, these two components provide the call routing function. Routing is primarily based on the caller's location, but it can also be affected by business rules or other policies in the NG9-1-1 network.

An IP-based PSAP contains an SIP proxy, an automatic call distributor, a conference server, and call-taker workstations. Automatic call distributor is a signaling component that routes incoming calls to a call-taker based on the PSAP's local policy. Voice and video streams are handled by the conference server. The IP-based PSAP provides the call termination function such as answering the emergency call, dispatching the responder to the incident location, and recording the call and incident information.

There are many types of call origination networks, but in the NG9-1-1 architecture, they share common responsibilities: identifying an emergency call, obtaining the location of the caller and making it available to the PSAP, and routing the call to the appropriate ESRP. For this purpose, the POC system introduced SIP border gateways, which are SIP outbound proxies that enable aforementioned functions for emergency calls that the end device cannot provide. Also, since the NG9-1-1 network is SIP-based, it is the call origination networks' responsibility to convert their signaling protocols to SIP. For example, a PSTN call is converted to an SIP call through a telephony gateway.

An example of an emergency call flow within the NG9-1-1 architecture is as follows. When the caller initiates an emergency call, the caller's device determines its civic or geospatial location and queries a LoST server to resolve the location to an ESRP URL. A SIP INVITE request with the location information is sent to the ESRP. The ESRP queries a LoST server to resolve the location to a PSAP URL and forwards the SIP INVITE request to the PSAP. Within the PSAP, the automatic call distributor selects a call-taker and forwards the call to the call-taker. The call-taker's status changes to busy and remains so until the end of the call. Meanwhile, the call-taker communicates with the caller to verify the location, determine the type of emergency, and dispatch help to the caller.

8.3.2.1 Hardware Components

Call Origination Hardware

Booz Allen Hamilton CNSI hosted the hardware for the call origination. To demonstrate the integration of various call origination sources in the NG9-1-1 system, we used different types of devices to place emergency calls. The call origination devices were a telephone, Cisco IP phone, IBM laptop with SIP client software, and a Samsung BlackJack PDA phone.

The Plain Old Telephone Service was used to simulate emergency calls from the PSTN device. For this device, CNSI provided the connectivity to the IP access network using a Cisco telephony gateway. The phone was connected to the telephony gateway using RJ11 interface. The gateway converted a PSTN call to an SIP call.

A Cisco IP phone was used as a hardware IP user agent. The IP phone was one of the call origination devices used to show a call coming in from an IP-enabled environment.

Another example of an IP-based call origination device is a laptop with an SIP client software installed. This device was able to place an emergency call with multimedia capability such as voice, video (with a webcam), and text. We connected a GPS "dongle" to the laptop to determine its location.

The PDA phone was used to test voice calls and SMS messages from cellular networks.

In addition to call origination devices, Booz Allen Hamilton CNSI housed other call origination network components such as the telephony gateway, the LoST server, and SIP border gateways.

Table 8.2 Call origination equipment list

Equipment	Purpose
AT&T corded phone	Legacy call origination device
Cisco IP phone 7960G	Hardware-based IP UA call origination device
IBM Lenovo T60 / Windows XP / Logitech QuickCam Messenger	Software-based IP UA call origination device
Samsung BlackJack SGH-i607 (AT&T)	Cellular / SMS call origination device
Cisco 2821 integrated services router / VIC2-2FXS two-port voice interface card	IP router / telephony gateway
Cisco Catalyst 3560-48TS switch	IP switch for network connectivity
Cisco ASA 5505 Adaptive Security	Firewall
Dell PowerEdge 1950 rack server / RHEL5 x64_64	SIP border gateways / LoST / DHCP / DNS
Hypermedia HG-1600 VoIP GSM gateway	SMS gateway

The Texas A&M laboratory had the PSTN VoIP gateway. The PSTN VoIP gateway was used to convert cellular calls to SIP. A telephone number was assigned to the PSTN VoIP gateway. The number acted as the simulated emergency number for cellular calls. For example, if a user placed a call to that number, the gateway recognized the call as an emergency call from a cellular network and forwarded the converted SIP message to the cellular SIP border gateway.

The Texas A&M laboratory also had the VoIP GSM gateway and the SMS server. Both were used to convert SMS messages to SIP MESSAGE requests and vice versa. The gateway had both GSM and IP network interfaces: the GSM interface was used to send or receive SMS messages while the IP network interface was connected to the SMS server via TCP. This part will be explained in more detail later.

The hardware used and their purpose in the call origination network are summarized in Table 8.2.

Call Routing Hardware

Columbia University IRT laboratory hosted five Dell PowerEdge servers used mostly as the routing components in the NG9-1-1 network. Two servers were used for ESRPs, one server for the business rule database used by ESRPs, one for the LoST server, and one for the video recorder.

A Snowshore IP Media server from Cantata, now owned by Dialogic, was also housed in Columbia University. In the POC system, we used one conference server for all PSAPs. However, in the real deployment, the conference server might be placed at each PSAP.

The laboratory was also equipped with a test PSAP system. Table 8.3 summarizes the network equipment list.

Call Termination Hardware

Call termination components were deployed in the five PSAPs listed above. Each PSAP system was composed of a Cisco router, an IP switch, a Dell server, and a call-taker workstation. The router and switch provided the IP infrastructure in the PSAP. The Dell server ran sipd and psapd: sipd is an SIP proxy, and psapd is an SIP-based automatic call distributor developed at Columbia University. The call-taker workstation was equipped with three monitors to

Table 8.3 NG9-1-1 network equipment list

Equipment	Purpose
Cisco 2811 router	IP router
NETGEAR ProSafe switch	IP switch for network connectivity
Dell PowerEdge 1950 rack server / RHEL5 x64_64	SIP border gateways / ESRPs / LoST / business DB / video recorder / test PSAP
Snowshore IP Media server	Conference server

display the call-taker software, the call information window, and the mapping software at the same time.

Table 8.4 summarizes the PSAP equipment list.

8.3.2.2 Software Components

Caller SIP Client

SIPc is a software SIP user agent developed by Columbia University [290]. With SIPc, users can make a call with multimedia support such as audio, video, and text.

In the POC system, SIPc is used as an IP user agent (IP UA) call origination software. To support emergency calls, SIPc determines its location, routes emergency calls to the proper ESRP with the help of LoST server, and sends emergency SIP INVITE requests.

The location of the caller can be determined by GPS, Dynamic Host Configuration Protocol (DHCP) [16, 26], Link Layer Discovery Protocol – Media Endpoint Discovery (LLDP-MED) [291], or manual entry. In the POC system, we used LLDP-MED as the primary location source, but also tested GPS, DHCP, and manual entry.

To support routing of emergency calls, SIPc sends a LoST query to the LoST server in order to resolve the destination ESRP URL after determining its location. This location determination and LoST query are performed before the user places an emergency call.

Users can place emergency calls by either dialing 9-1-1 or pressing the emergency button on SIPc. After identifying the emergency call, SIPc sends the emergency SIP INVITE request. The INVITE request contains the location information in its body as well as the SDP for media negotiation. The request URI and the SIP To header field are set with the emergency service URN, which is "urn:service:sos" [11]. This URN acts as an emergency call identifier

Table 8.4 PSAP equipment list

Equipment	Purpose
Cisco 2811 integrated services router	IP router
Cisco Catalyst Express switch	IP switch for network connectivity
Cisco ASA 5505 Adaptive Security	Firewall
Dell PowerEdge 1950 rack server / RHEL5 x64_64	SIP proxy / psapd
Dell Precision T3400 workstation / Windows XP / Dell 2208WFP LCD (×3)	Call-taker workstation

for subsequent entities in the routing path. The SIP Route header field is set with the ESRP URL that is obtained from the LoST query in the previous step.

The SIPc user interface and SIP stack were written in Tcl/Tk, a script language for quick prototyping and easy GUI development. The media processing was implemented by external programs. For example, the audio processing features such as capturing audio, encoding and decoding, transmitting RTP [174] packets, and playing audio were implemented using audio library from Columbia University. Vic [292], an open-source video conferencing application, was used for video conferencing. Real-time text (T.140) [227] was implemented using the t140 handler, developed at Columbia University. Those external modules were integrated with SIPc using Tcl/Tk.

LoST Server

In the NG9-1-1 architecture, the LoST server is used for the location-based routing. The LoST server maintains PSAP jurisdiction information in its database. Given the caller's location in civic or geospatial format, it returns the contact information of the PSAP that serves that area.

The LoST request contains the location of the caller and the emergency service identifier, which is "urn:service:sos" for emergency calls. The location information in the request could be in either civic or geodetic format. The LoST response contains the mapping information. The mapping information includes the PSAP URL, the PSAP display name, and the PSAP boundary.

In the POC system, we deployed two levels of LoST servers, namely the public and private LoST servers. The public LoST server maintained the state-level mapping information. The call origination devices used the public LoST server in order to get the ESRP URL. The private LoST server kept the county-level mapping information. It was accessed by ESRPs in the NG9-1-1 network to get the PSAP URL.

The LoST server was written in Java as a Web application. It ran on the Apache Tomcat Web application server. The PostgreSQL server [293] with the PostGIS extension [294] was used to store the mapping database in both civic and geodetic formats.

SIP Border Gateway

The SIP border gateway is an SIP proxy in the call origination network. It acts as an interface between the call origination device and the NG9-1-1 network. Each of the call origination networks in the POC system, such as the PSTN, IP network, and cellular and SMS networks, had its own SIP border gateway. The SIP border gateway fills the gap between what the call origination network needs to provide for emergency calls and what they can provide using the limited capability of their devices.

For example, the PSTN telephone does not have the capability to determine its own location. In this case, the PSTN SIP border gateway queries the Automatic Location Identification (ALI) database to retrieve the location of the PSTN telephone. Likewise, the cellular or SMS SIP border gateway interacts with the simulated Mobile Positioning Center (MPC) to get the location of the cellular phone. The IP UA SIP border gateway uses DHCP to determine the location of the hardware SIP phone. For all of these cases, the location information is added to the SIP INVITE request by the SIP border gateway.

The SIP border gateway was implemented using sipd and an SIP-CGI script [295]. Sipd is an SIP proxy and registrar within CINEMA [296], an SIP software suite developed at Columbia

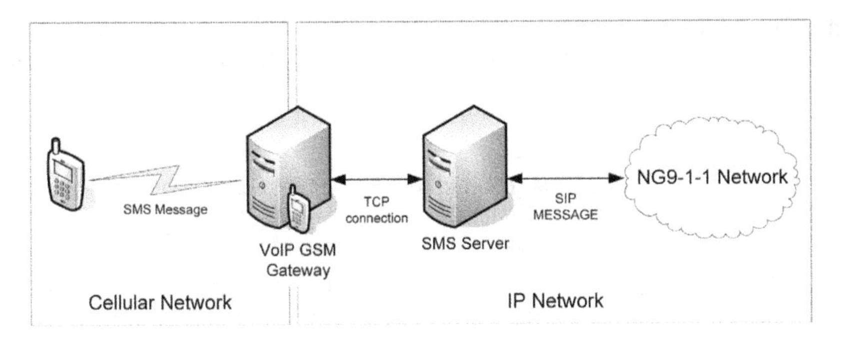

Figure 8.6 SMS server.

University. Emergency call-related functions were implemented in the SIP-CGI script. SIP-CGI is a "Common Gateway Interface (CGI) for service creation in an SIP environment" [295]. The SIP-CGI script was written in Perl. Sipd invokes the SIP-CGI script whenever it receives an SIP request message for emergency calls.

SMS Server

The SMS server, shown in Figure 8.6, interacts with the VoIP GSM gateway to convert SMS messages to SIP MESSAGE requests. The VoIP GSM gateway and the SMS server maintain a TCP connection to communicate with each other. The SMS server is located between the VoIP GSM gateway and the SMS SIP border gateway. The SMS server was implemented using Java and the JAIN SIP stack [297].

ESRP

The ESRP is an SIP proxy in the NG9-1-1 network. Its primary role is to route emergency SIP messages to the proper PSAP. The routing is based on the caller's location and/or the business rule of the NG9-1-1 network. In the POC system, the NG9-1-1 network business rule can override any routing decision that is based on the caller's location. For example, authorities may use business rules to redirect calls from a non-operational PSAP to its neighbor PSAP.

We deployed two ESRPs: ESRP West for the Western United States, and ESRP East for the Eastern United States. In the real deployment, there may be more ESRPs, reflecting the business relationship between groups of PSAPs.

We used sipd and an SIP-CGI script to implement the ESRP. Like the SIP border gateway, the SIP-CGI script implemented the emergency call-related functions and was written in Perl. The business rule was stored in the MySQL database.

The psapd Call Distributor

We used psapd as the automatic call distributor within PSAPs. It distributes incoming emergency calls to the call-taker depending on the call-taker's status as well as other PSAP policies such as language preference of call-takers. It also manages the call queue, logging, and conferencing.

We developed psapd using the CINEMA SIP library from Columbia University. The MySQL database was used to store the call information. The call distribution logic was written in a Tcl script for easy customization.

Video Recorder
In emergency communications, it is required to record all media streams so that people can replay them. In the POC system, the audio was recorded by the Snowshore conference server. However, it did not support video recording.

To support the video recording, we developed and deployed the video recorder in the POC system. The video recorder can relay the video stream in RTP, transcode it to MPEG-4, and store it into local files. The recording files can be accessed through the HTTP GET method later for replay.

The video recorder is transparent in that it records the video while forwarding it to the original destination. Here is how it works in the POC system. When initiating the emergency call session, psapd sends a "start recording" signal to the video recorder. Then it redirects the video stream from the caller to the video recorder by modifying the SDP sent to the caller. Originally, the conference server is supposed to receive the video stream for video conferencing. During the call, the video recorder records the incoming video from the caller while forwarding it to the conference server seamlessly. In this way, the caller's video is recorded without interrupting video conference.

The video recorder ran on a Linux platform and was implemented using the GNU ccRTP RTP library [298] and FFmpeg, a video conversion tool [299].

Call-Taker SIP Client
Call-takers in the POC system used SIPc to answer calls. Call-taker SIPc, shown in Figure 8.7, handles multimedia calls, displays emergency information such as the caller's location and the callback number, and allows the call-taker to enter incident details. It also uses Google Maps to display the caller's location on a map. Other functions include dynamic display of call scripts and standard operating procedures based on incident types, customizable one-click speed-dial buttons, and links to call supporting information.

The call-taker SIPc shares the code base with the caller SIPc, except that the user interface implements the extra functions needed by call-takers.

8.3.2.3 Features and Use Cases

The following sections describe four different aspects of the POC system, namely call origination from various devices, callback, conferencing and call transfer, and logging and recording.

Call Origination From Various Devices and Location Determination
The POC explored call origination use cases from five different communication service classes: IP UA such as laptops and IP phones, telephones, cellular phones, SMS, and telematics devices. Figure 8.8 shows this graphically for the different call origination networks.[2] Even

[2] While there are five communication classes, only four origination networks are shown in Figure 8.8, since cellular phones are used for voice and SMS.

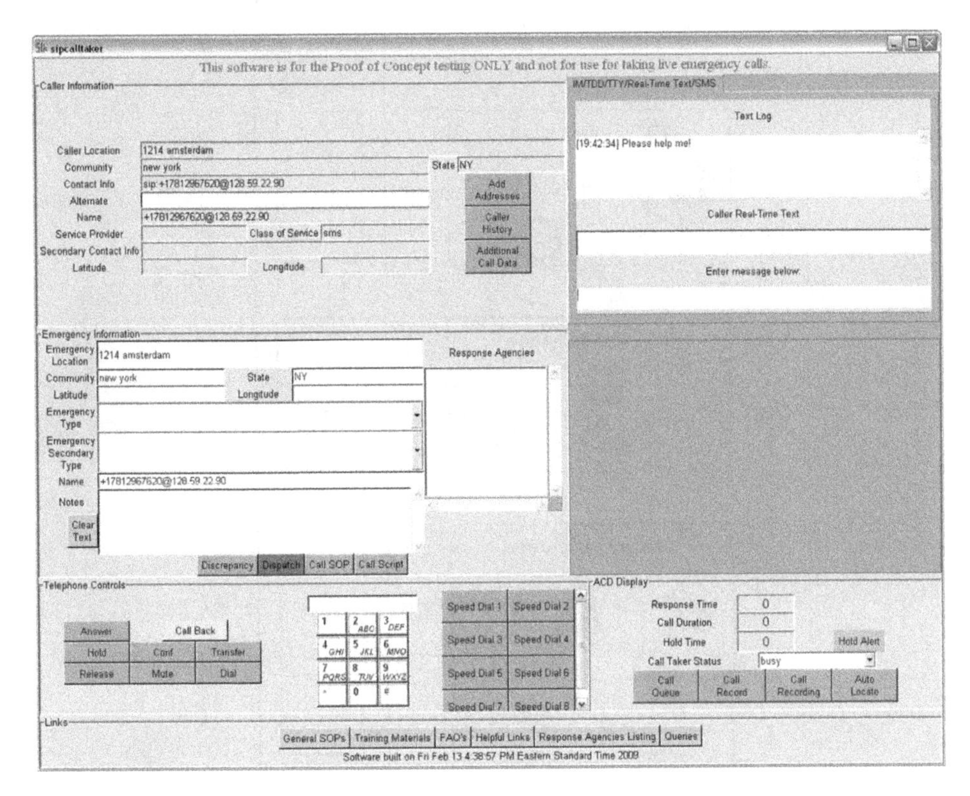

Figure 8.7 Screenshot of call-taker SIPc.

though these are five different services, emergency calls that originate from these services go through the similar call setup procedure. The procedure involves protocol conversion, location determination, and call routing. All calls are converted to SIP (for call signaling) and RTP (for media streams) if needed, and then forwarded to an SIP border gateway. The SIP border gateway is an SIP outbound proxy with emergency features such as determining the location of the calling device and querying a LoST server to route the call to an appropriate ESRP.

Location determination is necessary to forward the call to the correct ESRP and also for the call-takers to dispatch help to the correct location. The POC system places responsibility for determining location onto the call origination network. Once the location is determined, it is formatted into a PIDF-LO object [17], an XML document that carries the user's location, and added to the SIP INVITE request as a MIME attachment.

- *IP User Agents.* IP user agents (IP UAs) are devices that are used to communicate on IP-based services. The POC system tested two types of IP UA: a laptop with SIPc, and a Cisco IP phone. SIPc represents fully implemented NG9-1-1 capable IP UAs, while the Cisco IP phone represents SIP phones without NG9-1-1 capability. The POC system supported emergency calls from both IP UAs. Both IP UAs were SIP-based, so there was no need to convert the original protocol to SIP. Also, since both were SIP-based, the SIP

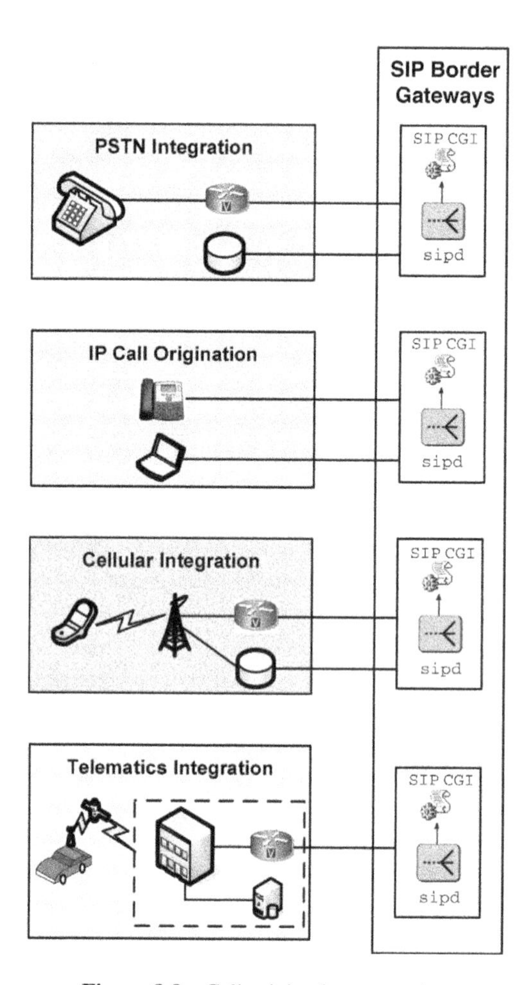

Figure 8.8 Call origination networks.

border gateway doubled as a registrar and IP UAs registered to the SIP border gateway. The following describes the call flow originating from SIPc and the Cisco IP phone.

SIPc tries to determine its location every time it launches. Location is determined automatically by using GPS or by querying a location-enabled DHCP server or an LLDP-MED switch. It can also be determined manually by the caller, who enters the street address or geodetic coordinates in the location configuration window.

With the location, SIPc queries a LoST server, which returns an appropriate ESRP URL for that location. In the case of an emergency, the ESRP URL is included in the SIP Route header field of the emergency SIP INVITE request. When the location changes, SIPc queries the LoST server again to find the appropriate ESRP URL for the updated location. This step is done before the caller places an emergency call.

SIPc recognizes an emergency call when the caller clicks on the emergency call button on its main window or when the caller dials an emergency number (e.g., 9-1-1 in the United

States, 1-1-2 in Europe). If the call is recognized as an emergency call, SIPc performs NG9-1-1 functions such as adding an SIP Geolocation header field to the SIP INVITE request, appending location information in PIDF-LO format to the SIP INVITE request as a MIME attachment, filling the SIP To header field and the request URI with "urn:service:sos," and sending the message to the SIP border gateway. The SIP border gateway does not have much to do except to forward it to the ESRP URL already included in the SIP Route header field.

Unlike SIPc, the Cisco IP phone does not distinguish an emergency call from a normal phone call. It simply forwards an SIP INVITE to the SIP border gateway. The SIP border gateway recognizes an emergency call by inspecting the SIP To header field. If 9-1-1 or "sos" is present in the SIP To header field, it is considered an emergency call. Once recognized as an emergency call, the SIP border gateway determines location information and queries the LoST server to find out where to forward this message.

The SIP border gateway determines the location of the Cisco IP phone by querying a simulated location database. It adds an SIP Geolocation header field to the SIP INVITE request and appends location information in PIDF-LO format to SIP INVITE request as a MIME attachment.

The SIP border gateway uses the location information to query a LoST server. The LoST server returns the appropriate ESRP URL. Then the SIP border gateway forwards the INVITE request to the ESRP.

- *Telephones.* A telephone was connected to a Cisco telephony gateway to demonstrate how the POC system can answer PSTN calls. The Cisco telephony gateway generated SIP messages and packetized voice into RTP streams, but was not NG9-1-1 capable.

  As with the Cisco IP phone, the telephone and the gateway do not distinguish between an emergency call and a non-emergency call, so the task of recognizing an emergency call is assigned to the PSTN SIP border gateway. The behavior of the SIP border gateway is identical to that in the Cisco IP phone use case, namely, it determines location information and queries the LoST server to find out where to forward this message. The difference is in how the SIP border gateway determines the telephone's location.

  The PSTN SIP border gateway determines the location of the telephone by querying a simulated ALI database. The ALI database maintains (telephone number, location information) tuples as the key, value pair. The SIP border gateway queries this database using the caller's telephone number and receives the caller's location. Then it appends the location information in PIDF-LO format to the SIP INVITE request as a MIME attachment and adds an SIP Geolocation header field to the SIP INVITE request.
- *Cellular Phones.* To support cellular phone calls, a PSTN VoIP gateway was introduced into the POC system. A telephone number was assigned to the PSTN VoIP gateway so that cellular phone callers can dial this number as the emergency number instead of dialing 9-1-1 on their cellular phones. When the call arrived at the PSTN VoIP gateway, it generated SIP messages and packetized voice into RTP streams. However, just like the Cisco telephony gateway, it was not NG9-1-1 capable.

  The PSTN VoIP gateway simply forwards a normal SIP INVITE request to the cellular SIP border gateway. It is the cellular SIP border gateway that recognizes an emergency call, determines and includes location, and routes the call to the correct ESRP.

  The cellular SIP border gateway determines the location of the cellular phone by querying a simulated Mobile Positioning Center (MPC) database. The database is identical to the ALI database: it uses the (cellular phone number, location information) tuple as the key, value pair. After the cellular SIP border gateway gets location information from this database, it appends location information in PIDF-LO format to SIP INVITE request as a MIME attachment and adds an SIP Geolocation header field to the SIP INVITE request. Then, as with other SIP border gateways, the cellular SIP border gateway queries the LoST server to resolve location to an ESRP URL.

- *Short Messaging Service.* The POC system explored the idea of using SMS as an emergency communication tool. To support this, a VoIP GSM gateway was paired with the SMS server. The VoIP GSM gateway was similar to the PSTN VoIP gateway. It had a telephone number assigned to it so that an SMS message can be sent to the gateway. The assigned number was used in lieu of 9-1-1. Once the SMS message reached the VoIP GSM gateway, the gateway converted it into a format that is readable by the SMS server. The SMS server's role was to generate an SIP MESSAGE request from the SMS content. The VoIP GSM gateway and the SMS server sent messages to each other over a persistent TCP connection. Both the VoIP GSM gateway and the SMS server were not NG9-1-1 capable.

  The SMS server simply forwards an SIP MESSAGE request to the SMS SIP border gateway. The behavior of the SMS SIP border gateway to support NG9-1-1 is identical to the cellular SIP border gateway, except that the forwarded message is an SIP MESSAGE request instead of an SIP INVITE request.
- *Telematics.* Telematics devices embedded in the car detect accidents, automatically dial an operator, and transmit relevant data. The important feature in Telematics is the delivery of crash data along with the call.

  The POC system emulated this feature with help from OnStar. A cellular phone was assigned as a Telematics call device. All calls from this cellular phone were converted to SIP by OnStar's gateway, and then forwarded to the Telematics SIP border gateway.

  OnStar used a simulated MPC database to get the caller's location, similar to the cellular phone use case. The location was then appended to the SIP INVITE request. Crash data were also included in the SIP INVITE request as an HTTP URL in the SIP Call-Info header field. Later, the call-taker was able to pull the crash data through HTTP GET method.

  Since OnStar provided the caller's location, the Telematics SIP border gateway simply routed the call to an ESRP after a LoST query.

PSAP Callback

Call-takers need to call back the emergency caller after calls are disconnected or to follow up on continuing incidents. The POC system implements a simple callback feature. The callback number is extracted from the SIP From header field in the caller's SIP INVITE request. The call-taker is able to dial back with one click on the callback button. To the NG9-1-1 network, which is an all-SIP network from ESRP to individual PSAPs, the callback is simply an SIP INVITE request from the call-taker to the caller.

When this SIP INVITE request comes to the call origination network, each network handles the request a little bit differently. In an IP UA network, the SIP INVITE request is forwarded to the caller without any modification. In a PSTN or cellular network, the SIP INVITE request is converted by gateways. The same happens with Telematics networks, as there is no data flowing from the call-taker to the caller. Only the voice is transmitted. In the SMS network, the SMS server converts the SIP MESSAGE request to a format readable by the VoIP GSM gateway, which in turn converts it into an SMS message. The SMS message is then sent to the caller's SMS device.

Conferencing and Call Transfer

During an emergency call, call-takers may need to add more participants or transfer the call. For example, the call-taker may need assistance from the supervisor or the call may have to be transferred to the local police station. Both functions are implemented in psapd and the call-taker software. The call-taker software provides the user interface to initiate these functions, while psapd provides the logic that makes these functions work.

In the POC system, all calls are conference calls. This means that all incoming calls from callers are automatically put into conference rooms where call-takers are ready to receive calls. In this way, it becomes easier to add new call-takers, administrators, or other emergency personnel into the call as needed.

Figure 8.9 shows psapd behavior when a new call comes in. It creates a conference session first, selects one of the available call-takers, then invites the call-taker to the conference. As a third-party call controller, psapd facilitates the media negotiations between the conference server and the call-taker software. After the negotiations are successful, the caller is added to the conference call. Once again, psapd facilitates media negotiations between the conference server and the caller software. If successful, the caller and the call-taker can talk to each other in the same conference room.

It is simple to add another party to this conference room. The call-taker enters the third party's SIP URL and clicks on the conference button in the call-taker software. The call-taker software constructs an SIP REFER request [232] with the third party's SIP URL in the SIP Refer-To header field and sends it to psapd. Using information in the REFER request, the psapd constructs and sends an SIP INVITE request to the third party. This begins media negotiations between the conference server and the third party's software. The third party is in the conference as soon as the negotiations are successful.

The call transfer is implemented by adding a third party to the conference room, as described above, and then the transfer-requesting call-taker voluntarily withdrawing from the conference room. This is due to the requirement that the transfer-requesting party has to make sure that the transfer is successful before dropping out of the call.

In both conferencing and call transfer, invited participants receive the caller's location. The psapd includes caller's location in the SIP INVITE request sent to invited participants. In call transfer, psapd also includes the PSAP database URL so that the invited party can pull relevant information such as incident information and comments from the original call-taker. The URL is conveyed in the SIP Call-Info header field.

Logging and Recording

The ESRP logs incoming SIP messages and outgoing SIP messages so that administrators can see routing decisions for each emergency call. The psapd logs include call logs, call distributions, and status of call queues. Call logs contain information such as timestamp, caller ID, caller location, call-taker who received this call, call duration, join and leave times of various parties, and call status. The call-taker software also logs information such as the call-taker's status, call hold time, incident details, and the call-taker's comments on incidents. Both the psapd and call-takers share a database to log such details. The ESRP has its own log repository.

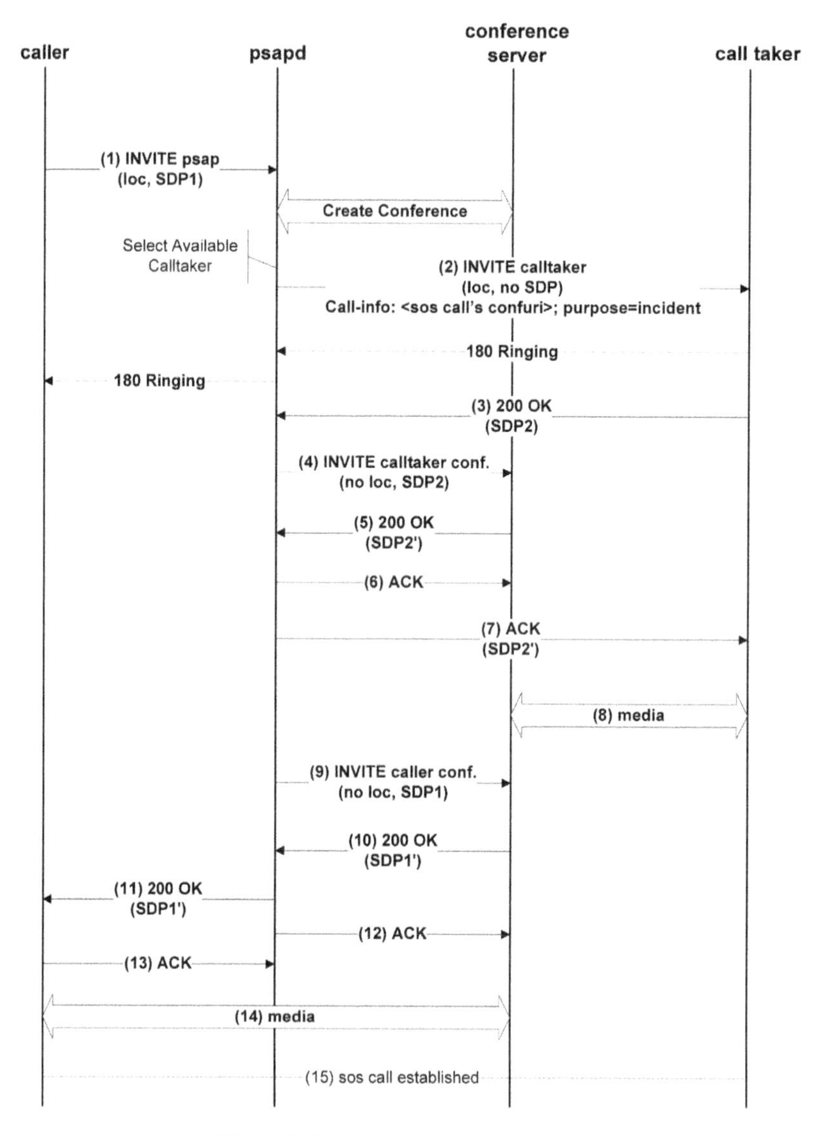

Figure 8.9 Basic emergency call flow.

Call recording is divided into three parts based on the type of media used. Voice is recorded and stored in the conference server. This is a natural choice because every call participant's voice goes through the conference server. Text is recorded by the call-taker software and stored in a database after call termination. Video is recorded by the video recorder, which intercepts video frames going to the conference server and records the frames while relaying it to the conference server.

The logs and the recordings are available to the PSAP administrator through the PSAP webpage.

8.3.3 *Testing*

The POC testing was completed by using the following process steps:

1. Develop testing plan
2. Develop training
3. Conduct training
4. Conduct testing
5. Document results

The entire testing process flow was built on the concept of "crawl, walk, run." The training and laboratory testing comprised the crawl phase, while everyone learned the new systems. The team was able to walk during the individual PSAP testing, and then run during the system and demonstration testing. This process was able to help the team members, including actual PSAP personnel, to develop a comfort level with the new systems.

8.3.3.1 Develop Testing Plan

The POC Test Plan defined the procedures and logistics necessary to test the functionality of equipment and applications associated with the NG9-1-1 POC. The plan was focused on the following objectives:

- Provide a formal testing process.
- Test each requirement identified in the NG9-1-1 system.
- Document the results of each test.
- Provide useable information from which to develop the final documents, such as final design, cost–benefit analysis, and transition plan.
- Provide verifiable documentation that the concept, as developed, is functional.

A test was developed for each selected requirement. For each requirement, a test script provided the following information:

- Test description – brief overview of the test.
- Test procedures – required test steps.
- Expected results – what was expected to happen.
- PASS/FAIL – general indicator of success or failure of the test.
- Results – documentation of the test outcome.

The scripts provided a set of test steps and expected results that were used as a measure of success for each requirement. There was considerable interrelationship among several of these requirements, and a single test call was used occasionally to test more than one requirement at a time.

The Test Plan was written before all system development had been completed. The test procedures did not always accurately reflect the proper operation of the system software, but in many cases the actual functional requirement was still met by the system. In these cases, the functional requirement received a PASS, and the details of how the requirement was met were added to the notes section of the test documentation.

8.3.3.2 Develop Training

Two sets of training were developed, one for the internal test team and one for the PSAP staff. The internal test team training was informal. This training consisted of review of the test plan, processes, and software associated with the testing. The development team went over the various systems and their functions.

The PSAP staff training was a formal, two to three hour block of instruction. This training covered the following aspects:

- Background information of the NG9-1-1 Initiative.
- Purpose of the Proof of Concept.
- Overview of the call-taker workstation and software.
- Hands-on practice.

8.3.3.3 Conduct Training

Internal test team training was held in conjunction with staff meetings at the Booz Allen Hamilton CNSI laboratory site. This allowed the team to get hands-on experience with the call origination and the call-taker equipment. The PSAP testing was conducted at each PSAP by two trainers and additional test team staff.

8.3.3.4 Conduct Testing

The testing proceeded in six phases as shown graphically in Figure 8.10. The three laboratories were tested as a group. The system components were located in the various laboratories but functioned as a single system. The Texas A&M and the Booz Allen Hamilton CNSI laboratories originated calls using each type of technology tested. Calls were answered using a

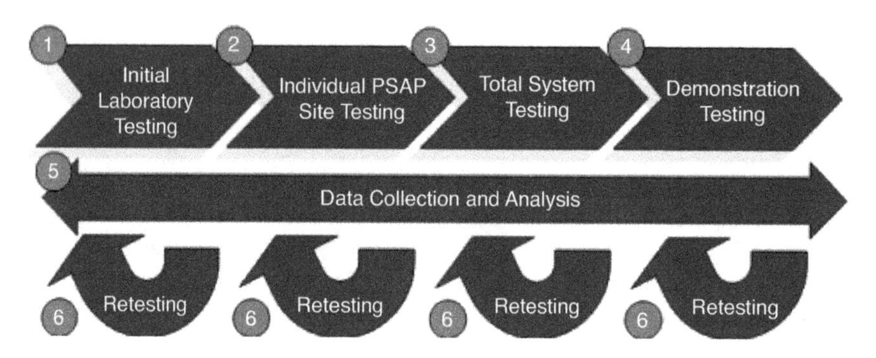

Figure 8.10 NG9-1-1 testing phases.

call-taker workstation located at the Booz Allen Hamilton CNSI laboratory, which simulated the infrastructure that was housed at each of the five PSAPs.

Each PSAP was tested individually. The equipment housed in the PSAPs was intended to work independently. The test team went to each PSAP and conducted the tests for a two-day period. The PSAP staff conducted the tests, in most cases operating the workstation while the test team of two personnel made sure that the scripts were followed and documented. A support team was also located at the laboratory sites as support for call origination and to resolve technical issues if they occurred. All tests used dedicated equipment, systems, and test data. At no time were live calls handled on the POC system or were test calls sent to live 9-1-1 systems.

At the completion of the PSAP testing, there was a total system test. This testing was completed by having all PSAPs operating at the same time. Transferring calls and changing call handling rules on the system were included in this testing.

The original testing scripts developed for the POC Test Plan were modified as a result of the actual testing experience with specific technologies. During the two-day test period, test team and PSAP personnel were present at each of the five PSAP locations, and the development and support team was located at the three laboratories. Seven tests were developed and tested. Each script tested a specific technology or function. The tests were as follows:

- Process Wireline Call
- Process Cellular Call
- Process Intelligent IP Call (Voice, Video, and Text)
- Process SMS Text Message
- Process Telematics Call
- Process Extended Transfer of Calls
- Business Rules Test

Each test was broken down into three scripts (except transfer and business rule, with two scripts), which used the same format as the other tests. The three test scripts for each test were: (1) Call Setup, (2) Call Handling, and (3) Data Handling. To test as many of the use cases as possible, the various use cases were integrated into each test. The use cases covered in each section of the test script are listed below.

1. Call Setup
   - Call Authentication
   - Recognize Originating Location
   - Recognize Call Type
   - Route Call to PSAP
2. Call Handling
   - Answer Call
   - Determine and Verify Location of Emergency
   - Determine Nature of Emergency
   - Identify Appropriate Responding Agency or Service
   - Establish Conference Call
   - Provide Network Bridging Services

3. Data Handling
   - Update Mobile Caller's Location Information (if caller is mobile)
   - Record Call
   - Obtain Supportive or Supplemental Data Post Call Delivery
   - Initiate Callback
   - Transfer Call Record

During the second day of total system testing, the system was used by all five PSAP locations as well as the three laboratory sites.

A demonstration was prepared, including animated slides describing the processes that each call source used. These slides were used to visualize and explain the processes that occur prior to and following placement of the test call. After the slides were reviewed, an actual test call was placed to the demonstration location.

Over a period of three weeks, seven demonstrations were conducted, one at each of the five PSAP locations, one at the Texas A&M Laboratory, and one at the Booz Allen Hamilton CNSI Laboratory. These demonstrations were held to allow input from a variety of 9-1-1 stakeholders regarding the processes and systems of the NG9-1-1 POC.

Throughout the testing process, data was collected to be evaluated on the functioning of the POC system. This data included data from dedicated test equipment used on the system as well as the test results and log records from the systems.

The POC was a research and development project, and the system was being constantly reviewed. A series of software versions and changes to the systems were developed and integrated during the testing process. To support the timelines, retesting occurred in an *ad hoc* manner during the testing process and whenever new features were released by the software development team.

8.3.3.5 Document Results

The test plan was developed to assist in the documentation of the testing. Each test script included two sections to document the testing results. Each test script conducted at each location was completed, and some scripts were retested and documented on the same script to show the retesting.

This resulted in a large number of tests being conducted during the testing session. Each test script was included in the test results document that was submitted to the US DOT.

The POC test report had the following summary of the testing at the laboratories and PSAPs:

> The three laboratory sites were tested at the beginning of the POC, prior to any PSAP-based testing. Although the three laboratories housed the equipment, logically they operated as a single system. Seven use cases were tested at the laboratories, and testing was completed by placing test calls to a workstation at the Booz Allen laboratory. During these initial tests, 47 requirements were tested, with 39 (83%) successfully passing. This lower pass rate can be attributed to the limited maturity of the POC software at the time of testing. The laboratory tests focused on the call setup and routing of calls, while the individual PSAP-based testing focused more on the call termination and handling. At the 5 PSAPs, 26 professional call-taker, dispatch, and supervisory personnel were trained to assist with the POC testing. Involving end users in the testing process provided a valuable benefit to the overall effort. These users were able to provide subjective

> feedback about functionality which someday they will come to depend on. Their input about their needs and the difficulties currently faced in today's 9-1-1 environment will help ensure that current problems are addressed by emerging NG9-1-1 technology. During the PSAP-based testing, 273 functional requirements were tested, with 241 (88.3%) successfully passing. While no industry benchmarks exist that gauge the success of emergency service network implementations, the team felt accomplished in having successfully demonstrated a significant portion of the NG9-1-1 concepts and use cases during the POC.

The total system testing went well. There were 114 total test scripts tested at the five PSAPs and the laboratory. Of those test scripts, only five were not tested (due to time constraints) and four failed. This is a 92% pass rate. No single test script failed at all sites. All failures were a result of individual issues at a single site.

During the second day of total system testing, all five PSAP locations, as well as the three laboratory sites, were tested at the same time. This did put stress on the systems, and some problems were experienced.

The network response time extended and call-takers began to report that the call-taker workstation software took a second or more to refresh the screen with the caller's information when they answered a call. They also noticed some of the other screens (map, call detail record) were slow to open.

Several SMS calls were not delivered or delivered to the default PSAP many hours later. During the demonstrations, a firewall was removed at one of the laboratories and this seemed to resolve the issue during those tests.

The POC demonstrations were conducted at the five PSAPs and two laboratory sites. There were approximately 200 attendees including representatives from media, elected officials, PSAP and 9-1-1 personnel, vendors, and members of the speech- and hearing-impaired communities.

After testing in the laboratory and PSAP, the total system was tested. Calls were processed through the complete system, recording the results. All tests used POC equipment, systems, and test data, and at no time were live 9-1-1 calls handled on the POC system or test calls sent to live 9-1-1 systems. Over a period of several months in mid-2008, a total of 116 functional requirements were tested using use case scenarios. In all, the project conducted 320 individual tests in the laboratories and PSAP facilities, with 280 (87.5%) successfully passing the test criteria. Table 8.5 summarizes the results of the POC.

8.3.4 Conclusion

The US DOT NG9-1-1 Initiative was one of the first federally funded studies to comprehensively define and document a future vision for 9-1-1 systems. As a result of this project, 9-1-1 stakeholders have begun to recognize that a fundamental transformation of the way 9-1-1 calls are originated, delivered, and handled is slowly under way. The NG9-1-1 Initiative's POC helped create action within the community to get more involved and to start discussing the issues. At the conclusion of the NG9-1-1 Initiative, the results transitioned to the E9-1-1 Implementation Coordination Office (ICO or National 9-1-1 Office). The ICO is a joint program between US DOT National Highway Traffic Safety Administration (NHTSA) and the Department of Commerce, National Telecommunications and Information Administration (NTIA).

Table 8.5 Proof of Concept findings

High-level functional component	Initial findings
Ability to send and receive voice, video, text (Instant Messaging (IM), Short Message Service (SMS)), and data	Successfully tested. Location data for SMS-based call is not currently commercially available. There is no method for guaranteed message delivery and some SMS failed to be delivered. Initially, issues arose with bandwidth and video streaming methods that caused video-based calls to fail
Increased deaf/hearing-impaired accessibility	Successfully tested. Upgrades and technological improvements will be needed at multiple levels
Caller's location identification	Successfully tested for wireline, wireless, and IP-based calls. A number of acquisition and identification processes were used to demonstrate functional capabilities to locate emergency callers
Call routing based on caller's location	Successfully tested use of an Emergency Services Routing Proxy (ESRP) and Location to Service Translation (LoST) servers, along with a Business Rules database
Transmitting Telematics data, including speed, vehicular rollover status, and crash velocity	Successfully tested. Demonstrated the ability to easily and automatically transfer important data associated with a vehicle crash
IP networking and security in an emergency communications environment	Successfully tested. Work is still needed to ensure the integrity of the network and system

During the POC trials, we implemented the NG9-1-1 system and deployed it in three laboratories and five real PSAPs. We were able to make emergency calls from devices such as telephones, software SIP clients, IP phones, and cellular phone. SMS was also tested as a potential emergency communication tool. Emergency calls were routed to the proper PSAP based on caller's location and policy in the NG9-1-1 network. It was also shown that data can be delivered along with the call. Examples of such data included the caller's location and crash data.

One of the greatest benefits of the NG9-1-1 Initiative was the sharing of information related to Next Generation 9-1-1. The POC testing report stated:

> It is imperative that this concept of information sharing continue to flourish and grow in the 9-1-1 community. There is definitely a need for a common location where stakeholders can collaborate and share information from past, present, and future NG9-1-1 efforts. Whether it be a repository, wiki, or combination thereof, all this information should continue to be open and available to all stakeholders. Providing a platform for communication will guarantee an increase in information sharing, leading to decreased cost and increased quality and interoperability of 9-1-1 systems. It will provide a place of recognition and scrutiny for all NG9-1-1 efforts and ensure that the 9-1-1 stakeholder community does not repeat past difficulties.

Several issues were identified during the processes that may benefit from further research and development work. These issues were broken into Operational and Technical. Additional details on these can be found in the test report from the POC located at the US DOT NG9-1-1 website [272].

The following operational issues that would benefit from additional development were identified during the POC:

- Process for handling abandoned, lost, and dropped calls.
- Call-taker interactions with SMS messaging.
- Concept of operations for business rules, policy-based routing, and NG9-1-1 system and software configuration.
- Integration of Telematics Automatic Crash Notification (ACN) data and criticality metric determination.
- Automatic third-party conferencing.
- Effective demonstration of sensor data integration into PSAP operations centers.
- Definition of a flexible, authoritative hierarchical governance and operation model for call handling and routing in NG9-1-1.
- Standards for integrating external systems and services into NG9-1-1.
- Flexible human–machine interface software architecture for taking in new data sets.
- Accreditation of NG9-1-1 systems to ensure interoperability.
- Operational model and technical feasibility study for authority-to-citizen communication.
- Interface with and transfer of NG9-1-1 information to other emergency services (fire, police, EMS).

The following technical issues that would benefit from additional development were identified during the POC:

- Importance of product selection and understanding the 9-1-1 vendor community.
- Improved network and system management of NG9-1-1 systems.
- Extensions of network monitoring, traffic generators, and packet sniffers for future interoperability, accreditation, and performance benchmarking of NG9-1-1 systems.
- Best practices guidelines in security for NG9-1-1.
- Building redundancy, reliability, and overflow into the NG9-1-1 network.
- Study and standardization of codecs for optimal voice and video transmission.
- Optimization of call routing based on call propagation timing.
- Location acquisition for all forms of emergency communication.
- Provision of imagery and additional supplemental data to the call-taker.
- Improved geospatial data fusion for the PSAP and call-takers.

9

Organizations

In earlier chapters, various standardization efforts from the IETF, 3GPP, OMA, and the IEEE have been described in some level of detail. These organizations have traditionally been involved in the standardization of communication protocols for telecommunication networks as well as for the Internet. Since these organizations are widely known even to those not active in standardization, we omit an overview of how they work. The emergency services ecosystem is larger, and other organizations with a focus on specific tasks and certain communities or specific regions are active as well. In this chapter we wanted to introduce you to a few of those organizations; the following sections are written by those who participate there or who belong to the leadership of the organizations.

The description aims to be short and should give you, as the reader, a broader perspective. The focus of the description is more on the structure of the organization and the main scope of the work rather than on the technical details. We asked every contributor to capture a set of questions about the membership of the organization, the goal of the work, and their vision for emergency services.

In section 9.1 Chantal Bonardi and Jean-Pierre Henninot describe the ETSI Special Committee EMTEL (Emergency Communications). Jean-Pierre is the Chairman of the group and Chantal has the role of the Secretary. Both of them have a long history in ETSI and are the best persons to write about the structure of ETSI EMTEL and the work the group is doing.

We asked Roger Hixson, the Technical Issues Director of NENA, to provide a write-up about NENA. We have been working with Roger for a long time in NENA Working Groups, in the preparation of emergency services workshops, and also on IETF-related matters. Roger also co-chaired the EENA NG112 Technical Committee for a few years. In section 9.2 Roger provides a short write-up about NENA. The length of his description understates the energy of the large number of members NENA has, and the high-quality work products it produces. We definitely encourage everyone to take a look at the NENA webpage and to read through some of their published standards, which are available for download for free. In sections 3.1 and 3.2 we have already described two NENA efforts, NENA i2 and NENA i3.

The European Emergency Number Association (EENA) is described in section 9.3. Gary Machado, the Executive Director of EENA, provides us with some insights about the organizational structure of EENA and work done in the different committees. EENA is seen by many as the European counterpart of NENA.

Internet Protocol-Based Emergency Services, First Edition. Hannes Tschofenig and Henning Schulzrinne.

John Elwell provided us with a description of Ecma in section 9.4. The contributions by Ecma are more focused on enterprise networks, and this was John's focus while he worked at Siemens.

Section 9.5 contains a description of the Alliance for Telecommunications Industry Services (ATIS). Martin Dawson, a regular ATIS participant, provided us with a description of this American forum.

In section 9.6 Carla Anderson describes the role of the NG9-1-1 Caucus and the NG9-1-1 Institute. Carla serves as the Deputy Executive Director of the E9-1-1 Institute.

Finally, in section 9.7 a short description of the European Commission's Communications Committee (COCOM) Expert Group on Emergency Access (EGEA) is provided. This write-up was provided with input from Martins Prieditis and reviews from Alain Van Gaever and Gyula Bara.

9.1 ETSI EMTEL

Chantal Bonardi[1] and Jean-Pierre Henninot[2]
[1]*European Telecommunications Standards Institute (ETSI)*
[2]*Ministère de l'Economie de l'Industrie et de l'Emploi (MEIE)*

9.1.1 Purpose of ETSI Special Committee EMTEL (Emergency Communications)

Emergency services are a "regalian" responsibility of a State toward its citizens. Efficient management of resources (equipment and human) rely more and more on the application of information and communication technologies (ICT). It is not surprising that ETSI decided to create a group dedicated to this topic.

9.1.2 Main Features of EMTEL

The ETSI EMTEL *ad hoc* group (OCG EMTEL) started its activity in 2002, initiated by the ETSI Board for Coordination of Emergency Telecommunication activities in ETSI. In 2005 it became a fully recognized ETSI Committee, Special Committee EMTEL. As such, EMTEL is one of the ETSI Technical Bodies (TBs), which number more than 30 in total. ETSI is a European standards organization active in all areas of ICT (including fixed, mobile, radio, broadcast, Internet, and several other areas) setting globally applicable standards. It is a not-for-profit organization, created in 1988. ETSI is not a certification body, but can develop specifications that may support certification.

The structure of SC EMTEL is similar to the other ETSI Technical Bodies. SC EMTEL elects a Chairman and Vice-Chairmen; elected from November 2010 are Jean-Pierre Henninot (MEIE, France) as Chair, and two Vice-Chairmen, Peter Sanders (one2many) and David Williams (Qualcomm, UK), and a Secretary, Chantal Bonardi (ETSI). The SC EMTEL work is to produce mainly Technical Specifications (TSs), Technical Reports (TRs) or Special

Reports (SRs). Only TSs are normative; the two other categories are only informative. As in other ETSI TBs, a new work item can be set up with the support of at least four ETSI members. Publications are freely available on the ETSI website [300].

9.1.3 *Scope of ETSI SC EMTEL Work*

SC EMTEL is responsible for identifying the operational and technical requirements of those involved in the provision of emergency communications, for conveying these requirements to other ETSI committees and for liaison with other organizations involved in this field.

The main responsibilities of SC EMTEL are as follows:

- To act as a key coordinator in collecting requirements on Emergency Communications from the different stakeholders (including National Authorities responsible for provisioning emergency communications, End Users, the European Commission, Communication Service Providers, network operators, manufacturers, and other interested parties), outside and inside ETSI. The scenarios to be considered include: communication of citizens with authorities, communication from authorities to citizens, communication between authorities, and communication among citizens. Generally agreed categories are to be considered in the provision of emergency communications for practically all types of scenario, including communications resilience and network preparedness.
- To elaborate, through coordination with other TBs the ETSI positions on emergency telecommunication issues. Although SC EMTEL was formed to specifically address public safety user requirements for Emergency Telecommunications, other Technical Bodies (TBs) within ETSI have been active for some time, such as a cooperating activity between 3GPP and ETSI TISPAN on the specification of a Mobile Location Positioning protocol for the delivery to the Emergency Authority the position of a caller to the Emergency Services.
- To continue to monitor the work being undertaken throughout ETSI that is relevant to emergency situations, to ensure a coordinated approach and the stimulation of further work. EMTEL thus works closely with the ETSI Technical Committee on Telecommunication and Internet converged Services and Protocols for Advanced Networking (TC TISPAN) and the Third Generation Partnership Project (3GPP) on emergency requirements in next-generation networks (NGNs), and with the ETSI Technical Committee on Mobile Standards Group (TC MSG) on the European Commission's eCall vehicle safety initiative.

  The definition of an SIP interface from the NGN system toward a PSAP may be under consideration. Clarification of the need for this so-called peer-to-peer SIP interface is sought from the EU Commission and PSAP Operators.

  Many standards related to EMTEL topics (more than 700) are developed by other ETSI bodies, that is, 3GPP, TC TISPAN, EP MESA, TC TETRA, and TC ERM.
- To take the lead with respect to emergency communications issues and be the interface between ETSI groups and other SDOs and organizations. EMTEL liaises with organizations outside ETSI such as the Telecommunications Standardisation sector of the International Telecommunication Union (ITU-T, e.g., Study Groups 2 and 13) and the Joint Technical Committee of the International Organisation for Standardisation and the International Electrotechnical Commission (ISO/IEC JTC). EMTEL continues to monitor discussions with the European Commission's COCOM group of experts, especially in relation to the

Common Alerting Protocol (CAP) and emergency calls in a VoIP environment, and is involved in discussions with the European Commission over the IP interface between all-IP networks and next-generation Public Safety Answering Points (PSAPs).

9.1.4 Operation and Activities of SC EMTEL

SC EMTEL meets three or four times per year all over Europe. Meetings are regularly attended by experts from telecommunication operators, equipment vendors, and administrations, and there is a significant representation of emergency organizations, which places the Committee in an ideal position to identify user needs. EMTEL's task is to distill user requirements into reports and specifications and to coordinate with other ETSI committees to ensure that the necessary standardization is achieved. As well as user needs, the Committee's mission implies consideration of topics linked to network architectures, network resilience, contingency planning, priority communications, priority access technologies and network management, national security, and Public Protection and Disaster Relief (PPDR).

EMTEL membership is about 100 delegates belonging to industry, PSAPs, governments, regulators, and others. Membership is global, as in other ETSI TBs. Non-members may participate in EMTEL meetings and be subscribed on the EMTEL exploder list, upon agreement of the TB Chairman, for six months to see if they have any interest in becoming and ETSI member. The membership fee varies according to the number of units (the "class"), which are estimated according to the member company's annual ICT ECRT (Electronics Communications Related Turnover). SMEs, Micro-Enterprises, user and trade associations, and university and public research bodies come under class 1.

As regards relations with other bodies, ETSI places emphasis on this in view of the increasing number of common subjects linked to the pervasive use of technologies.

A Memorandum of Understanding has been signed between ETSI and NENA (National Emergency Number Association), involving mainly SC EMTEL and TC TISPAN. There is also a mapping of activities between ETSI and IETF, which shows an active working relationship between SC EMTEL and IETF ECRIT.

EMTEL liaises with numerous bodies such as 3GPP, ITU-T and GSMA concerning Public Warning Systems (broadcasting national emergencies). These include the United States-led initiative CMAS (Commercial Mobile Alert Service) and the Japanese-led initiative ETWS (Earthquake and Tsunami Warning Service). As evoked above, EMTEL is involved in European Union-funded projects, such as the eCall project (in-vehicle automatic emergency call), a project required by the Commission to ETSI, in coordination with TC MSG (Mobile Standards Group) and 3GPP groups.

9.1.5 EMTEL Evolution and Strategy

Since its early beginnings, EMTEL has developed a portfolio of deliverables beyond the basic requirements (emergency call, communications for public authorities, information toward and between citizens). EMTEL is permanently aware of the need to keep pace with the changing technology and the possibility of new services.

A Special Report (SR 002 777 [301]) on call forwarding and the referral of emergency calls is being compiled. At present, in some countries, emergency calls have to be redialed if more than one service is required. The Committee is looking at ways in which the call can be forwarded, to save crucial time and to increase efficiency.

In view of the Telephony over IP development, ETSI TR 102 180 [302] (Report on "Basis of requirements for communication of individuals with authorities/organizations in case of distress") was revised. The EMTEL group is currently completing the work on "Total Conversation Access to Emergency Services" in [303] and [304].

Furthermore, the medium-term evolution of EMTEL will be based on a few guidelines, developing relationships, collaborating with public authorities, and promotion; EMTEL will also endeavor to maintain a vision of future emergency services.

9.1.5.1 European Public Warning System

In a major new initiative for the committee, EMTEL has taken a leading role with the output of an EC-funded project for a European Public Warning System. This project, which was supported by the governments of several European countries, has been the main focus of EMTEL's work during 2010. The aim of this work was to capture European-specific requirements for a Public Warning System (PWS) for broadcasting national emergencies using the Cell Broadcast Service into a Technical Specification. That Technical Specification (ETSI TS 102 900 [305, 306]) provided input to 3GPP for inclusion into the 3GPP PWS specification.

9.1.5.2 Closer Working Relationships With Public Safety Users

The Committee is making particular efforts to involve the public safety community more closely in its work and in the standardization process generally, and has adopted a strategy to encourage participation. Members of the Committee have made a number of presentations at relevant conferences, and EMTEL is investigating the enhancement of the current EMTEL website to provide a forum for the publication of documents, for the dissemination of information, and for consultation among the wider public safety community.

9.1.5.3 Contribution in Public Authority Instances

When the European Commission reactivated the EGEA expert group for emergency access [307], EMTEL proposed to present its activities and offered to collaborate, which was welcomed. In view of the execution of a new Mandate 493 on emergency caller location [308], SC EMTEL collaborated with ETSI TISPAN to prepare the corresponding deliverables.

9.1.6 Vision for Future Emergency Services

The ability to share information between all stakeholders in the provision of emergency services is vital to the effective response to day-to-day service requirements and major disasters, however that major disaster may occur or in which country.

Voice, data, image, and video technologies are everyday tools of the public, and the expectation is that the emergency services should be equally well equipped. There is therefore a need to provide common standards that enable the diverse public safety organizations to maintain, as far as practicable, their current business structures and practices while facilitating, when necessary, cross-service and cross-border cooperation.

The vision would be for the public safety and emergency services across Europe and globally to have available systems that operate in a globally harmonized spectrum allocation that is reserved solely for the use of public protection and disaster relief and that is capable of enabling the sharing of information that is up to date, accurate, and available when required. This spectrum should have sufficient bandwidth to enable the use of narrowband, wideband, and broadband data requirements, and should be a "publicly owned" asset and not be subject to auction as envisaged in some countries under what is commonly known as the "digital dividend."

With regard to emergency calls to designated three-digit numbers, it is recognized that there are a large number of such combinations in use throughout the world and the vision should be for technology to recognize any of the combinations in use, irrespective of the country in which the call is made, and route the call to the nearest public safety answering point (PSAP) in the country in which the call originated.

9.2 NENA

Roger Hixson
National Emergency Number Association (NENA)

The National Emergency Number Association's (NENA) mission is to foster the technological advancement, availability, and implementation of a universal emergency telephone number system (9-1-1). In carrying out its mission, NENA promotes research, planning, training, and education. The protection of human life, the preservation of property, and the maintenance of general community security are among NENA's objectives.

NENA, a non-profit corporation officially founded in 1982 to further the goal of "One Nation–One Number," has more than 7000 members and 48 chapters – a membership dedicated to saving lives. NENA works every day on a single, yet vital, mission: providing an effective and accessible 9-1-1 service for North America. NENA's membership is dedicated to making 9-1-1 and emergency communications work better, and they measure success in the lives that are saved by 9-1-1 each day.

Serving as a vital communications link in the delivery of emergency services, 9-1-1 has, throughout its evolution, become recognized as an asset of the North American public. And the NENA organization has been connected to 9-1-1 every step of the way. More and current information can be obtained from the NENA website (see [58]).

From assisting and promoting new system installations, through solving and defining standards and recommendations on technical and operational issues, to developing and clarifying 9-1-1 related policy issues, and educating managers on the latest technologies and business practices, NENA and its members have been intertwined with 9-1-1 during the growth and development of the 9-1-1 systems in North America. NENA is proud to share this heritage, and as an organization, we are uniquely positioned to take 9-1-1 to new heights and to meet new challenges. A major example is the initiation of what is now known as Next Generation 9-1-1 (NG9-1-1), beginning with concepts in 2001 and the start of technical development in 2003.

In parallel with continuing development, NENA's definitions of IP-based functions and architecture for NG9-1-1 were brought into the 2007–2008 US Department of Transportation's

NG9-1-1 R&D project and Proof of Concept demonstrations. NENA senior staff members were instrumental in this effort to showcase the capabilities of a technologically evolved 9-1-1 call processing and delivery process.

In cooperation and coordination with international standards organizations, and especially IETF, the results of NENA development of NG9-1-1 system standards were scheduled to be Beta tested and First Application trialed in 2010. Canada and the European Union are using NENA development content as a basis of their 9-1-1 and 1-1-2 development work. As the electronic communications revolution continues, NENA will continue work toward convergence of standards for quick and effective support of new sources of emergency calls, messages, and data of all types.

9.3 EENA

Gary Machado
European Emergency Number Association (EENA)

9.3.1 What Is EENA?

EENA, the European Emergency Number Association, is a Brussels-based NGO set up in 1999 dedicated to promoting high-quality emergency services reached by the number 112 throughout the European Union. EENA serves as a discussion platform for emergency services, public authorities, decision-makers, associations, and solution providers with a view to improving emergency response in accordance with citizens' requirements.

EENA is also promoting the establishment of an efficient system for alerting citizens about imminent or developing emergencies. EENA believes that European citizens have the fundamental right to know about the existence of the 112 number that can save their lives. EENA is also convinced that, when in distress, every citizen calling 112 within the European Union should get the appropriate help, as soon as possible, at the place of the emergency. Moreover, citizens in distress should be entitled to the same high-quality safety and security standards within the territory of the Member States and they should receive the same high-quality aftercare in case of accident or disaster. Citizens should also have the right to be informed as soon as possible about the behavior they need to adopt in case of an imminent or developing emergency or disaster, throughout the European Union.

In September 2011, the EENA membership included 550 emergency services and public authorities' representatives from 42 European countries, 25 solution providers, nine international associations or organizations, as well as 26 Members of the European Parliament.

9.3.2 What EENA Does?

EENA is involved in 112 issues through several activities.

EENA advocates to authorities the issues related to 112 and public warning based on citizens' requirements. The association also brings individuals and organizations together within the EENA Advisory Board, the 112 ESSN and the MEPs 112 Champions networks (as presented below).

Every year, EENA organizes several events: the 112 Awards Ceremony is held to congratulate individuals and organizations contributing to the improvement of 112 and emergency communications. EENA organizes round tables on 112 in European countries with a view to fostering the sharing of best practices and providing national stakeholders with a roadmap. EENA also gathers emergency services and stakeholders during yearly workshops on 112.

EENA works on technical and operational issues. Together with its members, EENA works on defining requirements and performance indicators for emergency services in the 112 operations and the Next Generation 112 (NG112) Committees.

The association also aims to raise awareness about 112. The association promotes 112 in the media. In collaboration with the EENA Advisory Board, EENA created the 112 Foundation, which focuses on educating citizens about the existence and use of 112. Moreover, EENA manages citizens' feedback and is at the disposal of those willing to report on their experience with 112. EENA makes sure that these feedbacks are considered by the relevant institutions at local, national, or European level.

EENA can serve as consultant. Being regularly consulted by governments and national authorities, EENA can participate in specific projects concerning the implementation and/or improvement of a 112 system in a country or region. EENA is also available to advise, review, and comment on feasibility studies by bringing its expertise to bear on the legal, operational, and technical aspects of 112 and emergency services. It should be noted that EENA will not respond to the call for tenders on its own but rather will advise respondents (e.g., consultancies). EENA's objective is to make sure that national and/or regional projects are in line with citizens' requirements, best practices, EU standards, and regulations.

9.3.3 What Are the EENA Memberships?

EENA brings individuals and organizations together. The EENA Advisory Board (section 9.3.3.1) gathers together politicians, associations, international organizations, and solution providers with the aim of improving the whole 112 chain. The 112 Emergency Services Staff Network – 112 ESSN (section 9.3.3.2) gathers individuals working for emergency services or relevant public authorities in order to foster sharing of experience and best practices. The MEPs 112 Champions (section 9.3.3.3) is a network of Members of the European Parliament committed to promoting better emergency communications for citizens.

9.3.3.1 The EENA Advisory Board

In 2008, following major political successes, such as the endorsement by a very large majority of the European Parliament of the Written Declaration on the European Emergency Number 112 and the Written Declaration on Early Warning Systems, EENA set up an Advisory Board in order to gather together all the major emergency telecommunications stakeholders, which include politicians, professional users' associations, international organizations and associations, as well as solution providers.

The EENA Advisory Board is structured in three committees in the following focus areas: technical, operations, and legal. The Advisory Board members hold regular conference calls (four per year) and communicate via emails. The committees hold between four and 20 conference calls per year, and have established charters and work plans for their activities.

The 112 Operations Committee is dedicated to defining requirements and enhancing the sharing of best practices on the whole 112 chain. The main tasks of this committee are to:

- create a set of requirements and recommendations for European emergency services;
- stimulate sharing of best practices among EU emergency services;
- review the different 112 models in Europe (PSAP representatives provide presentations about their national 112 system); and
- follow EU funding opportunities for emergency services.

The aim of the Next Generation 112 Committee is to develop European standards for the next-generation IP-based emergency services and to help the transition from legacy PSAPs. The main tasks of this committee are to:

- define NG112 requirements in collaboration with NENA's next-generation emergency services architecture;
- develop a NG112 architecture considering the different types of organization in the European Union (by reusing available emergency services standards as much as possible); and
- provide policy documents that can be presented to EU officials and national authorities with hints on how to implement IP-based emergency services and advantages (in collaboration with the legal committee).

The EENA Legal Committee has the aim of improving legislation pertaining to emergency communications in Europe and providing emergency services with the necessary means. The main tasks of this committee are to:

- follow relevant EU legislation (e.g., Universal Service Directive, Telecom Package, Cross-Border Health-Care Directives);
- propose amendments with a view to obtaining EU funding for emergency services and pilot projects (e.g., NG112 project);
- collaborate with NG112 and 112 Operations Committees in the preparation of policy documents;
- follow major updates in the national legislations; and
- advocate for high location accuracy of 112 calls.

In addition to these three well-established committees, the "EENA Network of Researchers" was launched during summer 2012. This international network will gather researchers in the fields of emergency communications and emergency services. It aims to foster the sharing of information and knowledge among researchers and between researchers and other EENA members from emergency services and industry.

9.3.3.2 The 112 ESSN – Emergency Services Staff Network

Aware of the non-existence of a platform that gathers emergency services staff involved with the European emergency number 112, EENA decided to set up the first Europe-wide 112

Emergency Services Staff Network (ESSN). Membership of the 112 ESSN is free and open to all (and only) emergency services staff involved directly or indirectly with 112, such as first responders, managers, call-takers, operators, technicians, managers, paramedics, ambulance staff, fire fighters, police officers, coastguards, search and rescue organizations, ministries, telecommunication regulators, and so on from the European Union as well as from non-European Union countries.

Members get several benefits from their participation in the 112 ESSN. They receive regular news on best practices in Europe, latest technological developments, EU funding opportunities, and EU regulations (with the possibility to comment on legislative proposals). They are also invited to participate in conferences and workshops organized in Europe. The mailing list of the network enables the sharing of experience and knowledge related to 112 among the members. Moreover, ESSN members get advance notification and discounts on all EENA organized events, such as the 112 Awards Ceremony and 112 educational round tables and workshops. EENA delivers to each member a Professional Certificate of Membership in the 112 ESSN, with unique membership number, to show support for the EENA Charter.

9.3.3.3 The MEPs 112 Champions

The MEPs 112 Champions network is a network of Members of the European Parliament (MEPs) committed to promote better emergency communications for citizens. As an MEP 112 Champion, the MEP agrees to advocate at the local, regional, and national level for the efficient use and functioning of the European emergency number 112. He or she also contributes to informing and educating the citizens of their region or country on the use of 112. Moreover, he or she raises awareness among decision-makers on the need to develop a proper emergency response to daily accidents as well as large-scale disaster.

More information can be found in [309].

9.4 Ecma International

John Elwell
Siemens Enterprise

9.4.1 Ecma International

Ecma International [310] is an industry association founded in 1961 and dedicated to the standardization of information and communication technology (ICT) and consumer electronics (CE). The aims of Ecma are as follows:

- To develop, in cooperation with the appropriate national, European, and international organizations, Standards and Technical Reports in order to facilitate and standardize the use of information and communication technology and consumer electronics.

- To encourage the correct use of Standards by influencing the environment in which they are applied.
- To publish these Standards and Technical Reports in electronic and printed form; the publications can be freely copied by all interested parties without restrictions.

For over 40 years Ecma has actively contributed to worldwide standardization in ICT and telecommunications. More than 390 Ecma Standards and 100 Technical Reports of high quality have been published, more than two-thirds of which have also been adopted as International Standards and/or Technical Reports.

The organizational structure of Ecma is a simple, flat one: the technical work is done by the Technical Committees, and the results of this work can be submitted twice a year to the General Assembly for approval for publication.

Ecma procedures consist of a set of Bylaws, Rules, and a Code of Conduct in Patent Matters. Five categories of members assure a wide participation of the industry and interested organizations in the work of Ecma.

9.4.2 Ecma Technical Committee TC32

Within Ecma, Technical Committee TC32 has responsibility for a number of standardization topics relating to business communications, and in particular corporate telecommunication networks. On such topics, Ecma TC32 complements the work of ITU-T, ETSI, and other organizations, which focus more on public telecommunications. Many of Ecma TC32's publications have been submitted to ISO/IEC JTC1 through the fast-track process for publication as International Standards or Technical Reports. A particular success story from Ecma TC32 in the past was the development of the series of Ecma and International Standards specifying the QSIG signaling system for use within private integrated services networks.

With the increasing migration of corporate telecommunications networks toward the use of VoIP in recent years, Task Group Ecma TC32-TG17 has conducted a number of investigations into the special needs of next-generation corporate networks (NGCNs). A series of Technical Reports on NGCNs has identified a number of recommendations for corporate networks, a number of requirements on public next-generation networks (NGNs) concerning interoperation with NGCNs, and a number of standardization gaps.

9.4.3 ECMA TR/101, Next Generation Corporate Networks (NGCN) – Emergency Calls

One Technical Report in the NGCN series, ECMA TR/101 [311], addresses the many aspects of support for emergency calls made by NGCN users to Emergency Response Centers (ERCs). Typically, particularly for smaller enterprises, the ERC will be a public ERC (i.e., a Public Safety Answering Point, PSAP). However, some enterprises with special needs use their own or third-party private ERCs. The focus is on the particular needs of emergency VoIP calls (as well as other media over IP), including the delivery of such calls to a public VoIP service provider (e.g., NGN) for onward delivery to a PSAP. Issues relating to the location of the caller, particularly for mobile or nomadic users outside enterprise premises, are given special consideration. The work of IETF ECRIT is used as the basis, but augmented to address the particular needs of corporate networks.

9.5 ATIS

Martin Dawson
CommScope

ATIS is the "Alliance for Telecommunications Industry Services." As its name suggests, it is an alliance of industry representatives, from more than 250 companies, that is focused on products and services in the telecommunications space. ATIS membership is not free, and companies typically pay a significant annual subscription fee. According to its website [312] ATIS has the following mission:

> **ATIS Mission**
> ATIS is committed to providing leadership for, and the rapid development and promotion of, worldwide technical and operations standards for information, entertainment and communications technologies using a pragmatic, flexible and open approach.

And, also as per the ATIS website, the following specific areas fall within its scope of activities:

- existing and next-generation IP-based infrastructures;
- reliable converged multimedia services, including IPTV;
- enhanced operations support systems and business support systems; and
- greater levels of service quality and performance.

ATIS is an American forum, accredited by the American National Standards Institute (ANSI), and many of the outputs generated by this organization take the form of American Standards documents and Technical Reports, the latter of which may also contain Standard Specifications.

The umbrella ATIS organization can be decomposed into a number of specialized committees, and one, in particular, is specifically focused on emergency services. This is the Emergency Services Interconnection Forum.

9.5.1 Emergency Services Interconnection Forum (ESIF)

The mission and scope of this committee is documented in the following way:

> **ESIF Mission**
> The mission of ESIF is to provide a forum to facilitate the identification and resolution of technical and/or operational issues related to the interconnection of wireline, wireless, cable, satellites, Internet, and emergency services networks.

ESIF Scope

ESIF is an open, technical/operational forum with the voluntary participation of interested parties to identify and resolve recognized interconnection issues. The interest of all members will be served by observing the principles of openness, fairness, consensus, and due process. ESIF will liaise with standards and governmental organizations to apprise them of its deliberations and decisions. Discussions will be focused on the FCC's Wireless Phase I and II mandates, and into other areas of emergency services interconnection. ESIF works closely with the National Emergency Number Association (NENA) [58], which currently manages the technical evolution of the 9-1-1 system and emergency communications process through its Future Path Plan [313]. A summary of the ESIF–NENA relationship can be found in [314], and two diagrams depicting the relationship can be found in the ESIF–NENA Relationship Diagrams [315].

As per the documented mission and scope of the committee, its focus is on ensuring that there are appropriately considered, recommended, and standardized ways in which different network elements that play a part in the provision of emergency services interconnect and communicate. As suggested by the statement of scope, particular priority has been given to issues arising from the US Federal Communications Commission's "Wireless Phase I and II" regulation. This regulation drove the industry standardization and implementation of systems to ensure that 9-1-1 emergency calls originating from cellular networks are appropriately routed based on the location of the point of origin of the call and that the call-related information, including the location, is transferred to the Public Safety Answering Point (PSAP) servicing the emergency call. The original emphasis of this regulation was on calls made using traditional circuit service cellular technology delivered over circuit-based trunks to the emergency network and associated PSAPs.

In more recent years, as the Internet has continued to displace the PSTN as the global communications network of choice, the focus of both the US regulator and committees such as ESIF has shifted to dealing with the issue of providing a reliable emergency service via Internet access networks whether fixed or mobile. ESIF conducts its work in separate subcommittees and, in response to this emerging focus, it established the Next-Generation Emergency Services (NGES) Subcommittee.

9.5.2 Next-Generation Emergency Services (NGES) Subcommittee

Consistent with the higher-level mission of ESIF, the NGES subcommittee concerns itself with network interconnectivity, the definition of appropriate standards, and ongoing consultation and collaboration with relevant external Standards Development Organizations (SDOs) such as NENA. NGES documents its mission as follows:

The ESIF NG Emergency Services (NGES) Subcommittee coordinates emergency services needs and issues with and among SDOs and industry forums/committees, within and outside ATIS, and develops emergency services (such as E9-1-1) standards, and other documentation related to advanced (i.e., Next Generation) emergency services architectures, functions, and interfaces for communications networks. The Subcommittee has an emphasis on standards development as it relates to North American communication networks, in coordination with the development of standards activities to include relative ATIS committees (e.g., PTSC), ITU, 3GPP, ETSI, and NENA.

A key focus of the NGES subcommittee has been in contributing to the development and elaboration of NENA's i3 standards [38, 39, 122]. Via the expertise and interests of its membership, NGES particularly seeks to identify gaps in the specifications from various SDOs when evaluated from the perspective of the manifold individual elements that constitute the details of the emergency and public carrier networks. For example, specific recommendations and specifications may be made to facilitate the interconnectivity of customer premises equipment (CPE) as used within the PSAP community to the larger emergency network infrastructure defined by i3 as a whole.

9.5.3 *Example ESIF Issues*

As is done within ESIF in general, particular work items are managed as unique "Issues" in accordance with procedures governing the Issue life-cycle defined by the organization at large. Examples of the Issues on which NGES has worked and is currently working include the following.

9.5.3.1 Issue 66 Ingress SIP Interface Specification Between IP Originating Networks and an ES-NGN

This Issue established the need for a specification for network-to-network interconnectivity between originating IP networks and the emergency services networks. The goal was to provide a consistent mechanism to route emergency calls from IP access networks to the correct PSAP.

The Issue was generated following a gap analysis, conducted as a prior separate issue, which analyzed the NENA i3 specification and identified the need for this document. This is typical of the way in which ESIF/NGES draws on the work of other forums and identifies and prioritizes work items. As such, this Issue is a good example of the way the forum works.

9.5.3.2 Issue 64 Conveying Access Subscription and Other Access Information to the PSAP in the Context of an IP Originated Emergency Call

When an emergency call arrives at the emergency services network, it is evident that it originated at some physical point of access to the Internet and, ideally, from a point within the service area of the PSAP to which it is routed. Beyond that, however, it is not necessarily clear what physical access network was used to originate the call or what its characteristics are. For example, is the network a public WiFi hotspot in a café, is it a residential broadband DSL connection, or is it a mobile 4G wireless connection?

This Issue was raised because of internal concerns of participants that some mechanism should be defined which allows the emergency service to "reach back" to the physical access network to determine this potentially useful information. As such, the Issue was not generated in response to a prior interaction with an external forum. Nevertheless, at the time of writing, the approach taken in addressing the Issue is to liaise with NENA to establish what consideration has been given to this concern and whether work is already under way to address it. Depending on the outcome, ESIF/NGES my leave the work to the other forum, assist in the work, or take on the development of a specification by itself. This sort of proactive initiative is also typical of the way that ESIF/NGES works.

9.5.3.3 Issue 63 Request for Assistance Interface (RFAI) Specification

This is one of the most recent Issues being worked on by ESIF/NGES at the time of writing, where the work is largely complete. The NENA i3 specification posits the existence of IP-based PSAPs, reachable via the Internet, and able to support emergency calls originating from IP devices attached to Internet access networks. The NENA i3 specification also acknowledges the transitional reality that traditional telephony devices attached to circuit service networks also exist. There is, similarly, recognition that a PSAP that has migrated to IP technology will have such traditional networks and devices active within their area of service. While the NENA specification acknowledges this transitional situation, the use cases associated with it are deemed to be out of scope. Discussion within ESIF/NGES subsequently determined that it was appropriate to address the scope within that forum.

This Issue focused on the definition of a standardized way in which calls delivered on traditional networks should be managed at the IP-based emergency network interface. It established the concept of an entity called an IP Selective Router (IPSR) in recognition of the fact that the Selective Router is the emergency network gateway entity seen by traditional networks at the traditional PSAP interface. It was determined that the Issue would deliver complete (Stages 1, 2, and 3) specifications related to this function in an ATIS standard. At the time of writing, this work was largely complete; a follow-up Issue had been further established to evolve the specification to an ANSI Standard. Taking cues from other forums such as NENA for important work items considered out of scope by the originating forum is a common way in which ESIF/NGES Issues are established.

9.5.3.4 Issue 58 Standardization of Location Parameter Conveyance Architecture and Protocols

A fundamental network element in the NG9-1-1 architecture as exemplified by the NENA i3 specification is the Location Information Server (LIS). This network element is responsible for providing the location of the calling device to both the device and the emergency services. The protocols and procedures used for providing the location information to these clients are well defined and specified by other forums (e.g., [31]). It is generally understood that the LIS can determine location via some interaction with the other elements of the physical serving network. What the specifications do not cover is how the LIS may interact with these physical network elements in order to obtain the information necessary to determine location. Through discussion, it was established that it would assist in the earlier implementation of effective NG9-1-1 services if LIS and network equipment vendors had a common protocol to facilitate the conveyance of what came to be labeled “Location Parameters” between network elements and the location server.

Work on this Issue led to the development of a conceptual framework that identified logical network entities called “Access Measurement Functions” (AMF) responsible for providing the parameters of interest to the location server. This further led to the development of the protocol for conveying these parameters between the location server and AMF instances. Since the protocol is intended to support all types of physical network characteristics, it included flexible semantics by which the location server and the AMF could have the "conversation." This protocol was named Flexible Location server to AMF Protocol (FLAP). The final output of the Issue was a Technical Report that describes and defines the FLAP protocol and general measurement architecture.

The subject of this Issue was originally identified in a prior Issue that was primarily concerned with the protocols used to provide the location service itself; the LIS protocols. Since this aspect fell outside the scope of that Issue, it was hived off and, when time allowed, was made the subject of its own Issue. Further, the effective evolution of this specification from a TR to a full ANSI Standard has been proposed as another follow-on Issue. This is another example of the way work items are identified and managed within ESIF/NGES.

9.5.4 Summary

ATIS is a telecommunications industry alliance with the key objective of establishing common standards to facilitate interoperability and cost-effective deployment of networking infrastructure and applications. Participation is generally company-based and available on a paid membership basis. ESIF is a committee within ATIS divided into several subcommittees, including NGES, which is focused on next-generation emergency services. ESIF deals with separate work items under the scope of what are termed "Issues." The subject and focus of Issues are debated and agreed by the subcommittee members and approved at the executive level of the governing committee. Issues can be generated from any relevant concern, but typical sources are liaisons or inferences from other forums, particularly NENA, and from the direct concerns of the participating membership of the subcommittees themselves.

9.6 The NG9-1-1 Caucus and the NG9-1-1 Institute

Carla Anderson
e-Copernicus

The 9-1-1 emergency call system is our citizen's crucial link to emergency service – whether it's an everyday crisis, natural disaster, or terrorist attack. Improving our nation's 9-1-1 system must become a national priority. On 25th February 2003 the Congressional E9-1-1 Caucus, now the NG9-1-1 Caucus, was formed to elevate the visibility of E9-1-1 at the Federal level, a welcome development to the thousands of public safety officials and first responders throughout the country. The NG9-1-1 Caucus is committed to making 9-1-1 services a national priority!

The Congressional NG9-1-1 Caucus is the only bipartisan, bicameral organization within the US Congress dedicated exclusively to emergency communications issues and to educating law-makers, constituents, and communities about the importance of citizen-activated emergency response systems. The Co-Chairs of the NG9-1-1 Caucus are: Senator Amy Klobucher (D-MN), Senator Richard Burr (R-NC), Congresswoman Anna Eshoo (D-CA), and Congressman John Shimkus (R-IL).

The Legislative Agenda of the Congressional NG9-1-1 Caucus is focused primarily on:

- securing funding for Federal 9-1-1 grants;
- ensuring that consumers can access 9-1-1 and E9-1-1, and NG 9-1-1 services over any communications device in legacy or future mobile and fixed broadband networks; and
- making 9-1-1 an essential component of the homeland security agenda of the United States.

The E9-1-1 Institute, now the NG9-1-1 Caucus, was established in 2003 as a not-for-profit organization that supports the work of the Congressional E9-1-1 Caucus. The NG9-1-1 Institute serves as the organizational structure for the Congressional NG9-1-1 Caucus, functioning as the think-tank on 9-1-1 issues, and provides administrative support to the NG9-1-1 Caucus events. It assists the NG9-1-1 Caucus in working with the private sector, providing valuable coordination, resources, and leadership for the public safety community, political leaders, industry, and the general public so the N9-1-1 Caucus get the critical infrastructure in place that will save lives.

Ensuring universal access to 9-1-1 capabilities to all people in all areas is important not only for the safety of individuals, but for our homeland security, and requires the cooperation and innovation of numerous stakeholders and policy-makers. This goal remains as the 9-1-1 system begins its migration to next-generation 9-1-1 services that will greatly enhance 9-1-1 access and capability.

The NG9-1-1 Institute provides an opportunity for public safety organizations, industry, and academia to come together and join the Congressional Caucus in promoting public education and discussion on these critical 9-1-1 issues. Specifically, the NG9-1-1 Institute:

- is a clearinghouse of information on 9-1-1 and emergency communications issues for the Congressional NG9-1-1 Caucus and the general public;
- promotes public education on 9-1-1 and emergency communications issues; and
- creates an open forum for discussion and development on 9-1-1 and emergency communications policy initiatives.

The objective of the NG9-1-1 Institute is to advance the best emergency communications system possible. Advancing the implementation of 9-1-1 systems, ensuring that 9-1-1 systems, networks, and operators are properly funded, and elevating emergency communications issues within all branches of government at the Federal, State, and local levels are just some of our goals.

Membership of the NG9-1-1 Institute is open to public safety organizations, industry, public interest organizations, and individuals who share in the goal of promoting 9-1-1 and emergency communications issues. The NG9-1-1 Institute has approximately 1500 members throughout the nation who are devoted to advancing 9-1-1 services. Members include public safety officials, telecommunication and industry professionals, policy-makers, academics, as well as concerned citizens.

The NG9-1-1 Institute urges all those involved in emergency communications to become engaged in the legislative, executive, and regulatory processes that control how our 9-1-1 services are implemented, upgraded, and funded. The easiest way to do this is by becoming a member of the NG9-1-1 Institute. Anyone can sign up online to receive our updates and invitations to upcoming NG9-1-1 Institute events.

The NG9-1-1 Institute is a not-for-profit organization that relies entirely on contributions and sponsorships received from its members and supporters. The purpose of the NG9-1-1 Institute is to promote public education and awareness of 9-1-1 and emergency communications issues and to serve the NG9-1-1 Congressional Caucus as a clearinghouse of information on E9-1-1 and emergency communications issues. The NG9-1-1 Institute [316] provides administrative and policy support to the Congressional NG9-1-1 Caucus. Contributions and participation is instrumental for the success of the NG9-1-1 Institute. Because of the generous support that the NG9-1-1 Institute has received we have:

- established an Annual Honors Awards Ceremony where we recognize government leaders and "citizens in action" whose work and life stories underscore the critical importance of 9-1-1 services;
- co-hosted "9-1-1 Goes To Washington" with NENA;
- hosted numerous round-table forums on Capitol Hill to educate Congress on the importance of 9-1-1 issues;
- hosted Hometown Security Forums in various states across the nation;
- hosted Membership Receptions at APCO and NENA annual events; and
- established Emergency Communication Technology Fairs held in Washington, DC, which provide industry professionals and public safety officials with an excellent opportunity to demonstrate the latest in E9-1-1 and emergency communications technologies to federal policy-makers.

We look forward to working with you on these issues and hope that you will join the NG9-1-1 Institute.

Gregory L. Rohde, former Administrator of the National Telecommunications and Information Administration (NTIA), serves as the NG9-1-1 Institute's Executive Director. Carla A. Anderson serves as the Deputy Executive Director of the E9-1-1 Institute. The NG9-1-1 Institute is governed by an elected Board of Directors, coming from the public safety community, industry, and not-for-profit organizations. The NG9-1-1 Institute also partners with several national public safety and industry organizations including APCO, CTIA, EENA, NASNA, NENA, and the 9-1-1 Industry Alliance. The current NG9-1-1 Institute Board of Directors includes four permanent seats and eleven elected seats.

9.7 COCOM EGEA*

Hannes Tschofenig
Nokia Siemens Networks

Section 7.2 provides background information for emergency services legislation in Europe explaining how the different institutions work together and what laws are in existence today. The European Commission established committees and Working Groups[1] to assist the Commission in its tasks and to create a harmonized implementation of the adopted regulatory

*The author would like to thank Martins Prieditis for his text contribution. Additionally, the author would like to thank Alain Van Gaever and Gyula Bara for their review comments. Any errors that remain are the sole responsibility of the author.

[1] For a list of these groups, see ref. [317].

frameworks. The Communications Committee (COCOM) [318], which was created in 2002, is the European Commission's comitology[2] committee in the field of electronic communications. The COCOM website [318] provides more information about the purpose and the membership and says:

> Made up of senior officials from the Member State authorities responsible for telecoms, the COCOM assists the Commission in carrying out its executive powers under the regulatory framework and also with regard to the .eu top level Internet domain. It exercises its functions through "advisory" and "regulatory with scrutiny" procedures in accordance with the Comitology Regulation.

The Body of European Regulators for Electronic Communications (BEREC) has a work scope that appears very similar to COCOM, as stated at [320]:

> The mission of BEREC is to assist the Commission and the national regulatory authorities (NRAs) in the implementation of the EU regulatory framework for electronic communications, to give advice on request and on its own initiative to the European institutions and to complement at European level the regulatory tasks performed at national level by the regulatory authorities.

BEREC, however, decided to leave emergency services-related activities to COCOM.

According to Article 26 of the Universal Service Directive (Directive 2002/22/EC) [224], Member States have an obligation to ensure the proper functioning of the European emergency number 112, including making sure that caller location information is available to emergency authorities. The number 112 is regularly on the COCOM agenda since the Commission uses this framework to gather information on its implementation in the Member States and to promote best practice in this respect. The most recent reports on the implementation of 112 are available at the following:

- 2011 Report [321] and its Annex [322];
- 2010 Report [323] and its Annex [324];
- 2009 Report [257] and its Annex [325];
- 2008 Report [326] and its Annex [327].

The above-listed COCOM reports were drafted based on the answers of the relevant national authorities on a questionnaire on the implementation of 112. In addition, the Commission is conducting Eurobarometer surveys on the awareness level of 112 in EU Member States. The COCOM reports and the surveys can be found on the Commission website [328].

On 11th February 2009, the European Commission, the European Parliament, and the Council approved the establishment of a "European 112 Day" to be celebrated on 11th February each year, in particular by the organization of awareness raising, experience sharing, and networking activities (see the Joint Tripartite Declaration [329]).

A subgroup of COCOM, the Expert Group on Emergency Access (EGEA) [307], is established to work on issues related to access to emergency services. The group functions on the

[2] "Comitology" in the European Union refers to the committee system that oversees the delegated acts implemented by the European Commission. For more information, see ref. [319].

basis of the "Terms of Reference" documents [330] providing (i) the scope of its work and (ii) information about the group's membership policy and on the purpose.

Regarding the membership policy of the group, the 2011 terms of reference document [331] states:

> Members of the Group are nominated by the Member States. They would typically represent emergency response organizations, Ministries responsible for civil protection issues and/or Ministries or regulatory authorities responsible for electronic communications. The members should accordingly be specialized and have knowledge of national emergency response structures (police, fire, ambulance . . .) and requirements regarding access to emergency services and electronic communications infrastructure in the area of emergency communications. Representatives of Candidate countries and EEA countries are also invited to participate. The Chair may invite external experts to the meetings of EGEA where appropriate.

[331] goes on to state that:

> The purpose of the Group is to deal with forward-looking and current aspects of access to emergency services in the European Union, in particular issues related to the application of new technologies for communication with emergency services and seeking practical solutions to problems experienced by emergency services in order to provide an efficient and effective service to all citizens, including the disabled. Traditionally, access to emergency services has been via the telephone, but new technologies like VoIP, SMS, MMS and instant messaging, may open up new possibilities for citizens to communicate with emergency services. The continuous deployment of next-generation networks (NGN) is expected to produce a significant impact on emergency services infrastructure. These developments give rise to various issues, such as how European citizens and emergency services can benefit from more effective ways of communication thanks to the development of new communication technologies and how to avoid the risk that the introduction of these new technologies does not give EU citizens the same level of access to emergency services as they enjoy with current technologies.

The precise work scope of the EGEA group changed over the years of its existence and was also influenced by the specific chairman of the group at the time. For most cases the group's main purpose was to share information, and all presentations and documents are publicly available [332]. In early 2011 the COCOM EGEA group contributed to the Mandate 493 [308] to

> prepare a coherent and complete set of specifications or standards containing the architecture, the interfaces and the protocols in support of the requirements set by article 26 of the amended Directive 2002/22/EC concerning the determination, transport and delivery of caller location information

by European standardization organizations (ESOs).[3] Since the release of the Mandate, ETSI has been active in investigating the state of the art in emergency services standardization and preparing new technical standards.

[3] As part of the assessment of the available standards, the EGEA group concluded that further work with technical standards on caller location are needed. This conclusion is not shared by everyone, as can be seen by the letter of the Internet Architecture Board sent to the European Commission in response to their Mandate [333].

10

Conclusion and Outlook

Those who have read through most or all of the chapters will agree with us that much has been accomplished in the standardization of IP-based emergency services and in the supporting ecosystem by involving a global stakeholder community. You may have had multiple reactions when you read through the different chapters: even for us some of the design decisions are sometimes hard to understand when we look back over the work produced in the last 10 years or more. There are also lots of acronyms and new terminology that one needs to become familiar with. The standardization process in the different groups, which is driven by the underlying communication system, business models, as well as regulatory requirements, leads to different results that are not always "perfectly aligned." A few trends can be observed: there are many developments in the area of the communication infrastructure – from IMS deployments to over-the-top application services (in form of Web services or smart-phone applications) – and an increasing penetration of the Internet Protocol into every corner of communication systems and emergency services networks.

While a great deal has been accomplished, there is no reason to stop working on improvements to the emergency services system. Ultimately, we want better emergency services offered to those in need of help. To accomplish this goal, there is still plenty of work to be done. We would like to share our views on what types of activities need to happen in the near future.

10.1 Location

In Chapter 2 we described location formats, encoding of location information, and the various location protocols. Clearly, there is no shortage of standardization work on location. Still, various new communication protocols (like WebRTC [8], which relies on the W3C Geolocation API [19] for location support) or older technologies like XMPP [9] lack even the most basic support for location conveyance [20] and [334]. To bring these new communication technologies up to par with the already standardized SIP-based emergency services support, additional work is needed. For geodetic location, the work done by the OGC, as explained in section 2.3, represents the best current practice for the entire industry. For civic location, some standardization work is nearing completion, for example, civic address extension [22]. However, there is still a lot of work ahead in various countries to develop profiles, in the same

Internet Protocol-Based Emergency Services, First Edition. Hannes Tschofenig and Henning Schulzrinne.

style as described in section 2.1 for the United States and Canada and as described in [25] for Austria. Without such civic location profiles, interoperability problems may arise. It is envisioned that this work will happen in organizations like EENA (for Europe) rather than in standards organizations like the IETF, 3GPP, or the IEEE.

Location protocols, on the other hand, are quite stable. IETF HELD, which is described in section 2.4, and OMA SUPL, which is described in section 2.5, allow an end device and a third-party entity, such as a PSAP, to obtain the location of the emergency caller's IP device. Only minor extensions are being added, such as the flow identity extensions [335] and device-provided location-related measurements [81]. For various enterprise deployments and fixed network environments, the use of the DHCP location extensions, described in Chapter 2 and section 2.2, and link-layer protocol extensions, see section 3.4, are viable choices.

With all these standards for accurately describing location information, and the ability to retrieve location information from dedicated location servers and to convey it to PSAPs, it may seem that high-quality location is only one step away from becoming a reality. It turns out that the regulatory requirements regarding location accuracy vary throughout the world; and while there are mandates for telecommunication operators in the United States to provide high-quality location, similar mandates are missing in European Member States. In Europe, location provided to emergency services is either the civic address of landline callers or location based on Cell-ID for mobile phone users. The quality of Cell-ID based positioning varies substantially, and can, in urban areas, lead to a poor accuracy of several kilometers [336, 337]. Martin Dawson explained the properties of different positioning techniques quite well in his April 2011 EENA workshop presentation (see slide 4-7 of [338]). With an increasing number of emergency calls made by cell phones (and the decline of fixed phone usage overall), the quality of location information provided to PSAPs today is effectively lower than it was 15 years ago. Issue 4 of the official NENA publication "The Call" [339] provides some data about the mobile phone usage for emergency services, and Barbara Jaeger, the NENA President, says:

> ... smart phones now represent 40 percent of all active mobile phones and approximately 600,000 wireless calls to 9-1-1 are placed each day – accounting for 70 to 80 percent of call volume in many areas.

10.2 Architectures

Architectures combine different protocols to form a complete system. Developing architectures is always difficult since there is always the risk of restricting future use when specifying interoperability points. Providing too much flexibility, on the other hand, leads to interoperability problems and walled gardens of proprietary systems.

Chapter 3 lists the dominating architectures, which differ from each other due to initial assumptions. There are often only minor protocol differences. The NENA i2 architecture, for example, assumes that VoIP systems interconnect with legacy PSAPs for emergency communication. This assumption emphasizes changes to certain elements, whereas an architecture like NENA i3 is not subject to these constraints.

Needless to say, one can create new architectural variants by introducing new design assumptions or by relaxing existing assumptions. In fact, we see evidence for interest in developing new architectures: The most recent example is the ETSI M493 group, which was established based on the EC mandate 493 [308]. The M493 group focuses on over-the-top VoIP providers, unlike many other standardization activities in ETSI that are primarily targeting telecommunication operators. While the work is still in its early stage at the time of writing, new baseline assumptions[1] are driving the work in a new direction. The main assumption of the work is that end hosts will not provide location information (or measurements data) because information that end hosts provide is untrusted and modifiable. In addition, the existing emergency services system is supposed to be left unmodified. Finally, since over-the-top VoIP providers are also considered untrusted, they will not be allowed to query a location server in the access network for location information of the emergency caller's device (for security and privacy reasons). Consequently, an over-the-top VoIP provider that receives an emergency call from one of its users will have difficulties routing the call to an appropriate PSAP.

Changing the assumptions leads easily to new systems. Since each country has different goals for integrating VoIP calls (and multimedia communication in general) into their existing emergency services deployment, many solution variants are possible.

Over the years, in our standardization work, we have noticed substantial differences in the treatment of location information – also noted in the M493 effort in the previous paragraph. Today's IP-enabled devices have access to more location data than ever, either from built-in GPS receivers or from third-party location databases (which are fed with the use of commercial location-based services). Still, in many parts of the world, this wealth of location information is not available to emergency services. There are two reasons for this development. First, many of the commercial location-based services fear liability problems and therefore prohibit the use of location data for emergency services purposes. Second, those working on the design of the emergency services system are aiming for the perfect system. A perfectly working system always provides the best possible location information in a verified form. Unfortunately, the desire to design such a system comes at a certain cost, as designers eschew available location data that does not match their ideals.

We fear that some of these architectural variants are incompatible with each other and will lead to interoperability problems. Our vision is that a random IP-enabled device with communication software from some application service provider is able to interact with emergency services using multiple media, not just voice. People roam freely with their devices and may need help outside their home country. Even in those cases, help from emergency services has to be available.

10.3 Deployments

Chapter 4 illustrates efforts in Canada, the United Kingdom, Indiana in the United States, and Sweden to extend their emergency services support to IP-based emergency calling. These groups have spent a lot of time in various stakeholder groups discussing new requirements, which extend beyond just technical considerations.

[1] The assumptions are not entirely new since they had been discussed in the IETF ECRIT Working Group under the topic of "location hiding" [128], but the industry-chosen solution for providing rough location to the end host (with appropriate security protection) [129] did not seem to meet the expectation of the European Commission or certain national regulators.

As countries upgrade their emergency infrastructure, the need for education and sharing of experience increases. Both NENA and EENA have included education tracks in their conference programs, and speakers from emergency services authorities share their views on what should or should not be done. Questions about the integration of smart-phone applications, support for real-time text and multimedia emergency calling, SMS support, and many other topics are popular.

As the technology we describe in this book is deployed on a nationwide scope (and even beyond), feedback will be incorporated into the standards process, new extensions will be developed, and specifications will be refined. This is the regular standards development process. We are looking forward to hearing about the next ambitious deployment effort and sharing it with the rest of the community.

10.4 Security and Privacy

Concerns about security and privacy are common with emergency services, and of course with the Internet in general. Unfortunately, the concerns are often unspecific and therefore hard to discuss and to dispel. In our experience, untangling the wide range of possible concerns is the first step to making progress on the topic. In Chapter 5 we approached the topic from a broad perspective and collected all threats that we came across in discussions. One specific concern relates to the ability to attack the limited resources of the emergency services authorities, namely call-takers and first responders. In the November 2011 issue of the "Communications of the ACM" magazine [340], we summarized the concerns related to swatting and caller identity spoofing. That security chapter offers a more detailed treatment of the topic, in addition to [131], since it closely relates to the concerns about fake location and SIM-less (or anonymous) emergency calls.

We see two areas in emergency services security where further work is required. First, many of the existing VoIP and application service offerings provide only a limited form of identity proofing, and their user authentication is often weak as well. Without the ability to link a specific call to a real-world person, it is difficult to punish misbehavior. Second, emergency services networks are IP-based networks and, if not operated correctly, updated frequently, and standardized, they quickly become vulnerable to various forms of attacks. The former challenge is part of a larger body of ongoing work to develop and deploy an identity layer for the Internet. An example effort can be found in the United States with the National Strategy for Trusted Identities in Cyberspace (NSTIC) [341]. The latter challenge requires guidance, such as offered by the NENA Security for Next-Generation 9-1-1 Standard (NG-SEC) [342], as well as certification and audits to ensure that equipment and services are indeed following the guidance.

10.5 Emergency Services for Persons With Disabilities

The capabilities of modern IP-based communication protocols, like SIP and XMPP, easily enable multimedia communication. Properties like capability and media type negotiation allow the two endpoints (e.g., the phone of the emergency caller and the phone of the PSAP call taker) to agree on the latest and best possible communication mechanisms to use. This is clearly important for extensibility and incremental deployment. Even at a higher level, LoST allows emergency services authorities to publish information about the protocols through which they are reachable via information encoded in the URIs returned by the LoST protocol.

As customers purchase more and increasingly capable devices, the ability to use multimedia to communicate increases. Transmitting instant messages, real-time text, pictures, and video is a common feature of applications running on smart phones and Internet tablets. Chapter 6 discusses the use of multimedia communication for persons with disabilities.

Unfortunately, more than just protocol support is needed. Support by emergency services for multimedia communication is still poor, if available at all. Fax has been used in several countries in Europe, and TTY in the United States. In the meantime, more and more countries are offering SMS to a broad range of consumer devices, including those that are not IP-capable, to reach emergency services. The European REACH112 pilot project showed that the standardized technology is available and is working. Proper integration of the text messaging system into the PSAP software environment is, however, often not available, and additional training of call-takers is necessary to ensure proper response times. From the REACH112 pilot and from various SMS-based deployments, it is known that only a few persons make use of the new capabilities. There are many reasons for this, such as a lack of awareness of the availability of such a service, unfamiliarity with multimedia emergency calls, and the smaller user base that needs such capabilities (compared to those who feel comfortable just making voice-based emergency calls).

Since the deployment of IP-enabled phones is increasing dramatically, we expect that more users, not only those with disabilities, will find the multimedia functionality attractive.

10.6 Regulation

The telecommunication sector has been subject to regulation for a very long time and, as the industry transitions from legacy Time Division Multiplexing (TDM) and Signaling System 7 (SS7) to all-IP technology, the interest in regulating various IP- and SIP-based communication infrastructures will increase as well.

Chapter 7 discusses the regulatory situation in Europe and in the United States. The regulatory situation continues to change. In Europe, the most recent activity by the European Commission was Mandate 493 [308] to support location-enhanced emergency call service specifically for pre-IMS and over-the-top VoIP deployments, which led to new standardization activities in ETSI.

The degree of regulation will very much depend on the expectations of end users, their communication behavior, and the trends in the industry over all. For the cellular operator sector, it remains to be seen how fast the deployment of LTE will lead to the deployment of IMS and the transition away from the Plain Old Telephone Service (POTS).

10.7 Research Projects and Pilots

Chapter 8 highlights the work done in three research and pilot projects. There are additional projects ongoing, such as the HeERO project [343], which focuses on in-vehicle emergency calls, and pilots of SMS emergency services in different countries. We did not include these projects in this book, since they are not utilizing any IP-based technologies.

Overall, the feedback from the various pilot efforts has been positive. The standardized technology works. Separate interoperability events have been created, such as the NENA Industry Collaboration Events (ICE) [344], to ensure that vendors have next-generation emergency services software that is compliant with the developed standards. This provides a good basis for any pilot project where additional concerns have to be taken into account, such as educating stakeholders, integrating the new equipment into the deployed infrastructure, and adjusting processes and procedures. Pilot projects often aim to keep the pilot system running after the end of the project, and thus business considerations need to be taken into account as well.

As the PEACE project has shown, the range of topics in emergency services for researchers to investigate is broad. While the PEACE project focused mostly on communication protocols research, the emergency services community now wants help from researchers to improve emergency center operations. This task involves analyzing data collected during daily operations. These investigations may provide insight into the efficiency and the performance of the service chain and may allow the comparison of the various countries' performances. One recent activity to involve researchers in emergency services work is the formation of the EENA Network of Researchers [345], which aims to foster the sharing of knowledge among researchers, with emergency services authorities as well as with industry.

At the time of writing, more government-funded research projects are being launched. These projects will contribute to the deployment of next-generation emergency services.

10.8 Funding

The emergency services value chain requires many stakeholders to contribute their equipment and resources to provide help to those who need it most. Quite naturally, this service costs money. This book has no separate chapter on the topic of funding emergency call handling. Funding is an important aspect but also fairly complex. The current funding models for the emergency services infrastructure vary considerably throughout the world.

As explained in section 7.2, the European Commission requires emergency services to be free of charge for the emergency caller as stated in the Universal Service Directive.

To illustrate an example from Europe, we use the United Kingdom.[2] The United Kingdom has a model with stage 1 and stage 2 PSAPs. British Telecom (BT) and Cable & Wireless (C&W) are the two main entities running the stage 1 PSAPs, and there is a large number of stage 2 PSAPs operated by various emergency services organizations, such as ambulance, fire brigade, police, and so on.

BT receives about 80% of the emergency calls in the United Kingdom at their stage 1 PSAP. BT obtains its funding for operating the stage 1 PSAP from other telecommunication operators, who pay for the emergency call service support. The technical requirements and the financial conditions are publicly available from BT, whereas those from C&W are not. According to [346] the current price for emergency call handling offered by BT is £0.56 for mobile calls, £0.66 for fixed line calls, and £0.79 for VoIP calls. The difference mainly reflects

[2] We would like to thank John Medland for providing us with the data. Further details about the wholesale emergency services call handling fees can be found from British Telecom's (BT) wholesale price list [346], and also from the Irish Commission for Communications Regulation (COMREG) [347], since BT also provides their services within Ireland.

the operator handling times associated with each call type, with VoIP calls being longest, as call-takers have to verbally confirm the caller's location before routing the call to a stage 2 PSAP. The SMS support is still at an early phase and therefore no prices are available. The cost for an emergency call is based on the call volume counted on the interconnection link rather than on the duration of the call.

The details of how European telecommunication operators recover costs for emergency services are unknown, but the money is ultimately provided by the end customer.

The stage 1 operator does not charge stage 2 PSAPs for making location information or other data about the emergency call available. However, money does not flow in the other direction either. Instead, stage 2 PSAPs obtain their financial support from local and national government funds. The details depend on each emergency service and vary for different regions of the UK. Stage 2 PSAPs use the financial support to operate their call centers, maintain their equipment (including phone lines and data connections), and pay their staff.

In the United States, the situation is different. Subscribers pay their 9-1-1 and enhanced 9-1-1 fees as part of their phone bill (with different fees collected based on region and phone type). This money is collected by the telecommunication operators and then transferred to a State or local government agency, which redistributes to pay for the 9-1-1 system. A NENA publication on "Funding 9-1-1 Into the Next Generation" provides a good overview of the topic [348], and the FCC also provides reports about the collection and the distribution of these surcharges [349–351].

We expect more discussions regarding emergency services funding as the relationship between the application service provider and applicable jurisdiction becomes more clouded.

References

1. MIT Technology Review. *Texting "SOS" to 911*, August 2011. URL: http://www.technologyreview.com/news/425079/texting-sos-to-911/.
2. PCMag.com. *112 Day: 74% of Europeans Don't Know What Emergency Number to Call When Travelling in the EU. New Campaign*, February 2012. URL: http://ec.europa.eu/commission_2010-2014/kallas/headlines/news/2012/02/112_day_en.htm.
3. PCMag.com. *Verizon First to Offer National Text-to-911 Service*, May 2012. URL: http://www.pcmag.com/article2/0,2817,2404036,00.asp.
4. IETF. *The Tao of IETF: A Novice's Guide to the Internet Engineering Task Force*, October 2011. URL: http://www.ietf.org/tao.html.
5. IETF. *61st IETF Meeting, Washington, DC USA: Emergency Context Resolution With Internet Technologies (ECRIT) BOF*, November 2004. URL: http://www.ietf.org/proceedings/61/ecrit.html.
6. Tschofenig, H. *SDO Emergency Services Coordination Workshop (ESW06)*, October 2006. URL: http://www.emergency-services-coordination.info/esw1.html.
7. Schulzrinne, H. and Marshall, R. *Requirements for Emergency Context Resolution With Internet Technologies*, January 2008. RFC 5012, Internet Engineering Task Force.
8. Alvestrand, H. *Overview: Real Time Protocols for Browser-Based Applications*, February 2013. draft-ietf-rtcweb-overview-06.txt (work in progress), Internet Engineering Task Force.
9. Saint-Andre, P. *Extensible Messaging and Presence Protocol (XMPP): Core*, March 2011. RFC 6120, Internet Engineering Task Force.
10. Carpenter, B. *Architectural Principles of the Internet*, June 1996. RFC 1958, Internet Engineering Task Force.
11. Schulzrinne, H. *A Uniform Resource Name (URN) for Emergency and Other Well-Known Services*, January 2008. RFC 5031, Internet Engineering Task Force.
12. EENA. *Public Safety Answering Points in Europe*, December 2011. URL: http://www.eena.org/ressource/static/files/psaps_in_europe_eena_publication_2011_abstract1.pdf.
13. Machado, G. *112 Models in Europe*, November 2009. URL: http://www.eena.org/ressource/static/files/National112Systems_models.pdf.
14. EENA. *Transnational Emergency Calls, Version 1.0*, January 2012. URL: http://www.eena.org/ressource/static/files/3-5-4-1_v1-0.pdf.
15. Tschofenig, H., Adrangi, F., Jones, M., Lior, A. and Aboba, B. *Carrying Location Objects in RADIUS and Diameter*, August 2009. RFC 5580, Internet Engineering Task Force.
16. Schulzrinne, H. *Dynamic Host Configuration Protocol (DHCPv4 and DHCPv6) Option for Civic Addresses Configuration Information*, November 2006. RFC 4776, Internet Engineering Task Force.
17. Peterson, J. *A Presence-Based GEOPRIV Location Object Format*, December 2005. RFC 4119, Internet Engineering Task Force.
18. Hildebrand, J. and Saint-Andre, P. *XEP-0080: User Location*, September 2009. URL: http://xmpp.org/extensions/xep-0080.html, Draft Standard of the XMPP Standards Foundation.

Internet Protocol-Based Emergency Services, First Edition. Hannes Tschofenig and Henning Schulzrinne.

19. Popescu, A. *W3C Geolocation API Specification*, September 2010. URL: http://www.w3.org/TR/2010/CR-geolocation-API-20100907/, W3C Candidate Recommendation.
20. Aboba, B. and Thomson, M. *Emergency Services Support in WebRTC*, June 2012. draft-aboba-rtcweb-ecrit-00 (work in progress), Internet Engineering Task Force.
21. Thomson, M. and Winterbottom, J. *Revised Civic Location Format for Presence Information Data Format Location Object (PIDF-LO)*, February 2008. RFC 5139, Internet Engineering Task Force.
22. Winterbottom, J., Thomson, M., Barnes, R., Rosen, B. and George, R. *Specifying Civic Address Extensions in PIDF-LO*, December 2012. draft-ietf-geopriv-local-civic-10.txt (work in progress), Internet Engineering Task Force.
23. Schulzrinne, H. and Tschofenig, H. *Location Types Registry*, July 2006. RFC 4589, Internet Engineering Task Force.
24. IANA. *Location Types Registry*. URL: http://www.iana.org/assignments/location-type-registry/.
25. Wolf, K. and Mayrhofer, A. *Considerations for Civic Addresses in the Presence Information Data Format Location Object (PIDF-LO): Guidelines and IANA Registry Definition*, March 2010. RFC 5774, Internet Engineering Task Force.
26. Polk, J., Schnizlein, J. and Linsner, M. *Dynamic Host Configuration Protocol Option for Coordinate-Based Location Configuration Information*, July 2004. RFC 3825, Internet Engineering Task Force.
27. Polk, J., Linsner, M., Thomson, M. and Aboba, B. *Dynamic Host Configuration Protocol Option for Coordinate-Based Location Configuration Information*, July 2011. RFC 6225, Internet Engineering Task Force.
28. Winterbottom, J., Thomson, M. and Tschofenig, H. *GEOPRIV Presence Information Data Format Location Object (PIDF-LO) Usage Clarification, Considerations, and Recommendations*, March 2009. RFC 5491, Internet Engineering Task Force.
29. Barnes, R., Lepinski, M., Cooper, A., Morris, J., Tschofenig, H. and Schulzrinne, H. *An Architecture for Location and Location Privacy in Internet Applications*, July 2011. RFC 6280, Internet Engineering Task Force.
30. Tschofenig, H. and Schulzrinne, H. *GEOPRIV Layer 7 Location Configuration Protocol: Problem Statement and Requirements*, March 2010. RFC 5687, Internet Engineering Task Force.
31. Barnes, M. *HTTP-Enabled Location Delivery (HELD)*, September 2010. RFC 5985, Internet Engineering Task Force.
32. Polk, J. and Rosen, B. *Location Conveyance for the Session Initiation Protocol*, December 2011. RFC 6442, Internet Engineering Task Force.
33. Roach, A. *Session Initiation Protocol (SIP)-Specific Event Notification*, June 2002. RFC 3265, Internet Engineering Task Force.
34. Mahy, R., Rosen, B. and Tschofenig, H. *Filtering Location Notifications in the Session Initiation Protocol (SIP)*, January 2012. RFC 6447, Internet Engineering Task Force.
35. Winterbottom, J., Tschofenig, H. and Barnes, R. *Use of Device Identity in HTTP-Enabled Location Delivery (HELD)*, March 2011. RFC 6155, Internet Engineering Task Force.
36. US Federal Geographic Data Committee. *United States Thoroughfare, Landmark, and Postal Address Data Standard, FGDC-STD-016-2011*, February 2011. URL: http://www.fgdc.gov/standards/standards_publications/.
37. Sugano, H., Fujimoto, S., Klyne, G., Bateman, A., Carr, W. and Peterson, J. *Presence Information Data Format (PIDF)*, August 2004. RFC 3863, Internet Engineering Task Force.
38. NENA. *NENA Functional and Interface Standards for Next Generation 9-1-1 Version 1.0 (i3), Standard Number: 08-002 v1*, December 2007. URL: http://www.nena.org/standards/technical/voip/functional-interface-NG911-i3.
39. NENA. *Detailed Functional and Interface Standards for the NENA i3 Solution Version 1.0, Standard Number: 08-003 v1*, June 2011. URL: http://www.nena.org/standards/technical/i3-solution.
40. Rosen, B., Schulzrinne, H., Polk, J. and Newton, A. *Framework for Emergency Calling Using Internet Multimedia*, December 2011. RFC 6443, Internet Engineering Task Force.
41. FGDC. *Standards Publications – Federal Geographic Data Committee*, September 2012. URL: http://www.fgdc.gov/standards/standards_publications/.
42. International Standards Organization. *ISO 3166-1: Codes for the Representation of Names of Countries and Their Subdivisions – Part 1: Country Codes*, 1997. URL: http://www.iso.org/iso/country_codes/iso_3166_code_lists/country_names_and_code_elements.
43. US Postal Service. *Postal Addressing Standards, Publication 28*, April 2010. URL: http://pe.usps.gov/cpim/ftp/pubs/Pub28/Pub28.pdf.

44. US Census Bureau. *INCITS 31:2009, Codes for the Identification of Counties and Equivalent Entities of the United States, its Possessions, and Insular Areas*, February 2011. URL: http://www.census.gov/geo/www/ansi/ansi.html.
45. Rosen, B. *Interior Location in the Presence Information Data Format – Location Object*, March 2010. draft-rosen-geopriv-pidf-interior-01 (expired), Internet Engineering Task Force.
46. *IETF Geographic Location/Privacy (Geopriv) Working Group*. URL: http://datatracker.ietf.org/wg/geopriv/charter/.
47. Droms, R. *Dynamic Host Configuration Protocol*, March 1997. RFC 2131, Internet Engineering Task Force.
48. Droms, R., Bound, J., Volz, B., Lemon, T., Perkins, C. and Carney, M. *Dynamic Host Configuration Protocol for IPv6 (DHCPv6)*, July 2003. RFC 3315, Internet Engineering Task Force.
49. Patrick, M. *DHCP Relay Agent Information Option*, January 2001. RFC 3046, Internet Engineering Task Force.
50. IEEE. *IEEE 802.11k-2008, IEEE Standard for Information Technology – Local and Metropolitan Area Networks – Specific Requirements – Part 11: Wireless LAN Medium Access Control (MAC) and Physical Layer (PHY) Specifications. Amendment 1: Radio Resource Measurement of Wireless LANs*, June 2011.
51. IEEE. *IEEE 802.11-2007, IEEE Standard for Information Technology – Telecommunications and Information Exchange Between Systems – Local and Metropolitan Area Networks – Specific Requirements – Part 11: Wireless LAN Medium Access Control (MAC) and Physical Layer (PHY) Specifications*, June 2007.
52. IEEE. *IEEE-802.11y, IEEE Standard for Information Technology – Telecommunications and Information Exchange Between Systems – Local and Metropolitan Area Networks – Specific Requirements – Part 11: Wireless LAN Medium Access Control (MAC) and Physical Layer (PHY) Specifications. Amendment 3: 3650–3700 MHz Operation in USA*, November 2008.
53. International Association of Oil and Gas Producers (OGP). *Geodesy Resources, Geomatics Committee*. URL: http://info.ogp.org.uk/geodesy/.
54. US National Imagery and Mapping Agency. *Department of Defense (DoD) World Geodetic System 1984 (WGS 84), Third Edition, Amendment 1, NIMA TR8350.2*, January 2000. URL: http://earth-info.nga.mil/GandG/publications/tr8350.2/wgs84fin.pdf.
55. International Organization for Standardization (ISO). URL: http://www.iso.org.
56. Organization for the Advancement of Structured Information Standards (OASIS). URL: http://www.oasis-open.org.
57. Internet Engineering Task Force (IETF). URL: http://www.ietf.org.
58. National Emergency Number Association (NENA). URL: http://www.nena.org.
59. Open Mobile Alliance (OMA). URL: http://www.openmobilealliance.org.
60. OGC. *OGC Sensor Web Enablement*. URL: http://www.ogcnetwork.net/SWE.
61. Wikipedia. *Coordinate Reference Systems and Positioning*. URL: http://en.wikibooks.org/wiki/Coordinate_Reference_Systems_and_Positioning.
62. Lott, R. *OGC Abstract Specification Topic 2, Spatial Referencing by Coordinates*, August 2008. URL: http://portal.opengeospatial.org/files/?artifact_id=6716.
63. Reed, C. and Thomson, M. *GML 3.1.1 PIDF-LO Shape Application Schema for Use by the Internet Engineering Task Force (IETF)*, April 2007. URL: http://portal.opengeospatial.org/files/?artifact_id=21630.
64. Lake, R. and Reed, C. *GML Point Profile, Version 0.4*, July 2005. Open Geospatial Consortium, URL: http://portal.opengeospatial.org/files/?artifact_id=11606.
65. Vretanos, P. *GML Simple Features Profile, Version 1.0*, April 2006. Open Geospatial Consortium, URL: http://portal.opengeospatial.org/files/?artifact_id=15201.
66. Vretanos, P. *OpenGIS Web Feature Service 2.0 Interface Standard (also ISO 19142), Document #09-025r1*, November 2010. Open Geospatial Consortium, URL: http://www.opengeospatial.org/standards/wfs.
67. OGC. *GML Application Schemas and Profiles*. URL: http://www.ogcnetwork.net/node/210.
68. Hardie, T., Newton, A., Schulzrinne, H. and Tschofenig, H. *LoST: A Location-to-Service Translation Protocol*, August 2008. RFC 5222, Internet Engineering Task Force.
69. Global Disaster Alert and Coordination System. URL: http://www.gdacs.org/.
70. CityGML. URL: http://www.citygmlwiki.org/index.php/Basic_Information.
71. Thomson, M. and Winterbottom, J. *Discovering the Local Location Information Server (LIS)*, September 2010. RFC 5986, Internet Engineering Task Force.
72. Daigle, L. *Domain-Based Application Service Location Using URIs and the Dynamic Delegation Discovery Service (DDDS)*, April 2007. RFC 4848, Internet Engineering Task Force.

73. Mockapetris, P. *Domain Names – Implementation and Specification*, November 1987. RFC 1035, Internet Engineering Task Force.
74. Thomson, M. and Bellis, R. *Location Information Server (LIS) Discovery Using IP Address and Reverse DNS*, April 2013. draft-ietf-geopriv-res-gw-lis-discovery-05 (work in progress), Internet Engineering Task Force.
75. NENA. *Recommended Method(s) for Location Determination to Support IP-Based Emergency Services, Technical Information Document (TID)*, 2006. NENA 08-505, Issue-1.
76. Dawson,M., Winterbottom, J. and Thomson, M. *IP Location*, November 2006. McGraw-Hill Osbourne.
77. Rosenberg, J., Mahy, R., Matthews, P. and Wing, D. *Session Traversal Utilities for NAT (STUN)*, October 2008. RFC 5389, Internet Engineering Task Force.
78. IANA. *PIDF-LO Method Tokens Registry*, November 2004. URL: http://www.iana.org/assignments/method-tokens/method-tokens.xhtml.
79. IANA. *Geopriv HTTP Enabled Location Delivery (HELD) Parameters Registry*, September 2009. URL: http://www.iana.org/assignments/held-parameters/held-parameters.xhtml.
80. Schulzrinne, H. *The tel URI for Telephone Numbers*, December 2004. RFC 3966, Internet Engineering Task Force.
81. Thomson, M. and Winterbottom, J. *Using Device-Provided Location-Related Measurements in Location Configuration Protocols*, April 2013. draft-ietf-geopriv-held-measurements-07 (work in progress), Internet Engineering Task Force.
82. Thomson, M. and Winterbottom, J. *Location Measurements for IEEE 802.16e Devices*, July 2010. draft-thomson-geopriv-wimax-measurements-04 (expired), Internet Engineering Task Force.
83. Winterbottom, J., Tschofenig, H., Schulzrinne, H. and Thomson, M. *A Location Dereference Protocol Using HTTP-Enabled Location Delivery (HELD)*, October 2012. RFC 6753, Internet Engineering Task Force.
84. Marshall, R. *Requirements for a Location-by-Reference Mechanism*, May 2010. RFC 5808, Internet Engineering Task Force.
85. Barnes, R., Thomson, M., Winterbottom, J. and Tschofenig, H. *Location Configuration Extensions for Policy Management*, October 2012. draft-ietf-geopriv-policy-uri-07 (work in progress), Internet Engineering Task Force.
86. Schulzrinne, H., Tschofenig, H., Morris, J., Cuellar, J., Polk, J. and Rosenberg, J. *Common Policy: A Document Format for Expressing Privacy Preferences*, February 2007. RFC 4745, Internet Engineering Task Force.
87. Thomson, M., Winterbottom, J. and Barnes, M. *Device Capability Negotiation for Device-Based Location Determination and Location Measurements in HELD*, March 2011. draft-thomson-geopriv-held-capabilities-09 (expired), Internet Engineering Task Force.
88. Open Mobile Alliance. *Secure User Plane Location Architecture, Version 1.0*, June 2007. URL: http://www.openmobilealliance.org/technical/release_program/supl_v1_0.aspx.
89. Open Mobile Alliance. *Secure User Plane Location Architecture, Version 2.0*, April 2012. URL: http://www.openmobilealliance.org/technical/release_program/supl_v2_0.aspx.
90. Open Mobile Alliance. *Secure User Plane Location Architecture, Version 3.0*, September 2011. URL: http://www.openmobilealliance.org/technical/release_program/supl_v3_0.aspx.
91. 3GPP. *TS 23.271: Functional Stage 2 Description of Location Services (LCS), Release 10, Version 10.2.0*, March 2011.
92. 3GPP. *TS 25.305: Stage 2 Functional Specification of User Equipment (UE) Positioning in UTRAN, Release 10, Version 10.0.0*, October 2010.
93. 3GPP. *TS 36.331: Evolved Universal Terrestrial Radio Access (E-UTRA); Radio Resource Control (RRC); Protocol Specification, Version 11.0.0*, July 2012.
94. 3GPP2. *3GPP2 C.S0084-006-0: Connection Control Plane for Ultra Mobile Broadband (UMB) Air Interface Specification, Version 2*, August 2007.
95. Wimax Forum. *WiMAX Forum Network Architecture, Stage 2: Architecture Tenets, Reference Model and Reference Points, Release 1, Version 2.0*, January 2008.
96. 3GPP. *TS 44.031: Location Services (LCS); Mobile Station (MS) – Serving Mobile Location Centre (SMLC) Radio Resource LCS Protocol (RRLP), Version 10.0.0*, June 2011.
97. 3GPP. *TS 25.331: Radio Resource Control (RRC); Protocol Specification, Version 11.2.0*, July 2012.
98. 3GPP2. *3GPP2 C.S0022-0: Position Determination Service Standard for Dual Mode Spread Spectrum Systems, Version 3*, February 2001.

99. 3GPP. *TS 36.355: Evolved Universal Terrestrial Radio Access (E-UTRA); LTE Positioning Protocol (LPP), Version 11.0.0*, September 2012.
100. 3GPP. *TS 23.167: IP Multimedia Subsystem (IMS) Emergency Sessions, Version 9.3.0*, December 2009.
101. Open Mobile Alliance. *OMA Mobile Location Service, Version 1.0*, July 2011. URL: http://www.openmobilealliance.org/technical/release_program/mls_v1_0.aspx.
102. Open Mobile Alliance. *OMA Mobile Location Service, Version 1.1*, January 2011. URL: http://www.openmobilealliance.org/technical/release_program/mls_v1_1.aspx.
103. Open Mobile Alliance. *OMA Mobile Location Service, Version 1.2*, November 2011. URL: http://www.openmobilealliance.org/technical/release_program/mls_v1_2.aspx.
104. Open Mobile Alliance. *Mobile Location Protocol Specification, LIF TS101, Version 3.0.0*, June 2002. URL: http://www.openmobilealliance.org/tech/affiliates/LicenseAgreement.asp?DocName=/lif/LIF-TS-101-v3.0.0.zip.
105. Open Mobile Alliance. *Mobile Location Protocol (MLP), Version 3.1*, September 2011. URL: http://www.openmobilealliance.org/technical/release_program/mlp_v31.aspx.
106. Canadian Radio Television and Telecommunications Commission. *Wireless E9-1-1 Phase 2 Stage 1 Technical Specification Recommendation Version 1.4*, May 2009. URL: http://www.crtc.gc.ca/public/cisc/items/ESRE0047.pdf.
107. Open Mobile Alliance. *Presence Simple Architecture, Version 2.0*, December 2010. URL: http://www.openmobilealliance.org/technical/release_program/Presence_simple_archive.aspx.
108. Open Mobile Alliance. *Global Permissions Management Architecture, Version 1.0*, November 2011. URL: http://www.openmobilealliance.org/Technical/release_program/gpm_archive.aspx.
109. Open Mobile Alliance. *OMA XML Document Management, Version 2.1*, December 2010. URL: http://www.openmobilealliance.org/technical/release_program/XDM_archive.aspx.
110. Niemi, A., Kiss, K. and Loreto, S. *Session Initiation Protocol (SIP) Event Notification Extension for Notification Rate Control*, January 2012. RFC 6446, Internet Engineering Task Force.
111. 3GPP. *TS 22.071: Location Services (LCS); Service Description; Stage 1, Version 9.0.0*, December 2009.
112. 3GPP. *TS 36.305: Stage 2 Functional Specification of User Equipment (UE) Positioning in E-UTRAN, Release 10, Version 10.3.0*, October 2011.
113. 3GPP. *TS 43.059: Functional Stage 2 Description of Location Services (LCS) in GERAN, Release 10, Version 10.0.0*, April 2011.
114. 3GPP. *TS 24.229: IP Multimedia Call Control Protocol Based on Session Initiation Protocol (SIP) and Session Description Protocol (SDP), Stage 3, Release 11, Version 11.4.0*, June 2012.
115. Garcia-Martin, M., Henrikson, E. and Mills, D. *Private Header (P-Header) Extensions to the Session Initiation Protocol (SIP) for the 3rd-Generation Partnership Project (3GPP*, January 2003. RFC 3455, Internet Engineering Task Force.
116. ETSI. *NGN Functional Architecture; Network Attachment Sub-System (NASS), Version 3.4.1*, March 2010.
117. EENA. *Next Generation 112 Long Term Definition, Version 1.0*, April 2012. URL: http://www.eena.org/ressource/static/files/eena_ng112_ltd_v1-0_final.pdf.
118. Broadband Forum. URL: http://www.broadband-forum.org/.
119. CableLabs. URL: http://www.cablelabs.com/.
120. NENA. *ESQK Guidelines for VoIP to E9-1-1 Connectivity, NENA 03-507, Version 1*, 9 March 2009.
121. IETF. *IETF Emergency Context Resolution With Internet Technologies (ECRIT) Working Group*. URL: http://datatracker.ietf.org/wg/ecrit/charter/.
122. NENA. *NENA i3 Technical Requirements Document, Standard Number: 08-751 v1*, August 2006. URL: http://www.nena.org/standards/technical/voip/i3-requirements.
123. Rosen, B. and Polk, J. *Best Current Practice for Communications Services in Support of Emergency Calling*, March 2013. RFC 6881, Internet Engineering Task Force.
124. Rosenberg, J., Schulzrinne, H., Camarillo, G., Johnston, A., Peterson, J., Sparks, R., Handley, M. and Schooler, E. *SIP: Session Initiation Protocol*, June 2002. RFC 3261, Internet Engineering Task Force.
125. Schulzrinne, H., McCann, S., Bajko, G., Tschofenig, H. and Kroeselberg, D. *Extensions to the Emergency Services Architecture for Dealing With Unauthenticated and Unauthorized Devices*, September 2012. draft-ietf-ecrit-unauthenticated-access-05 (work in progress), Internet Engineering Task Force.
126. Polk, J. *Dynamic Host Configuration Protocol (DHCP) IPv4 and IPv6 Option for a Location Uniform Resource Identifier (URI)*, February 2013. draft-ietf-geopriv-dhcp-lbyr-uri-option-19 (work in progress), Internet Engineering Task Force.

127. Schulzrinne, H. and Tschofenig, H. *Synchronizing Location-to-Service Translation (LoST) Protocol Based Service Boundaries and Mapping Elements*, October 2012. RFC 6739, Internet Engineering Task Force.
128. Schulzrinne, H., Liess, L., Tschofenig, H., Stark, B. and Kuett, A. *Location Hiding: Problem Statement and Requirements*, January 2012. RFC 6444, Internet Engineering Task Force.
129. Barnes, R. and Lepinski, M. *Using Imprecise Location for Emergency Context Resolution*, July 2012. draft-ietf-ecrit-rough-loc-05 (work in progress), Internet Engineering Task Force.
130. Jennings, C., Peterson, J. and Watson, M. *Private Extensions to the Session Initiation Protocol (SIP) for Asserted Identity Within Trusted Networks*, November 2002. RFC 3325, Internet Engineering Task Force.
131. Tschofenig, H., Schulzrinne, H. and Aboba, B. *Trustworthy Location Information*, March 2013. draft-ietf-ecrit-trustworthy-location-05.txt (work in progress), Internet Engineering Task Force.
132. Schulzrinne, H. *Location-to-URL Mapping Architecture and Framework*, September 2009. RFC 5582, Internet Engineering Task Force.
133. Schulzrinne, H., Polk, J. and Tschofenig, H. *Discovering Location-to-Service Translation (LoST) Servers Using the Dynamic Host Configuration Protocol (DHCP)*, August 2008. RFC 5223, Internet Engineering Task Force.
134. Rosen, B. *Dial String Parameter for the Session Initiation Protocol Uniform Resource Identifier*, July 2007. RFC 4967 (Proposed Standard).
135. IEEE. *IEEE 802.11-2011, Wireless LAN Medium Access Control (MAC) and Physical Layer (PHY) Specifications*, June 2011.
136. IEEE. *IEEE 802.16-2009, IEEE Standard for Local and Metropolitan Area Networks, Part 16: Air Interface for Fixed and Mobile Broadband Wireless Access Systems, Amendment for Physical and Medium Access Control Layers for Combined Fixed and Mobile Operation in Licensed Bands*, June 2009.
137. WiMAX Forum. *WiMAX Forum Network Architecture, Stage 3, Base Specification, Release 1.5 V01*, November 2009.
138. WiMAX Forum. URL: http://www.wimaxforum.org.
139. 3GPP. *TS 23.228: Technical Specification Group Services and System Aspects; IP Multimedia Subsystem (IMS); Stage 2 (Release 7), Version 7.13.0*, July 2008.
140. WiMAX Forum. *WiMAX Forum Network Architecture, Stage 2, Base Specification, Release 1.5 V01*, November 2009.
141. Aboba, B., Blunk, L., Vollbrecht, J., Carlson, J. and Levkowetz, H. *Extensible Authentication Protocol (EAP)*, June 2004. RFC 3748, Internet Engineering Task Force.
142. Rigney, C., Willens, S., Rubens, A. and Simpson, W. *Remote Authentication Dial In User Service (RADIUS)*, June 2000. RFC 2865, Internet Engineering Task Force.
143. Calhoun, P., Loughney, J., Guttman, E., Zorn, G. and Arkko, J. *Diameter Base Protocol*, September 2003. RFC 3588, Internet Engineering Task Force.
144. WiMAX Forum. *WiMAX Forum Network Architecture; Architecture, Detailed Protocols and Procedures; IP Multimedia Subsystem (IMS) Interworking*, November 2009.
145. WiMAX Forum. *WiMAX Forum Network Architecture; Architecture, Detailed Protocols and Procedures; Emergency Services Support*, November 2009.
146. Adoba, B., Beadless, M., Arkko, J. and Eronen, P. *The Network Access Identifier*, December 2005. RFC 4282, Internet Engineering Task Force.
147. Aboba, B. and Simon, D. *PPP EAP TLS Authentication Protocol*, October 1999. RFC 2716, Internet Engineering Task Force.
148. WiMAX Forum. *WiMAX Forum Network Architecture – X.509 Device Certificate Profile Draft Specification, Version 1.1.0*, June 2009. URL: http://www.wimaxforum.org/sites/wimaxforum.org/files/page/2009/12/WiMAX_Forum_X.509_Device_Certificate_Profile.pdf.
149. WiMAX Forum. *WiMAX Forum PKI (X.509)*. URL: http://www.wimaxforum.org/resources/pki.
150. WiMAX Forum. *WiMAX Network Protocols and Architecture for Location Based Services, Release 1.5 V01*, November 2009.
151. WiMAX Forum. *Universal Services Interface. An Architecture for Internet+ Service Mode, Release 1.5 V01*, November 2009.
152. Open Mobile Alliance. *Mobile Location Protocol, Version 3.3*, June 2008. URL: http://member.openmobilealliance.org/ftp/Public_documents/LOC/Permanent_documents/OMA-TS-MLP-V3_3-20080627-C.zip.
153. Open Mobile Alliance. *Secure User Plane Location, Version 2.0*, July 2008. URL: http://member.openmobilealliance.org/ftp/Public_documents/LOC/Permanent_documents/OMA-ERP-SUPL-V2_0-20080627-C.zip.

154. Open Mobile Alliance. *UserPlane Location Protocol, Version 2.0*, June 2006. URL: http://member.openmobilealliance.org/ftp/Public_documents/LOC/Permanent_documents/OMA-TS-ULP-V2_0-20080627.
155. Eronen, P. and Tschofenig, H. *Pre-Shared Key Ciphersuites for Transport Layer Security (TLS)*, December 2005. RFC 4279, Internet Engineering Task Force.
156. 3GPP. *TS 22.101: Service Aspects; Service Principles, Version 9.6.0*, December 2009.
157. 3GPP. *TR 22.871: Study on Non-Voice Emergency Services, Release 11, Version 11.3.0*, October 2011.
158. 3GPP. *TS 22.004: General on Supplementary Services (Release 9), Version 9.0.0*, December 2009.
159. 3GPP. *TS 22.173: IP Multimedia Core Network Subsystem (IMS) Multimedia Telephony Service and Supplementary Services; Stage 1, Version 9.4.0*, December 2009.
160. Schulzrinne, H., Tschofenig, H., Holmberg, C. and Patel, M. *Public Safety Answering Point (PSAP) Callback*, March 2013. draft-ietf-ecrit-psap-callback-09.txt (work in progress), Internet Engineering Task Force.
161. 3GPP. *TS 23.018: Basic Call Handling; Technical Realization, Version 9.0.0*, December 2009.
162. 3GPP. *TS 24.008: Mobile Radio Interface Layer 3 Specification; Core Network Protocols; Stage 3, Version 9.1.0*, December 2009.
163. 3GPP. *eCall Specifications Set to Save Lives*, March 2009. URL: http://www.3gpp.org/eCall.
164. 3GPP. *TS 43.022: Functions Related to Mobile Station (MS) in Idle Mode and Group Receive Mode, Version 9.0.0*, December 2009.
165. 3GPP. *TS 25.304: User Equipment (UE) Procedures in Idle Mode and Procedures for Cell Reselection in Connected Mode, Version 8.7.0*, December 2009.
166. 3GPP. *TS 36.304: Evolved Universal Terrestrial Radio Access (E-UTRA); User Equipment (UE) Procedures in Idle Mode, Version 9.0.0*, December 2009.
167. 3GPP. *TS 24.301: Non-Access-Stratum (NAS) Protocol for Evolved Packet Services (EPS); Stage 3, Version 9.1.0*, December 2009.
168. 3GPP. *TS 23.221: Architectural Requirements, Version 9.2.0*, December 2009.
169. 3GPP. *TS 31.102: Characteristics of the Universal Subscriber Identity Module (USIM) Application, Version 9.1.0*, December 2009.
170. Emergency Response Center Administration Finland. *Wireless E9-1-1 Phase 2 Stage 1 Technical Specification Recommendation Version 1.3*. URL: http://www.112.fi.
171. Berners-Lee, T., Fielding, R. and Masinter, L. *Uniform Resource Identifier (URI): Generic Syntax*, January 2005. RFC 3986, Internet Engineering Task Force.
172. Rosenberg, J. and Schulzrinne, H. *An Offer/Answer Model With the Session Description Protocol (SDP)*, June 2002. RFC 3264, Internet Engineering Task Force.
173. Rosenberg, J. *The Session Initiation Protocol (SIP) UPDATE Method*, September 2002. RFC 3311, Internet Engineering Task Force.
174. Schulzrinne, H., Casner, S., Frederick, R. and Jacobson, V. *RTP: A Transport Protocol for Real-Time Applications*, July 2003. RFC 3550, Internet Engineering Task Force.
175. Camarillo, G., Marshall, W. and Rosenberg, J. *Integration of Resource Management and Session Initiation Protocol (SIP)*, October 2002. RFC 3312, Internet Engineering Task Force.
176. Camarillo, G. and Kyzivat, P. *Update to the Session Initiation Protocol (SIP) Preconditions Framework*, March 2005. RFC 4032, Internet Engineering Task Force.
177. Rosenberg, J. and Schulzrinne, H. *Session Initiation Protocol (SIP): Locating SIP Servers*, June 2002. RFC 3263, Internet Engineering Task Force.
178. Schulzrinne, H. and Casner, S. *RTP Profile for Audio and Video Conferences With Minimal Control*, July 2003. RFC 3551, Internet Engineering Task Force.
179. Sjoberg, J., Westerlund, M., Lakaniemi, A. and Xie, Q. *RTP Payload Format and File Storage Format for the Adaptive Multi-Rate (AMR) and Adaptive Multi-Rate Wideband (AMR-WB) Audio Codecs*, April 2007. RFC 4867, Internet Engineering Task Force.
180. Network Interoperability Consultative Committee (NICC) Standards Limited. URL: http://www.niccstandards.org.uk/.
181. NICC. *VOIP – Location for Emergency Calls (Architecture), NICC ND 1638 Issue 1.1.2*, March 2010. URL: http://www.niccstandards.org.uk/files/current/ND1638%20V1.1.2.pdf.
182. Metz, C. *Canadian Toddler Dies After VOIP 911 call; Ambulance Dispatched to Wrong City*, May 2008. URL: http://www.theregister.co.uk/2008/05/06/crtc_investigates_failed_911_call/.
183. NENA. *NENA Interim VoIP Architecture for Enhanced 9-1-1 Services (i2), Standard 08-001*, December 2005.
184. CRTC. *CRTC Communications Monitoring Report*, August 2009. URL: http://www.crtc.gc.ca/eng/publications/reports/policymonitoring/2009/cmr.htm.

185. CRTC. *CRTC Communications Monitoring Report*, September 2012. URL: http://www.crtc.gc.ca/eng/publications/reports/policymonitoring/2012/cmr.htm.
186. CRTC. *Telecom Notice of Consultation CRTC 2009-194*, April 2009. URL: http://www.crtc.gc.ca/eng/archive/2009/2009-194.htm.
187. CRTC. *Telecom Decision CRTC 97-8*, May 1997. URL: http://crtc.gc.ca/eng/archive/1997%5CDT97-8.htm.
188. CRTC. *Telecom Decision CRTC 2003-53*, August 2003. URL: http://www.crtc.gc.ca/eng/archive/2003/dt2003-53.htm.
189. CRTC. *Telecom Decision CRTC 2009-40*, February 2009. URL: http://www.crtc.gc.ca/eng/archive/2009/2009-40.htm.
190. CRTC. *Telecom Decision CRTC 2008-8*, June 2008. URL: http://www.crtc.gc.ca/eng/archive/2008/pt2008-8.htm.
191. CRTC. *Telecom Decision CRTC 2005-21*, April 2005. URL: http://www.crtc.gc.ca/eng/archive/2005/dt2005-21.htm.
192. CRTC. *Telecom Decision CRTC 2010-387*, June 2010. URL: http://www.crtc.gc.ca/eng/archive/2010/2010-387.htm.
193. CRTC. *Telecom Public Notice CRTC 2004-2*, April 2004. URL: http://www.crtc.gc.ca/eng/archive/2004/pt2004-2.htm.
194. CRTC. *Telecom Decision CRTC 2005-61*, October 2005. URL: http://www.crtc.gc.ca/eng/archive/2005/dt2005-61.htm.
195. CRTC. *Telecom Decision CRTC 2006-60*, September 2006. URL: http://www.crtc.gc.ca/eng/archive/2006/dt2006-60.htm.
196. CRTC. *Telecom Decision CRTC 2007-44*, June 2007. URL: http://www.crtc.gc.ca/eng/archive/2007/dt2007-44.htm.
197. CRTC. *Telecom Decision CRTC 2007-125*, December 2007. URL: http://www.crtc.gc.ca/eng/archive/2007/dt2007-125.htm.
198. CRTC. *Telecom Decision CRTC 2011-426*, July 2011. URL: http://www.crtc.gc.ca/eng/archive/2011/2011-426.htm.
199. CRTC. *CRTC Three-Year Plan 2012–2015*. URL: http://www.crtc.gc.ca/eng/backgrnd/plan2012.htm.
200. Indiana General Assembly. *Indiana Code IC Title 36, Article 8, Chapter 16.5 Enhanced Wireless Emergency Telephone Service*, 2010. URL: http://www.in.gov/legislative/ic/2010/title36/ar8/ch16.5.html.
201. Wolf, K.H. *Mozilla SIP Client Zap With Emergency Services Extensions*, September 2011. URL: http://ecrit.labs.nic.at.
202. Polk, J. *IANA Registering a SIP Resource Priority Header Field Namespace for Local Emergency Communications*, February 2013. draft-polk-local-emergency-rph-namespace-05 (work in progress). Internet Engineering Task Force.
203. Mayville, C. *911 Swatters Cost Thousands, Endanger Lives*, February 2009. URL: http://www.govtech.com/security/911-Swatters-Cost-Thousands-Endanger-Lives.html.
204. FBI. *Don't Make the Call – The New Phenomenon of "Swatting"*, February 2008. URL: http://www.fbi.gov/page2/feb08/swatting020408.html.
205. EENA. *False Emergency Calls, EENA Operations Document, Version 1.1*, May 2011. URL: http://www.eena.org/ressource/static/files/2012_05_04-3.1.2.fc_v1.1.pdf.
206. Droms, R. and Arbaugh, W. *Authentication for DHCP Messages*, June 2001. RFC 3118, Internet Engineering Task Force.
207. Rescorla, E. *HTTP Over TLS*, May 2000. RFC 2818, Internet Engineering Task Force.
208. Arends, R., Austein, R., Larson, M., Massey, D. and Rose, S. *DNS Security Introduction and Requirements*, March 2005. RFC 4033, Internet Engineering Task Force.
209. Daigle, L. and Newton, A. *Domain-Based Application Service Location Using SRV RRs and the Dynamic Delegation Discovery Service (DDDS)*, January 2005. RFC 3958, Internet Engineering Task Force.
210. Atkins, D. and Austein, R. *Threat Analysis of the Domain Name System (DNS)*, August 2004. RFC 3833, Internet Engineering Task Force.
211. Handley, M., Jacobson, V. and Perkins, C. *SDP: Session Description Protocol*, July 2006. RFC 4566, Internet Engineering Task Force.
212. Peterson, J. and Jennings, C. *Enhancements for Authenticated Identity Management in the Session Initiation Protocol (SIP)*, August 2006. RFC 4474, Internet Engineering Task Force.
213. Rescorla, E. and Modadugu, N. *Datagram Transport Layer Security*, April 2006. RFC 4347, Internet Engineering Task Force.

214. Lazzaro, J. *Framing Real-Time Transport Protocol (RTP) and RTP Control Protocol (RTCP) Packets Over Connection-Oriented Transport*, July 2006. RFC 4571, Internet Engineering Task Force.
215. Yon, D. and Camarillo, G. *TCP-Based Media Transport in the Session Description Protocol (SDP)*, September 2005. RFC 4145, Internet Engineering Task Force.
216. Lennox, J. *Connection-Oriented Media Transport Over the Transport Layer Security (TLS) Protocol in the Session Description Protocol (SDP)*, July 2006. RFC 4572, Internet Engineering Task Force.
217. Baugher, M., McGrew, D., Naslund, M., Carrara, E. and Norrman, K. *The Secure Real-Time Transport Protocol (SRTP)*, March 2004. RFC 3711, Internet Engineering Task Force.
218. Wing, D., Fries, S., Tschofenig, H. and Audet, F. *Requirements and Analysis of Media Security Management Protocols*, April 2009. RFC 5479, Internet Engineering Task Force.
219. Fischl, J., Tschofenig, H. and Rescorla, E. *Framework for Establishing a Secure Real-Time Transport Protocol (SRTP) Security Context Using Datagram Transport Layer Security (DTLS)*, May 2010. RFC 5763, Internet Engineering Task Force.
220. McGrew, D. and Rescorla, E. *Datagram Transport Layer Security (DTLS) Extension to Establish Keys for the Secure Real-Time Transport Protocol (SRTP)*, May 2010. RFC 5764, Internet Engineering Task Force.
221. Andreasen, F., Baugher, M. and Wing, D. *Session Description Protocol (SDP) Security Descriptions for Media Streams*, July 2006. RFC 4568, Internet Engineering Task Force.
222. Franks, J., Hallam-Baker, P., Hostetler, J., Lawrence, S., Leach, P., Luotonen, A. and Stewart, L. *HTTP Authentication: Basic and Digest Access Authentication*, June 1999. RFC 2617, Internet Engineering Task Force.
223. Dierks, T. and Rescorla, E. *The Transport Layer Security (TLS) Protocol – Version 1.2*, August 2008. RFC 5246, Internet Engineering Task Force.
224. The European Parliament and the Council of the European Union. *Directive 2002/22/EC of the European Parliament and of the Council of 7 March 2002 on Universal Service and Users' Rights Relating to Electronic Communications Networks and Services (Universal Service Directive)*, April 2002. Official Journal of the European Communities, L 108/51, URL: http://eur-lex.europa.eu/LexUriServ/LexUriServ.do?uri=OJ:L:2002:108:0051:0051:EN:PDF.
225. U.S. Department of Justice. *Americans With Disabilities Act – Access for 9-1-1 and Telephone Emergency Services*, July 1998. URL: http://www.ada.gov/911ta.htm.
226. FCC. *Ten-Digit Numbering and Emergency Call Handling Procedures for Internet-Based TRS*. URL: http://www.fcc.gov/guides/ten-digit-numbering-and-emergency-call-handling-procedures-internet-based-trs.
227. Hellstrom, G. and Jones, P. *RTP Payload for Text Conversation*, June 2005. RFC 4103, Internet Engineering Task Force.
228. Campbell, B., Rosenberg, J., Schulzrinne, H., Huitema, C. and Gurle, D. *Session Initiation Protocol (SIP) Extension for Instant Messaging*, December 2002. RFC 3428, Internet Engineering Task Force.
229. Campbell, B., Mahy, R. and Jennings, C. *The Message Session Relay Protocol (MSRP)*, September 2007. RFC 4975, Internet Engineering Task Force.
230. van Wijk, A. and Gybels, G. *Framework for Real-Time Text Over IP Using the Session Initiation Protocol (SIP)*, June 2008. RFC 5194, Internet Engineering Task Force.
231. Wenger, S., Hannuksela, M.M., Stockhammer, T., Westerlund, M. and Singer, D. *RTP Payload Format for H.264 Video*, February 2005. RFC 6280, Internet Engineering Task Force.
232. Sparks, R. *The Session Initiation Protocol (SIP) Refer Method*, April 2003. RFC 3515, Internet Engineering Task Force.
233. Johnston, A., Sparks, R., Cunningham, C., Donovan, S. and Summers, K. *Session Initiation Protocol Service Examples*, October 2008. RFC 5359, Internet Engineering Task Force.
234. Rosenberg, J., Schulzrinne, H. and Kyzivat, P. *Caller Preferences for the Session Initiation Protocol (SIP)*, August 2004. RFC 3841, Internet Engineering Task Force.
235. NENA. *9-1-1 Origin and History*. URL: http://www.nena.org/?page=911overviewfacts, last accessed September 2012.
236. Verizon. *Verizon Global Wholesale – E9-1-1*. URL: http://www22.verizon.com/wholesale/solutions/solution/e911.html, last accessed September 2012.
237. FCC. *Fact Sheet: FCC WIRELESS 911 REQUIREMENTS*, January 2001. URL: http://www.fcc.gov/pshs/services/911-services/enhanced911/archives/factsheet_requirements_012001.pdf.
238. Crawford, S. *The Ambulance, the Squad Car, and the Internet*, February 2006. Berkeley Technology Law Journal, available at SSRN: http://ssrn.com/abstract=885582.
239. Carroll, K. *One Fine E911 Mess*, August 2001. http://connectedplanetonline.com/mag/telecom_one_fine_mess/.

240. Hatfield, D.N. *A Report on Technical and Operational Issues Impacting the Provision of Wireless Enhanced 911 Services, Prepared for the Federal Communications Commission*, October 2002. http://gullfoss2.fcc.gov/prod/ecfs/retrieve.cgi?native_or_pdf=pdf&id_document=6513296239.
241. Cauley, L. *FCC Cut Study Finding 911 Flaws; Indoor Wireless Calls Often Can't Be Tracked*, March 2007. http://usatoday30.usatoday.com/printedition/money/20070314/1b_wireless14.art.htm.
242. FCC. *Voice-over-Internet Protocol*, August 2005. http://transition.fcc.gov/voip/.
243. Duncan, G. *VOIP Providers Largely Miss E911 Deadline*, November 2005. http://www.digitaltrends.com/lifestyle/voip-providers-largely-miss-e911-deadline/.
244. U.S. Congress. *New and Emerging Technologies 911 Improvement Act of 2008 (NET 911 Act), Pub. L. No. 110-283, 122 Stat. 2620*, 2008. www.gpo.gov/fdsys/pkg/BILLS-110hr3403enr/pdf/BILLS-110hr3403enr.pdf.
245. The Council of the European Communities. *Council Decision of 29 July 1991 on the Introduction of a Single European Emergency Call Number (91/396/EEC)*, August 1991. Official Journal of the European Communities, L 217, 0031–0032, URL: http://eur-lex.europa.eu/LexUriServ/LexUriServ.do?uri=CELEX:31991D0396:EN:HTML.
246. The European Parliament and the Council of the European Union. *Directive 97/13/EC of the European Parliament and of the Council of 10 April 1997 on a Common Framework for General Authorizations and Individual Licences in the Field of Telecommunications Services*, May 1997. Official Journal of the European Communities, L 117/15, URL: http://eur-lex.europa.eu/LexUriServ/LexUriServ.do?uri=OJ:L:1997:117:0015:0027:EN:PDF.
247. The European Parliament and the Council of the European Union. *Directive 98/10/EC of the European Parliament and of the Council of 26 February 1998 on the Application of Open Network Provision (ONP) to Voice Telephony and on Universal Service for Telecommunications in a Competitive Environment*, April 1998. Official Journal of the European Communities, L 101/24, URL: http://eur-lex.europa.eu/LexUriServ/LexUriServ.do?uri=CELEX:31998L0010:EN:PDF.
248. Michalski, W. *Technical and Regulatory Issues of Emergency Call*, March 2009. Journal of Telecommunications and Information Technology, URL: http://www.nit.eu/czasopisma/JTIT/2009/3/123.pdf.
249. The European Parliament and the Council of the European Union. *Directive 2002/21/EC of the European Parliament and of the Council of 7 March 2002 on a Common Regulatory Framework for Electronic Communications Networks and Services (Framework Directive)*, April 2002. Official Journal of the European Communities, L 108/33, URL: http://eur-lex.europa.eu/LexUriServ/LexUriServ.do?uri=OJ:L:2002:108:0033:0033:EN:PDF.
250. The European Parliament and the Council of the European Union. *Directive 2002/19/EC of the European Parliament and of the Council of 7 March 2002 on Access to, and Interconnection of, Electronic Communications Networks and Associated Facilities (Access Directive)*, April 2002. Official Journal of the European Communities, L 108/7, URL: http://eur-lex.europa.eu/LexUriServ/LexUriServ.do?uri=CELEX:32002L0019:EN:PDF.
251. The European Parliament and the Council of the European Union. *Directive 2002/20/EC of the European Parliament and of the Council of 7 March 2002 on the Authorisation of Electronic Communications Networks and Services (Authorisation Directive)*, April 2002. Official Journal of the European Communities, L 108/21, URL: http://eur-lex.europa.eu/LexUriServ/LexUriServ.do?uri=OJ:L:2002:108:0021:0021:EN:PDF.
252. The European Parliament and the Council of the European Union. *Directive 2002/58/EC of the European Parliament and of the Council of 12 July 2002 Concerning the Processing of Personal Data and the Protection of Privacy in the Electronic Communications Sector (Directive on Privacy and Electronic Communications)*, July 2002. Official Journal of the European Communities, L 201/37, URL: http://eur-lex.europa.eu/LexUriServ/LexUriServ.do?uri=OJ:L:2002:201:0037:0037:EN:PDF.
253. The European Parliament and the Council of the European Union. *Directive 2002/77/EC of the European Parliament and of the Council of 16 September 2002 on Competition in the Markets for Electronic Communications Networks and Services (Text With EEA Relevance)*, September 2002. Official Journal of the European Communities, L 249/21, URL: http://eur-lex.europa.eu/LexUriServ/LexUriServ.do?uri=OJ:L:2002:249:0021:0021:EN:PDF.
254. The European Parliament and the Council of the European Union. *Regulation (EC) No 2887/2000 of the European Parliament and of the Council of 18 December 2000 on Unbundled Access to the Local Loop (Text With EEA Relevance)*, December 2000. Official Journal of the European Communities, L 336/4, URL: http://eur-lex.europa.eu/LexUriServ/LexUriServ.do?uri=OJ:L:2000:336:0004:0004:EN:PDF.
255. Bundesministerium der Justiz. *Telekommunikationsgesetz (TKG)*, June 2004. URL: http://www.gesetze-im-internet.de/tkg_2004/index.html.
256. Bundesnetzagentur. *Verordnung über Notrufverbindungen (NotrufV)*, March 2009. URL: http://www.gesetze-im-internet.de/notrufv/index.html.

257. European Commission Communications Committee. *Implementation of the European Emergency Number 112 – Results of the Second Data-Gathering Round (January 2009)*, March 2009. URL: http://circa.europa.eu/Public/irc/infso/cocom1/library?l=/public_documents_2009/cocom09-11_final/_EN_1.0_&a=d.
258. Garrie, D. and Wong, R. *Regulating Voice over Internet Protocol: An EU/US Comparative Approach*, 2007. American University International Law Review, 22(4), 549–481, URL: http://ssrn.com/abstract=997592.
259. Parrillo, G. *The Current Status of VoIP Regulation in Italy*, 2006. Computer and Telecommunications Law Review, 12(4), 111.
260. Ofcom. *Regulation of VoIP Services: Access to the Emergency Services: Statement and Publication of a Statutory Notification Under Section 48(1) of the Communications Act 2003 Modifying General Condition 4*, December 2007. URL: http://stakeholders.ofcom.org.uk/binaries/consultations/voip/statement/voipstatement.pdf.
261. CGALIES (Coordination Group on Access to Location Information by Emergency Services). *Report on Implementation Issues Related to Access to Location Information by Emergency Services (E112) in the European Union: Final Report*, January 2002. URL: http://www.emtel.etsi.org/Workshop/Non-presented_papers/Cgalies%20final%20V1.0%20Jan%20M.doc.
262. Commission of the European Communities. *Commission Recommendation of 25 July 2003 on the Processing of Caller Location Information in Electronic Communication Networks for the Purpose of Location-Enhanced Emergency Call Services (Notified Under Document Number C(2003) 2657)*, July 2003. Official Journal of the European Communities, L 189/49, URL: http://eur-lex.europa.eu/LexUriServ/LexUriServ.do?uri=OJ:L:2003:189:0049:0051:EN:PDF.
263. Press Releases RAPID. *Telecoms: Italy Urged to Take Action on 112 Caller Location to Comply with Court Judgement, IP/09/774*, May 2009. URL: http://europa.eu/rapid/pressReleasesAction.do?reference=IP/09/774.
264. European Commission. *Europe's Information Society – Thematic Portal: The Telecoms Reform*. URL: http://ec.europa.eu/information_society/policy/ecomm/tomorrow/index_en.htm.
265. The European Parliament and the Council of the European Union. *Regulation (EC) No 1211/2009 of the European Parliament and of the Council of 25 November 2009 Establishing the Body of European Regulators for Electronic Communications (BEREC) and the Office*, December 2009. Official Journal of the European Communities, L 201/37, URL: http://eur-lex.europa.eu/LexUriServ/LexUriServ.do?uri=OJ:L:2009:337:0001:0010:EN:PDF.
266. The European Parliament and the Council of the European Union. *Directive 2009/136/EC of the European Parliament and of the Council of 25 November 2009 Amending Directive 2002/22/EC on Universal Service and UsersâĂŹ Rights Relating to Electronic Communications Networks and Services, Directive 2002/58/EC Concerning the Processing of Personal Data and the Protection of Privacy in the Electronic Communications Sector and Regulation (EC) No 2006/2004 on Cooperation Between National Authorities Responsible for the Enforcement of Consumer Protection Laws*, December 2009. Official Journal of the European Communities, L 337/11, URL: http://eur-lex.europa.eu/LexUriServ/LexUriServ.do?uri=OJ:L:2009:337:0011:0036:En:PDF.
267. The European Parliament and the Council of the European Union. *Directive 2009/140/EC of the European Parliament and of the Council of 25 November 2009 Amending Directives 2002/21/EC on a Common Regulatory Framework for Electronic Communications Networks and Services, 2002/19/EC on Access to, and Interconnection of, Electronic Communications Networks and Associated Facilities, and 2002/20/EC on the Authorisation of Electronic Communications Networks and Services*, December 2009. Official Journal of the European Communities, L 337/37, URL: http://eur-lex.europa.eu/LexUriServ/LexUriServ.do?uri=OJ:L:2009:337:0037:0069:EN:PDF.
268. European Parliament. *European Parliament Legislative Resolution of 6 May 2009 on the Common Position Adopted by the Council With a View to the Adoption of a Directive of the European Parliament and of the Council Amending Directive 2002/22/EC on Universal Service and Users' Rights Relating to Electronic Communications Networks, Directive 2002/58/EC Concerning the Processing of Personal Data and the Protection of Privacy in the Electronic Communications Sector and Regulation (EC) No 2006/2004 on Cooperation Between National Authorities Responsible for the Enforcement of Consumer Protection Laws (16497/1/2008 – C6-0068/2009–2007/0248(COD))*, May 2009. URL: http://eur-lex.europa.eu/LexUriServ/LexUriServ.do?uri=OJ:L:2009:337:0037:0069:EN:PDF.
269. REACH112. *REsponding to All Citizens needing Help (REACH112) Project Website*. URL: http://www.reach112.eu.
270. Tschofenig, H., Eggert, L. and Sarker, Z. *Report from the IAB/IRTF Workshop on Congestion Control for Interactive Real-Time Communication*, April 2013. draft-iab-cc-workshop-report-01 (work in progress), Internet Engineering Task Force.

271. PEACE. *IP-Based Emergency Applications and ServiCes for Next Generation Networks (PEACE) Project Webpage*. URL: http://www.ict-peace.eu.
272. US DOT. *US Department of Transportation Next Generation 9-1-1 Project Website*. URL: http://www.its.dot.gov/ng911/.
273. Bradner, S., Conroy, L. and Fujiwara, K. *The E.164 to Uniform Resource Identifiers (URI) Dynamic Delegation Discovery System (DDDS) Application (ENUM)*, March 2011. RFC 6116, Internet Engineering Task Force.
274. Galbraith, R., Turnbull, J., Dearnaley, R., Hart, J. and Nolde, K. *Do We Need Quality of Service in the Customer Environment?* 2005. BT Technology Journal, 23, 146–159, URL: http://dx.doi.org/10.1007/s10550-005-0013-6.
275. 3GPP. *3rd Generation Partnership Project*. URL: http://www.3gpp.org/.
276. IETF. *IETF Session Initiation Protocol Core (Sipcore) Working Group*. URL: http://datatracker.ietf.org/wg/sipcore/charter/.
277. Open Mobile Alliance. *OMA Location in SIP/IP Core, Version 1.0*, November 2010. URL: http://member.openmobilealliance.org/ftp/Public_documents/LOC/Permanent_documents/OMA-TS-LOCSIP-V1_0-20101125-C.zip.
278. Taylor, T., Tschofenig, H., Schulzrinne, H. and Shanmugam, M. *Security Threats and Requirements for Emergency Call Marking and Mapping*, January 2008. RFC 5069, Internet Engineering Task Force.
279. Fraunhofer FOKUS. *Open IMS Core Home Page*. URL: http://openimscore.org/.
280. Fraunhofer FOKUS. *Installation Guide for the Emergency Branch of the Open IMS Core*. URL: http://openimscore.org/emergency_installation_guide.
281. Fraunhofer FOKUS. *myMONSTER (Multimedia Open InterNet Services and Telecommunication EnviRonment)*. URL: http://www.monster-the-client.org/.
282. NMEA. *National Marine Electronics Association (NMEA) Standard 0183 V 3.01*, January 2002. URL: http://www.nmea.org/content/nmea_standards/nmea_083_v_400.asp.
283. Google. *Google Maps API*. URL: http://code.google.com/apis/maps/.
284. Google. *Geocoding Service from Google Maps API*. URL: http://code.google.com/apis/maps/documentation/geocoding/index.html.
285. Rosenberg, J. *A Presence Event Package for the Session Initiation Protocol (SIP)*, August 2004. RFC 3856, Internet Engineering Task Force.
286. 3GPP. *TS 29.214: Policy and Charging Control over Rx Reference Point; Stage 3, Version 7.11.0*, January 2011.
287. 3GPP. *TS 29.273: Evolved Packet System (EPS); 3GPP EPS AAA Interfaces, Version 8.8.0*, June 2011.
288. Fraunhofer FOKUS. *Open Evolved Packet Core*. URL: http://www.openepc.net/.
289. Farinacci, D., Li, T., Hanks, S., Meyer, D. and Traina, P. *Generic Routing Encapsulation (GRE)*, March 2000. RFC 2784, Internet Engineering Task Force.
290. Wu, X. and Schulzrinne, H. *Sipc, a Multi-Function SIP User Agent*. In 7th IFIP/IEEE International Conference, Management of Multimedia Networks and Services (MMNS, pp. 269–281. IFIP/IEEE, Springer, 2004.
291. ANSI/TIA. *Telecommunications: IP Telephony Infrastructure: Link Layer Discovery Protocol for Media Endpoint Devices, ANSI/TIA-1057-2006*, April 2006.
292. *VIC – Video Conferencing Tool*. URL: http://ee.lbl.gov/vic/.
293. *PostgreSQL*. URL: http://www.postgresql.org/.
294. *PostGIS*. URL: http://postgis.refractions.net/.
295. Lennox, J., Schulzrinne, H. and Rosenberg, J. *Common Gateway Interface for SIP*, January 2001. RFC 3050, Internet Engineering Task Force.
296. *CINEMA: Columbia InterNet Extensible Multimedia Architecture*. URL: http://www.cs.columbia.edu/irt/cinema/.
297. *JAIN SIP – Java API for SIP Signaling*. URL: https://jain-sip.dev.java.net/.
298. *GNU ccRTP*. URL: http://www.gnu.org/software/ccrtp/.
299. *FFmpeg*. URL: http://ffmpeg.org/.
300. ETSI. *ETSI Publications Download Area*. URL: http://pda.etsi.org/pda/queryform.asp.
301. EMTEL. *Emergency Communications (EMTEL); Test/Verification Procedure for Emergency Calls, Special Report ETSI 002 777 V1.1.1*, June 2010. URL: http://www.etsi.org/deliver/etsi_sr/002700_002799/002777/01.01.01_60/sr_002777v010101p.pdf.
302. EMTEL. *Emergency Communications (EMTEL); Basis of Requirements for Communication of Individuals With Authorities/Organizations in Case of Distress (Emergency Call Handling), ETSI TR 102 180 V1.3.1*, September 2011. URL: http://www.etsi.org/deliver/etsi_tr/102100_102199/102180/01.03.01_60/tr_102180v010301p.pdf.

303. EMTEL. *Emergency Communications (EMTEL); Total Conversation Access to Emergency Services, ETSI TS 101 470*, June 2012. URL: http://webapp.etsi.org/workprogram/Frame_WorkItemList.asp?qETSI_NUMBER=101+470.
304. EMTEL. *Emergency Communications (EMTEL); Total Conversation Access to Emergency Services, ETSI TR 103 170*, June 2012. URL: http://webapp.etsi.org/workprogram/Frame_WorkItemList.asp?qETSI_NUMBER=103+170.
305. EMTEL. *Emergency Communications (EMTEL); European Public Warning System (EU-ALERT) Using the Cell Broadcast Service, ETSI TS 102 900 Ver. 1.1.1*, October 2010. URL: http://webapp.etsi.org/workprogram/Frame_WorkItemList.asp?qETSI_NUMBER=102+900.
306. EMTEL. *Emergency Communications (EMTEL); European Public Warning System (EU-ALERT) Using the Cell Broadcast Service, ETSI TS 102 900 Ver. 1.2.1*, January 2012. URL: http://webapp.etsi.org/workprogram/Frame_WorkItemList.asp?qETSI_NUMBER=102+900.
307. European Commission. *Communications Committee (COCOM) Expert Group on Emergency Access (EGEA)*. URL: http://circa.europa.eu/Public/irc/infso/egea/home.
308. European Commission. *Standardisation Mandate to the European Standards Organizations (ESO) in Support of the Location Enhanced Emergency Call Service, M/493*, May 2011. URL: http://circa.europa.eu/Public/irc/infso/egea/library?l=/standardisation/emergency_m493pdf_1/_EN_1.0_&a=d.
309. *European Emergency Number Association (EENA)*. URL: http://www.eena.org.
310. *European Computer Manufacturers Association (ECMA)*. URL: http://www.ecma-international.org.
311. ECMA. *Technical Report TR/101, Next Generation Corporate Networks (NGCN) – Emergency Calls, 2nd Edition*, December 2010. URL: http://www.ecma-international.org/publications/techreports/E-TR-101.htm.
312. ATIS. *Alliance for Telecommunications Industry Services (ATIS)*. URL: http://www.atis.org.
313. NENA. *NENA Future Path Plan – FPP*. URL: http://www.nena.org/future-path-plan.
314. ATIS. *The ESIF–NENA Relationship*, November 2006. URL: http://www.atis.org/ESIF/Docs/Final-Documents/ESIF-NENA-Relationship.doc.
315. ATIS. *ESIF–NENA Relationships – Emergency Services Problem Solving*. URL: http://www.atis.org/esif/Docs/Final-Documents/ESIF-NENA-Relationship-Diagrams.ppt.
316. NG9-1-1 Institute. URL: http://www.e911institute.org.
317. European Commission. *Committees and Working Groups of the European Commission*. URL: http://ec.europa.eu/information_society/policy/ecomm/committees_working_groups/index_en.htm.
318. European Commission. *Communications Committee (COCOM)*. URL: http://ec.europa.eu/information_society/policy/ecomm/implementation_enforcement/comm_committee/index_en.htm.
319. Wikipedia. *Wikipedia Entry for "Comitology"*. URL: http://en.wikipedia.org/wiki/Comitology.
320. European Commission. *Body of European Regulators for Electronic Communications (BEREC) and the Office*. URL: http://ec.europa.eu/information_society/policy/ecomm/implementation_enforcement/berec/index_en.htm.
321. European Commission Communications Committee. *Implementation of the European Emergency Number 112 – Results of the Fourth Data-Gathering Round)*, May 2011. URL: http://circa.europa.eu/Public/irc/infso/cocom1/library?l=/public_documents_2011/cocom10-38_finalpdf/_EN_1.0_&a=d.
322. European Commission Communications Committee. *Annex to Document COCOM10-38 "Report on the Implementation of 112"*, May 2011. URL: http://circa.europa.eu/Public/irc/infso/cocom1/library?l=/public_documents_2011/cocom10-38_annexpdf/_EN_1.0_&a=d.
323. European Commission Communications Committee. *Implementation of the European Emergency Number 112 – Results of the Third Data-Gathering Round)*, February 2010. URL: http://www.eena.org/ressource/static/files/cocom_report2010.pdf.
324. European Commission Communications Committee. *Annex to Document COCOM10-09 "Report on the Implementation of 112"*, February 2010. URL: http://www.eena.org/ressource/static/files/cocom_annex2010.pdf.
325. European Commission Communications Committee. *Annex to Document COCOM09-11 Final "Report on the Implementation of 112"*, March 2009. URL: http://circa.europa.eu/Public/irc/infso/cocom1/library?l=/public_documents_2009/cocom09-11_final_1/_EN_1.0_&a=d.
326. European Commission Communications Committee. *Implementation of the European Emergency Number 112 – Summary Report*, July 2008. URL: http://circa.europa.eu/Public/irc/infso/cocom1/library?l=/public_documents_2008/cocom08-17_final_1/_EN_1.0_&a=d.
327. European Commission Communications Committee. *Annex to Document COCOM08-17 Final "Implementation of the European Emergency Number 112 – Summary Report"*, July 2008. URL: http://circa.europa.eu/Public/irc/infso/cocom1/library?l=/public_documents_2008/cocom08-17_final/_EN_1.0_&a=d.

328. European Commission. *112 – The European Emergency Number*. URL: http://www.112.eu.
329. European Commission. *Joint Tripartite Declaration Establishing a "European 112 Day"*, February 2009. URL: http://ec.europa.eu/information_society/activities/112/docs/decl/en.pdf.
330. European Commission. *EGEA Terms of Reference*. URL: http://circa.europa.eu/Public/irc/infso/egea/library?l=/terms_reference&vm=detailed&sb=Title.
331. European Commission Communications Committee. *Terms of Reference of the Working Group on Emergency Access (EGEA) for 2011*, January 2011. URL: http://circa.europa.eu/Public/irc/infso/egea/library?l=/terms_reference/cocom10-39_2011pdf/_EN_1.0_&a=d.
332. European Commission. *Communications Committee (COCOM) Expert Group on Emergency Access (EGEA) Presentation and Document Archive*. URL: http://circa.europa.eu/Public/irc/infso/egea/library.
333. Internet Architecture Board. *Letter to the European Commission on Global Interoperability in Emergency Services*. URL: http://www.iab.org/documents/correspondence-reports-documents/2011-2/letter-to-the-european-commission-on-global-interoperability-in-emergency-services/.
334. Tschofenig, H. *Emergency Services Functionality With the Extensible Messaging and Presence Protocol (XMPP)*, March 2012. draft-tschofenig-ecrit-xmpp-es-00 (work in progress), Internet Engineering Task Force.
335. Bellis, R. *Flow Identity Extension for HTTP-Enabled Location Delivery (HELD)*, April 2013. RFC 6915, Internet Engineering Task Force.
336. Trevisani, E. and Vitaletti, A. *Cell-ID Location Technique, Limits and Benefits: An Experimental Study*. In WMCSA'04: Proceedings of the Sixth IEEE Workshop on Mobile Computing Systems and Applications (WMCSA'04), pp. 51–60, 2004.
337. EENA. *Caller Location in Support of Emergency Services, EENA Operations Document, Version 1.3*, May 2011. URL: http://www.eena.org/ressource/static/files/2011_05_27_2.2.2.cl_v1.3.pdf.
338. Dawson, M. *Presentation About the "The Standard Mobile Emergency Service" at the EENA*, April 2011. URL: http://www.eena.org/ressource/static/files/commscope-eena-mobilelocation-april-2011.pdf.
339. NENA. *The Call, Issue No. 4, The Official Publication of NENA*, August 2012. URL: http://www.nena.org/?page=TheCallMagazine.
340. Tschofenig, H. *Security Risks in Next-Generation Emergency Services*, 2011. Communications of the ACM, 54(11), 23–25, URL: http://doi.acm.org/10.1145/2018396.2018405.
341. NIST. *National Strategy for Trusted Identities in Cyberspace*. URL: http://www.nist.gov/nstic/.
342. NENA. *NENA Security for Next-Generation 9-1-1 Standard (NG-SEC), NENA 75-001, Version 1*, February 2010. URL: http://nena9-1-1.org/sites/default/files/75-001_NG-Security.pdf.
343. Rooke, A. and Paris, J. *HeERO – Harmonised eCall European Pilot*. URL: http://www.heero-pilot.eu/view/en/index.html.
344. NENA. *NG9-1-1 ICE – Industry Collaboration Events*. URL: http://www.nena.org/?page=NG911_ICE.
345. EENA. *EENA Network of Researchers*. URL: http://www.eena.org/view/en/Memberships/eena_network_of_researchers.html.
346. British Telecom. *British Telecommunications Carrier Price List, Ancillary Service – Emergency Services*, March 2010. URL: https://www.btwholesale.com/shared/document/CPL/NCCN_2010_jan/nccn_998a.rtf.
347. Irish Commission for Communications Regulation. *Emergency Call Answering Service: Call Handling Fee Review 2012/2013, Reference Number: 11/81*, November 2011. URL: http://www.comreg.ie/publications/emergency_call_answering_service__call_handling_fee_review_2012_2013.596.103967.p.html.
348. NENA. *Funding 9-1-1 Into the Next Generation: An Overview of NG9-1-1 Funding Model Options for Consideration*, March 2007. URL: http://www.nena.org/resource/resmgr/NGPP/NGFundingReport.pdf.
349. FCC. *911 Fees, Congressional Report*, November 2011. URL: http://www.fcc.gov/document/911-fees-congressional-report.
350. FCC. *Second Annual Report to Congress on the State Collection and Distribution of 911 and Enhanced 911 Fees and Charges*, August 2010. URL: http://hraunfoss.fcc.gov/edocs_public/attachmatch/DOC-300946A1.pdf.
351. FCC. *3rd Report to Congress on State 911 Fees and Charges*, November 2011. URL: http://www.fcc.gov/document/3rd-report-congress-state-911-fees-and-charges.

Index

Internet Protocol-Based Emergency Services, First Edition. Hannes Tschofenig and Henning Schulzrinne.

www.ingramcontent.com/pod-product-compliance
Lightning Source LLC
Chambersburg PA
CBHW082045280125
20788CB00044B/47
9780470689769